SOFT X-RAY BAND SPECTRA

AND
THE ELECTRONIC STRUCTURE
OF METALS AND MATERIALS

*Conference on Soft X-Ray Spectrometry
" and the Band Structure of
Metals and Alloys*

SOFT X-RAY BAND SPECTRA

AND
THE ELECTRONIC STRUCTURE
OF METALS AND MATERIALS

Edited by

Derek J. Fabian

Department of Metallurgy, University of Strathclyde, Glasgow, Scotland

Based on the Proceedings of a Conference held at the University of Strathclyde, Scotland, 18–21 September 1967 with support from The Royal Society European Programme and the University of Strathclyde

1968  **Academic Press**
London and New York

ACADEMIC PRESS INC. (LONDON) LTD
Berkeley Square House
Berkeley Square
London, W.1

U.S. Edition Published by
ACADEMIC PRESS INC.
111 Fifth Avenue
New York, New York 10003

Library of Congress Catalog Card Number: 68–9111

FILMSET BY
KEYSPOOLS LTD., GOLBORNE, WARRINGTON, LANCS.

PRINTED IN GREAT BRITAIN BY
C. TINLING & CO. LTD., LIVERPOOL, LONDON AND PRESCOT

Contributors

Participants at the Conference on "Soft X-Ray Spectrometry and the Band Structure of Metals and Alloys" who have contributed to this book.

ABELÈS, F., *Institut d'Optique et Faculté des Sciences de Paris, 3 Boulevard Pasteur, Paris 15, France* (p. 191)

ALTMANN, S. L., *Department of Metallurgy, University of Oxford, Parks Road, Oxford, U.K.* (pp. 265, 366)

ASHCROFT, N. W., *Laboratories of Atomic and Solid State Physics, Cornell University, Ithaca, New York, U.S.A.* (p. 249)

BERGERSEN, B., *Institute of Theoretical Physics, Chalmers Tekniska Hojskola, Sven Hultins Gata, Göteborg S, Sweden* (p. 351)

BONNELLE, Mme. C., *Laboratoire de Chimie Physique, 11 Rue Pierre-Curie, Paris 5, France* (p. 163)

BROUERS, F., *Université de Liège, Faculté des Sciences, 1a quai F. Roosevelt, Liège, Belgium* (p. 329)

CAUCHOIS, Mlle. Y., *Laboratoire de Chimie Physique, 11 Rue Pierre-Curie, Paris 5, France* (p. 73)

CURRY, C., *Department of Physics, The University, Leeds 2, U.K.* (p. 173)

CUTHILL, J. R., *Alloy Physics Section, U.S. Department of Commerce, National Bureau of Standards, Washington D.C. 20234, U.S.A.* (p. 151)

DIMOND, R. K., *Department of Metallurgy, University of Strathclyde, Royal College, George Street, Glasgow, U.K.* (p. 46)

FABIAN, D. J., *Department of Metallurgy, University of Strathclyde, Royal College, George Street, Glasgow, U.K.* (pp. 46, 215)

FAESSLER, A., *Sektion Physik der Universität, Geschwister-Scholl-Platz 1, 8 München 22, West Germany* (p. 93)

GLICK, A. J., *Faculté des Sciences, Service de Physique des Solids, Orsay (S. et O.), France* (p. 319)

HARRISON, W. A., *Stanford University, Microwave Laboratory, Stanford, California 94305, U.S.A.* (p. 227)

HEDIN, L., *Institute of Theoretical Physics, Chalmers Tekniska Hojskola, Sven Hultins Gata, Göteborg S, Sweden* (p. 337)

HOLLIDAY, J. E., *U.S. Steel Corporation, Edgar C. Bain Laboratories, Research Center M.S. 59, Monroeville, Philadelphia 15146, U.S.A.* (p. 101)

JACOB, L., *Department of Natural Philosophy, University of Strathclyde, George Street, Glasgow, U.K.* (p. 31)

LIEFELD, R. J., *Department of Physics, New Mexico State University, Las Cruces, New Mexico, U.S.A.* (p. 153)

LONGE, P., *Université de Liège, Faculté des Sciences, 1a quai F. Roosevelt, Liège, Belgium* (p. 329)

MARCH, N. H., *Department of Physics, University of Sheffield, Sheffield 10, U.K.* (pp. 224, 283)

ROOKE, G. A., *Department of Metallurgy, University of Strathclyde, Royal College, George Street, Glasgow, U.K.* (pp. 3, 185)

SAGAWA, T., *Department of Physics, Tohoku University, Japan* (p. 29)

STOTT, M. J., *Department of Physics, The University, Sheffield 10, U.K.* (pp. 283, 303)

WATSON, L. M., *Department of Metallurgy, University of Strathclyde, Royal College, George Street, Glasgow, U.K.* (p. 46)

WIECH, G., *Sektion Physik der Universität, Geschwister-Scholl-Platz 1, 8 München 22, West Germany* (p. 59)

Foreword

This volume seems to me to represent an important trend which is now becoming more manifest. An enormous amount of information in science and technology is made available today, but it is not easily possible for the specialist reader to select those parts which are of most significance to himself. In the Conference that led to this publication, the participants were encouraged to think without inhibition on the present position and the future progress of their subject, and since they were themselves chosen as representatives of those making significant contributions to the field, this volume serves as a very useful guide to current thought in this expanding area of study and research.

The Conference participants formed a good cross-section of workers in the field including both theoretical and experimental scientists, physicists and metallurgists, and no doubt this resulted in the cross-fertilization that was clearly apparent. It is a good many years since very fundamental observations were made by a few physicists such as the late H. W. B. Skinner, whom I had the pleasure of knowing personally during his time at Harwell. It was appreciated then that their findings were very important to the proper development of the science but such contributions were subsequently insufficiently pursued in adjoining sectors. With the increasing insight available today, the moment is ripe for a substantial advancement and it is hoped that this book will prove valuable in expanding interest in and knowledge of the field.

May, 1968

S. C. CURRAN
Principal and Vice-Chancellor
University of Strathclyde

Preface

The purpose of this book is to collect under one cover the thoughts and works of a number of widely located research groups, particularly concerned with soft x-ray spectrometry and with the determination of band-structure information for solids. To this end the Conference on which this book is founded was planned and organized. It was the first international conference specially to combine these fields, and the book therefore provides an opportunity to bring focus to the research and to attempt a consistent terminology, which are inevitably missing when a subject of this kind arises from several sources in different parts of the world.

The Conference was designed to bring together the theoretician and the experimentalist and to promote among them useful discussion and interchange of ideas. In this respect the meeting was voted an outstanding success by all participants, and this book is similarly aimed at stimulating scientists interested in related fields.

The meeting was also prompted by new developments in nearly all aspects of the combined subject: both experimental advances and recent considerations given to theoretical interpretation. In short, the field has run through the phases that are common to many scientific subjects—oversimplification followed by theoretical awareness and experimental "doldrums"—now to emerge on the threshold of the exciting, a point at which we pay tribute to the foundations, both experimental and theoretical, solidly laid by the late H. W. B. Skinner and by H. Jones and Sir Neville F. Mott.

In shaping the book, I have chosen a broad arrangement, with introductory interpretation and experimental results in Parts 1 and 2, which respectively discuss the light metals and the heavier metals and alloys, and electronic-structure calculations in Parts 3 and 4, which cover one-electron band theory followed by detailed effects of electron–electron interaction.

Discussion as such has not been reproduced. This was announced policy, chosen to avoid the inhibiting effect that anticipated publication often has on the discussion itself. The policy succeeded. For this reason contributors have been invited to express, in a few instances, some of their more tentative ideas not necessarily thrashed out completely but constituting important material that arose from discussion. I believe that the book therefore gains considerably from being not just a reprinting of material as presented at the meeting but instead a carefully ordered and edited arrangement of text written in the "light" of the conference.

The success of a specialist symposium of this kind depends, to a very large extent, upon the participants and particularly upon the contributed material. It is a pleasure to thank all who have helped, often spontaneously, in this respect and in shaping this book. I should like in addition to express gratitude to: the Science Research Council for a grant in support of experimental work that led us to plan and organize the symposium; The Royal Society for financial support of the meeting; the University Principal, Dr. S. C. Curran, for sponsoring the conference and for his guidance; The University Court for financial assistance; and Professor E. C. Ellwood for his unstinting support and encouragement.

Finally, I should like to record the invaluable assistance and enthusiasm of my friends and colleagues L. M. Watson, R. K. Dimond and G. M. Lindsay, in organizing the conference; and the help and encouragement in editing the following material given by G. A. Rooke and C. A. W. Marshall.

July, 1968   DEREK J. FABIAN
Strathclyde

Contents

Part I

LIGHT-METAL SPECTRA
AND THEIR INTERPRETATION

The Interpretation of X-Ray Band Spectra

G. A. ROOKE

Solid State Physics Division, A.E.R.E., Harwell, England [†]

ABSTRACT

X-Ray band-spectra are discussed with the purpose of determining the accessible information of interest in the theory of metals. Plasmon satellites are used to estimate plasmon energies, and arguments are presented to show that the cut-off parameter β is larger than is normally expected and that exchange and correlation effects are often negligible. This allows emission-bandwidths to be favourably compared with theoretical values and estimates of effective potentials to be made. It is shown that the derivation of densities of states from the spectra is not possible, but that intensity distributions may be useful as a yardstick against which to compare theoretical calculations. Finally, emission spectra of light metals are discussed individually to illustrate the extent to which each is understood. It is concluded that their plasmon satellites have been correctly identified, that the bandwidths are qualitatively well explained, and that in most cases the band shapes show semi-quantitative agreement with theoretical values.

1. INTRODUCTION

Soft x-ray spectroscopy made a spectacular start with the first relative-intensity measurements reported by Skinner in 1934. Spectra showing Fermi-edges, van Hove discontinuities and nearly free-electron-like bandwidths were observed, in addition to predicted shapes for the bottom (low-energy) regions of the emission-bands. During the "Skinner" period considerable consolidation occurred. Emission and absorption spectra from a number of metals were recorded, and the low-energy tails and high-energy satellites were explained and the L_1–L_3 lines identified. The work done during this period is well summarized in the review papers by Skinner (1940) and Tomboulian (1957).

Unfortunately, in the last decade, confidence in these interpretations has been partly destroyed by a series of small but important events. The shape of the lithium spectrum was shown to be anomalous (Crisp and Williams, 1960a). Pines (1955) commented that electron-correlation causes the apparently fortuitous agreement between theoretical and experimental bandwidths to be lost. The possibility that the core-vacancy perturbs the valence band has also tended to bias some judgements.

[†] Present address: Metallurgy Department, University of Strathclyde, Glasgow, Scotland.

The effect of such doubts was to divert attention to the Fermi-surface experiments, which are well explained theoretically and have provided adequate data with which to compare theoretical results. Computers have made band-structure calculations available with sufficient accuracy for comparison with Fermi-surface results, but a theory which would predict binding energies, and hence crystal structure of metals and alloys, has yet to be formulated. The latter requires exact knowledge of the behaviour of *all* electrons, and not just those at the Fermi-surface. Since most filled states lie below the Fermi-surface, experimental information from these states is only directly obtained by some form of optical-property measurement, of which x-ray spectroscopy may prove to be the best.

In this paper we discuss generalizations that can be made from calculations of the expected shape of the aluminium band-emission spectrum. Included in the general aim of the article, which is an attempt to bring experiment and theory into closer agreement, are two specific objectives. One is to indicate lines of experimental research worthy of more detailed attention; the other is to show that it would be valuable for theoreticians in the course of band-structure calculation to estimate transition probabilities. Several different approximations may be used, depending on the nature of the information available. In most cases the interpretation of band-emission spectra is qualitatively understood, and therefore spectra from alloys should provide important information which is at present overlooked by theoreticians.

The scope of this article has been restricted by some important limitations. One restriction is that absorption spectra will not be discussed: first because the experiments are harder to carry out due to the difficulty of obtaining suitable x-ray sources consequently the range of published results is limited; second because the processes involved are not as well understood as those for emission spectra. This second reason makes the interpretation more interesting from the point of view of those who study x-ray processes, but may not rapidly lead to the use of x-ray absorption-spectra in the study of metals and alloys.

Another restriction is that much of the theory to be discussed is only applicable to the light metals. This is because the spectra of the heavy metals are obtained mostly from those metals that have a definite *d*-band in the filled part of their valence band. To date, the spectra from these metals show only a limited number of features which are easily interpreted.

In order to provide a consistent and coherent description of spectral shape, the electron density N_0 is expressed in terms of a characteristic radius r_s which is normally in atomic units

$$\frac{4}{3}\pi r_s^3 a_0^3 = \frac{1}{N_0} = \frac{1}{NZ} \qquad (1)$$

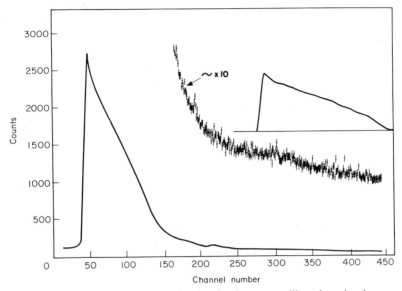

FIG. 1. Sodium $L_{2,3}$-emission spectrum showing the plasmon satellite enlarged and compared against the parent band, shifted 5·9 ev towards lower energies.

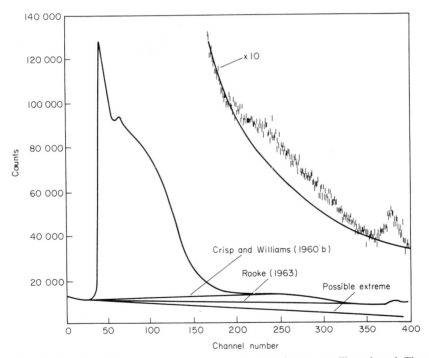

FIG. 2. Aluminium $L_{2,3}$-emission spectrum showing the plasmon satellite enlarged. Three different possible estimates of the background are also shown.

where a_0 is the Bohr radius, N is the atom density, N_0 is the electron density and Z is the valency.

2. PLASMON SATELLITES

A. Satellite Position

If, with the emission of the photon, the electron system is left in oscillation, the photon will appear with an energy loss equal to the amount taken up by the oscillation. The oscillations are subject to quantum conditions and the energy losses are discrete and approximately given by

$$\hbar\omega_p = 47 \cdot 11/r_s^{3/2} \tag{2}$$

These oscillations are called plasma oscillations, or plasmons, and have been discussed by Pines (1955) and more recently by Raimes (1963) and

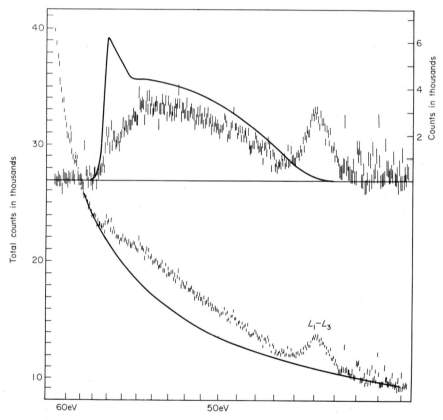

Fig. 3. Aluminium plasmon satellite showing the extrapolated tail of the parent spectrum and, above, the plasmon satellite with this tail subtracted, and compared against the parent band shifted 15·3 ev towards lower energies. (Reproduced with permission from Rooke, 1963.)

Pines again (1964). Ferrell (1957) predicted that satellites would be seen on the low-energy side of band-emission spectra. He predicted that the high-energy edge of the satellite would be lower in energy than the high-energy edge of the parent band-spectrum by the minimum energy-loss of approximately $\hbar\omega_p$. These plasmon satellites were first observed experimentally by Rooke (1963). The raw data for sodium and aluminium satellites are shown together with raw data for the parent spectra in Figs. 1 and 2, and modified data for the satellites of aluminium and magnesium are reproduced in detail in Figs. 3 and 4. The plasmon satellite for beryllium has also been recorded along with that for magnesium by Watson, Dimond and Fabian (this Volume, p. 56). Table I shows the comparison of the measured energy-

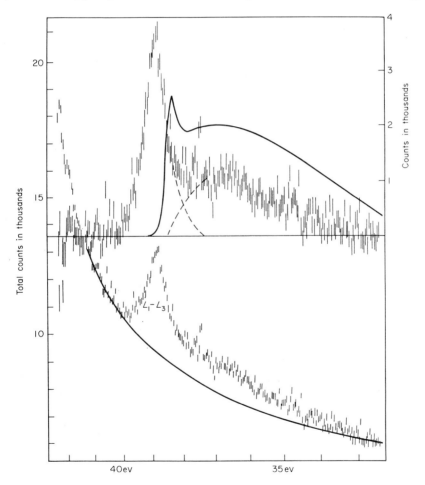

FIG. 4. Magnesium plasmon satellite showing the extrapolated tail of the parent spectrum and, above, the plasmon satellite with this tail subtracted, and compared against the parent band shifted 10·6 ev towards lower energies. (Reproduced with permission from Rooke, 1963.)

loss values determined by various methods with $\hbar\omega_p$. We can see that the agreement is good, indicating that these satellite emission-bands from the pure metal are indeed plasmon satellites.

TABLE I. Comparison of parameters used to interpret x-ray band emission-spectra

	Li	Be	Na	Mg	Al	K
$r_s^{(1)}$	3·2662	1·8916	3·993	2·6556	2·0737	4·84
$\hbar\omega_p$ (Theoretical)[2]	7·979	18·104	5·90	10·884	15·772	4·4
ΔE (Elect. energy loss)	7·12[3]	18·7[4]	5·85[5]	10·6[6]	15·3[7]	4·05[5]
ΔE (Opt. properties)	8[8]	18·4[9]	5·49[10]	10·34[11]	14·9[12]	4·25[10]
ΔE (Plasmon satellites)	—	18·2±0·5[13]	5·5±0·2[14]	10·6±0·5[14]	14·9±0·1[14]	—
β (Ferrell)[15]	0·84	0·71	0·95	0·81	0·74	1·02
β (Bohm and Pines)[16]	0·63	0·48	0·70	0·57	0·50	0·77
β (Elect. energy loss)	0·8±0·2[3]	0·67±0·05[4]	0·67±0·2[3]	0·8±0·1[4]	0·86±0·15[17]	1·2±0·2[3]
β (Plasmon satellites)	—	—	<1[14]	0·9±0·1[18]	0·8±0·04[18]	—
β (Auger-broadening)	—	—	0·82[19]	—	0·57[20]	—
$[E_F - E_0]$ (Free electron)[21]	4·691	13·987	3·139	7·097	11·638	2·14
$[E_F - E_0]$ (Band structure)	3·42[22]	11·38[23]	3·31[22]	7·03[24]	11·28[25]	2·20[22]
$[E_F - E_0]$ (Pseudo-potential)[26]	3·95	10·94	3·14	7·03	11·19	2·16
$[E_F - E_0]$ (Soft x-rays)	3·0±0·1[27]	13·8±1·0[28]	2·6±0·3[29]	6·84±0·05[13]	11·3±0·05[30]	1·62±0·04[31]
$[E_F - E_0]$ (Pines)[32]	3·9	12·1	4·2	8·9	14·2	—
$[E_F - E_0]$ (Ferrell)[33]	2·7	9·9	2·9	7·8	12·2	1·76

[1] Values derived from the lattice spacings given by ASTM (1967) and using $\alpha_0 = 0·52912$ Å.
[2] From Eq. (2).
[3] Kunz (1965).
[4] Sueoka (1965).
[5] Swan (1964). For alternative values see Kunz (1965).
[6] Powell and Swan (1959a).
[7] Powell and Swan (1959b).
[8] Wood and Lukens (1938).
[9] La Villa and Mendlowitz (1965).
[10] Mayer and Hietel (1965).
[11] Hunter (1964).
[12] Arakawa et al. (1965).
[13] Watson, Dimond and Fabian, this Volume, p. 56.

[14] These results have been measured from the spectra for this paper and were not directly determined in the earlier publication (Rooke, 1963).
[15] These results have been determined using formulae given by Ferrell (1957).
[16] These results have been determined using formulae given by Pines (1955).
[17] Watanabe (1956).
[18] These results have been estimated by Brouers and Deltour (1964) from the data by Rooke (1963).
[19] Values used by Blokhin and Sachenko (1960).

[20] Value used by Rooke (1968a)
[21] Using Eq. (6).
[22] Ham (1962).
[23] Terrell (1966).
[24] Falicov (1962).
[25] Segall (1961).
[26] Weaire (1967).
[27] Crisp and Williams (1960a).
[28] Skinner (1940).
[29] Crisp and Williams (1961).
[30] Rooke (1968b).
[31] Crisp (1960).
[32] These widths are the band-structure widths added to the Bohm and Pines (Pines, 1955) corrections.
[33] These widths are as for (32) but using corrections by Ferrell (1957).

B. Satellite Shape

In the free-electron approximation, plasmons with wave vector $k = 0$ have a quantum of energy of $\hbar\omega_p$, but plasmons with non-zero wave vectors k have a larger quantum of energy which is related to $\hbar\omega_p$ by (see Ferrell, 1957)

$$\hbar\omega(K) = \hbar\omega_p[1 + \tfrac{6}{5}\gamma^2 K^2 + (\tfrac{1}{2}\gamma^2 - \tfrac{6}{35}\gamma^4)K^4 + \ldots] \tag{3}$$

where $K|k| = k/k_F$ (k_F = the magnitude of the wave vector of an electron at the Fermi-surface) and $\gamma = (E_F - E_0)/\hbar\omega_p$, and $E_F - E_0$ is the free electron bandwidth. $\omega(K)$ is almost constant for small K but increases slowly for larger K as shown in Fig. 5. In the electron gas (which, because of the electron-interactions, is probably better described as a "sea" of electrons) the life-time of a plasmon depends on its ability to screen out a disturbance. A crude estimate of the variation of the life-times with respect to K can be made by considering the phase velocities $v = \omega(K)/K$. For the plasmons, assuming $\omega(K) = \omega_p$, $v_p \propto 1/K$ and v_p decreases as K increases. For electrons, $\omega(K) \propto K^2$, and $v_e \propto K$. Hence v_e is less than v_p for small K but becomes equal to it when the plasmon dispersion curve intersects the band of possible one-electron excitations, as shown in Fig. 5; we define β as the magnitude of K when this occurs. Since the velocity of electrons at small K is less than v_p, the electrons are normally slow to screen out a plasmon but become more effective at higher K, until at β only single-electron excitations are possible.

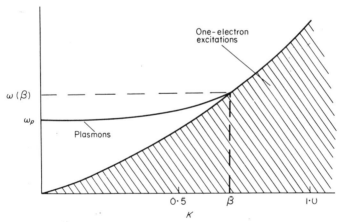

FIG. 5. Plasmon dispersion relation, shown with the band of possible one-electron excitations.

Considering only the formation of plasmons with $K = 0$, it is easily seen that the contribution to the plasmon satellite will form a low-intensity band which is approximately the same shape as the parent band but displaced to lower energies by an amount $\hbar\omega_p$. Plasmons with small K will result in an additional contribution which has a similar shape but is displaced by an energy slightly greater than $\hbar\omega_p$, and with a lower intensity due to the shorter plasmon life-time. As K increases, each incremental addition to the total plasmon satellite is displaced by a slightly greater energy and has a decreasing intensity until when $K = \beta$ the intensity is practically zero. Therefore the plasmon satellite does not have a sharp edge, like the parent emission-band, but has instead a slow rise from $\hbar\omega_p$ to somewhere in the region of an energy $\hbar\omega(\beta)$ below the parent band (see Brouers, 1964, 1965).

More detailed theory, which will be discussed in Part 4 of this Volume, has been developed by Glick and Longe (1965); the calculated curve does not show as sharp a maximum as the experimental spectrum. Hedin (1967, and this Volume, p. 337) has predicted that hole–plasmon interactions (plasmarons) should produce a low-energy fall-off near the bottom of the satellite band. The first experimental evidence of this feature is reported by Cuthill and co-authors (this Volume, p. 159). Another useful experimental result that can be compared with theory is the ratio of the satellite-intensity to the intensity of the high-energy edge of the parent band or to the intensity of the tail of the parent band at the high-energy end of the satellite. This comparison requires knowledge of the spectrometer response, and a carefully thought out estimate of the background intensity. Some of the problems of estimating background-shapes and intensities are discussed by Liefeld later in this Volume (p. 133).

C. Estimation of β

The Bohm and Pines theory (discussed by Pines, 1955) introduces the concept of β as a sharp definite value, but this is an artificial feature of the theory. However, it is useful to retain the parameter in order to present a unified approach to the description of experimental spectra, keeping in mind the fact that any sharp features associated with β may not appear as such in the experimental spectra. Brouers (1965) has shown that values of β can be obtained that make it possible to predict the correct shapes of satellites. Using the maximum in the intensity of the satellite as a lower limit for the determination of $\hbar\omega(\beta)$, estimates of β have been made from experiment and are shown in Table I.

The correlated motion of the electrons, which causes the plasma oscillation, also allows the sea of electrons to screen any electric field; the screened interaction between any two electrons is approximately given by

$$V = \frac{e}{r}\exp\left(-\mathbf{k}_c \cdot \mathbf{r}\right) \tag{4}$$

where $\mathbf{k}_c = \beta k_F$. As we shall see later, this affects the Auger broadening. Estimated values of β which have been used to predict the shapes of emission-spectra are given in Table I: β (Auger-broadening). Measurements of the scattering-angle of electrons that have suffered energy-loss have also been used to obtain estimates of β (Watanabe, 1956; Kunz, 1965). It may be seen that because of the diffuseness of associated features, the experimental values are difficult to estimate and lie somewhere between the two respective theoretical extremes estimated by Bohm and Pines (Pines, 1955) and by Ferrell (1957). Larger values of β are also required to permit a simple interpretation of the parent bandwidths. However, it must be emphasized

that β is not an intrinsic physical parameter but only a mathematical convenience. The differences between the values of β estimated from different
experiments may only be due to the associated mathematical approximations, or they may be real. Many conclusions have been drawn using definite
values of β, and these have been referred to recently by Falicov (1962) and
by Terrell (1966) who both imply that there exists considerable uncertainty
in the determination of experimental bandwidths. These uncertainties are
not necessarily well-founded, and the impression should perhaps be
countered.

3. PARENT EMISSION-BANDS

A. Basic Theory

In order to discuss the interpretation of parent spectra it is useful to
summarize the basic theoretical background. This is done briefly following
the stages of development of the theory.

(i) *Free-electron Theory*

For free-electrons the dispersion relation may be written

$$E - E_0 = \frac{50 \cdot 05}{r_s^2} K^2 \tag{5}$$

where $E - E_0$ is the band-energy of an electron in state K; energies are
given in electron volts (ev), the unit most convenient for experimentalists.
The bandwidths are given by

$$(E_F - E_0)_{FE} = \frac{50 \cdot 05}{r_s^2} \tag{6}$$

Some of the values for these bandwidths are given in Table I, as $(E_F - E_0)$
(free electron). The density of states is parabolic.

(ii) *Nearly-free-electron Theory*

The dispersion relation in the nearly-free-electron case is

$$E - E_0 = \frac{m}{m^*} \frac{50 \cdot 05}{r_s^2} K^2 \pm D(V_G, K) \tag{7}$$

Where V_G is the Gth Fourier-coefficient of the effective potential, m^*/m is
the band effective-mass ratio, and $D(V_G, K)$ is the deviation from a parabola
of the dispersion curve. The introduction of $D(V_G, K)$ is associated with the
appearance of band-energy gaps at $K = G/2$.

From Eq. 7 we obtain for the bandwidth

$$(E_F - E_0)_{NFE} = \frac{m}{m^*} \frac{50 \cdot 05}{r_s^2} + \varepsilon(D) \tag{8}$$

Where $\varepsilon(D)$ in this equation will be very small, unless a large energy-gap occurs at the Fermi-surface; Ashcroft (1963) has shown that $\varepsilon(D)$ is negligible in the case of aluminium. Values of some bandwidths calculated for nearly-free-electron-like metals by various authors are given in Table I as $(E_F - E_0)$ (band-structure). The density of states is parabolic near the bottom of the band and is a distorted parabola elsewhere. The deviations from a parabola are caused by van Hove singularities, and it has been shown (Rooke, 1968a) that for aluminium the size of these may be small. Ashcroft (this Volume, p. 249) discusses this method of determining the density of states in more detail.

(iii) Exchange and Correlation Theory

The Hartree-Fock correction assumes that all electrons with parallel spins repel one another and with a potential given by

$$V = \frac{e^2}{r}$$

This would cause them to remain highly separated in the crystal. However, the theory of electron-correlation shows that the sea of electrons provides dielectric screening, so that the potential for interaction is approximately given by Eq. (4). This reduces the long-range interactions $(r > k_c^{-1})$ and the electrons are not so rigidly kept apart. Hence we may expect any band-energy corrections introduced by the Hartree-Fock theory to be almost cancelled by dielectric screening.

According to Pines (1955), the dispersion relation when exchange and correlation are introduced will be

$$E - E_0 = \frac{m}{m^*}\frac{50\cdot05}{r_s^2}K^2 - \frac{8\cdot310}{r_s}\left(\frac{1-K^2}{K}\ln\frac{1+K}{1-K} - 2\right)$$

when $0 \leqslant K \leqslant 1 - \beta$, and

$$E - E_0 = \frac{m}{m^*}\frac{50\cdot05}{r_s^2}K^2 - \frac{8\cdot310}{r_s}\left(\frac{1-K^2}{K}\ln\frac{1+K}{\beta} - 1 + 2 + \frac{\beta^2 + 3K^2 - 1}{2K}\right) \qquad (9)$$

when $1 - \beta \leqslant K \leqslant 1$.

The correction for exchange and correlation slightly increases the band-width by an amount Δ given by

$$\Delta = \frac{8\cdot310}{r_s}\left(2 - 2\beta - \frac{\beta^2}{2}\right) + (0\cdot366 + 0\cdot852\ln\beta - 0\cdot029\beta^2) \qquad (10)$$

The factor 2 in the first term provides the correction due to exchange, and the remaining part of the first term (the factor, $-2\beta - \beta^2/2$) subtracts the effect of long-range interactions. The second term is a crude estimate of the correction due to the change in the short-range interaction, and is always small. The values of Δ given by Pines (1955) are not always small, and give densities of states similar to those obtained by Raimes (1954). In Table I, the values of bandwidths $(E_F - E_0)$ (Pines) are compared with the values $(E_F - E_0)$ (Ferrell) which are predicted using the calculations given by Ferrell (1957).

The best agreement with the experimental bandwidths is obtained for valence bandwidths derived either from the values of β calculated by Ferrell or from nearly-free-electron calculations uncorrected for exchange and correlation; because of the doubt that β is unique, the correction calculations for exchange and correlation effects can contain uncertainties which are of the same order as the correction itself, and the effects are therefore better neglected.

B. Bandwidths

(i) *Shape at the Bottom of the Emission Bands*

A theoretical density-of-states curve corrected for exchange and correlation is shown in Fig. 6. A discontinuity occurs at an energy E given by

$$E - E_0' = (1 - \beta)^2 (E_F - E_0') \tag{11}$$

which arises from the two-part nature of Eq. (9). The curves were calculated

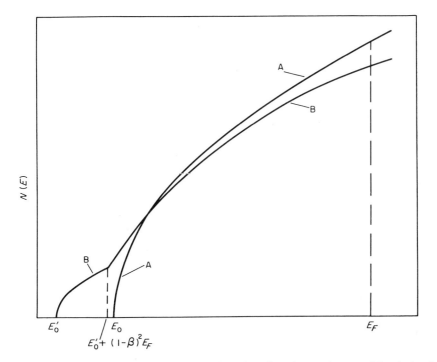

FIG. 6. Density of states for sodium: A, calculated from free-electron theory and B, calculated from free-electron theory allowing for exchange and correlation effects.

by Raimes (1954) and he has suggested that a linear rather than a parabolic extrapolation gives a better method of finding the energy of the bottom of the band. However, if larger values of β are used (typically, referring to Table I, values of 0·7 and 0·8 instead of 0·5 and 0·6) the values of $(1-\beta)^2$ are greatly reduced (to $\sim 1/16$ for the example given). This greatly reduces Δ and leaves the bottom of the density-of-states distribution almost parabolic. Density-of-states extrapolations of the $E^{1/2}$-type and $E^{3/2}$-type are therefore the more correct.

It is also possible that soft x-ray spectra do not distinguish the tails as predicted by Raimes, and this may lead to underestimation of the bandwidth. However, the errors will be considerably smaller than Raimes suggests and parabolic-type extrapolations will be more correct.

(ii) Auger Effect

If it were not for the dielectric screening, the scattering of an electron by a hole in the valence band would be enormous. Calculations of the tailing for sodium have been made by Landsberg (1949) and by Blokhin and Sachenko (1960), and for aluminium by Rooke (1968a); the results all suggest that β should be larger than the values estimated by Pines. The problems of estimating the background make comparison between theory and experiment difficult. Three different backgrounds are shown in Fig. 2. The first was used by Crisp and Williams (1960b) before the existence of the plasmon satellite was confirmed. The second was used by Rooke (1968b) on the assumption that the intensity of the plasmon satellite would fall off at lower-energies (this in fact may not occur: Hedin, 1967). The third corresponds to the opposite extreme, that the plasmon satellite is as large as possible at its low-energy end.

Rooke (1968b) has shown that the use of the parabolic-type extrapolation for finding the bottom of emission bands which have been broadened by the Auger effect, probably introduces an error of less than 0·05 ev for L-spectra of the light metals. The error may be larger for K-spectra. If the effects of the Auger broadening could be removed, the error might be reduced. Figure 7 shows the manner in which the broadening for aluminium varies with energy. Because of this variation the usual debroadening techniques cannot be applied, and some new approximate method should be sought.

(iii) Measurement of Effective Mass Ratios from Experiment

If the effects of exchange and correlation are neglected, it can be seen that bandwidths for the light metals can be estimated to an accuracy of $\sim 0·1$ ev. Hence the experimental band effective-mass ratios

$$\frac{m^*}{m} = \frac{(E_F - E_0)_{FE}}{(E_F - E_0)_{exp}} \tag{12}$$

can be estimated to within 1–2%. Most of the error comes from the estimation

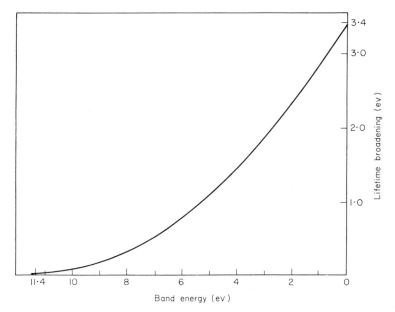

Band energy (ev)

FIG. 7. Widths of the Auger-broadening for aluminium. (Reproduced with permission from Rooke, 1968a.)

of the energy of the bottom of the band. Theoretical values for the effective-mass-ratios of pure metals have been estimated using pseudopotential theory by Weaire (1967); and bandwidths, $(E_F - E_0)$ (pseudopotential), derived using these values are also given in Table I.

(iv) *Conclusion Regarding Bandwidths*

Experimental bandwidths agree fairly well with those predicted from a nearly-free-electron model, provided that the exchange and correlation corrections are not applied. This agreement suggests that the combined effects of exchange and correlation are small for the light metals.

C. Van Hove Discontinuities

(i) *Size and Prominence*

Van Hove singularities should show up as discontinuities in emission-spectra. However, until recently, fewer discontinuities than expected have been observed in emission spectra from polyvalent metals. It has been shown (Rooke, 1968b) that discontinuities in the density of states for aluminium may be small, and this indicates that discontinuities in the band spectra could also be small.

Transition probabilities will cause many of these discontinuities to be even smaller. This can be shown by the following argument in which we refer to the Brillouin zone for aluminium (Fig. 8). Group theory suggests that the wavefunction at the point Γ should be s-like for most metals, and this has led Jones *et al.* (1934) to postulate that the bottoms of L-emission bands will fall sharply as $(E-E_0)^{1/2}$, forming a discontinuity with the background. However, in a K spectrum the emission will "see" only p-like wavefunctions, and the intensity can be shown to rise as $(E-E_0)^{3/2}$. This has zero gradient at $E = E_0$; hence no discontinuity will be observed.

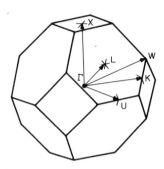

FIG. 8. Brillouin zone for a face-centred cubic metal.

At the point L, group theory predicts that one of the two states L_1 and L'_2, will be s-like and d-like, and the other p-like. Band-structure calculations (Segall, 1961) indicate that the p-state L'_2 has the lower energy. Hence the discontinuity expected at L'_2, like the discontinuity at the bottom of the band, will probably be too small to be seen in an L spectrum, particularly when Auger-broadening has occurred. The wavefunction at L_1 is s-like and d-like, and a moderate-size discontinuity is expected. At X the same situation occurs: X'_4 can be expected to be very small, whilst X_1 should be more prominent. Figure 9 shows the experimental spectra with these features indicated.

(ii) *Energies of the Discontinuities*

As seen from Fig. 9 there is good agreement between the energies at which experimental discontinuities occur and the energies (given by the labelled vertical lines) that correspond to the theoretical singularities, uncorrected for exchange and correlation, taken from Segall's band-structure calculations.

The extent of agreement between discontinuities in the magnesium emission spectrum and singularities from band-structure calculations is discussed by Watson, Dimond and Fabian in this Volume, p. 45; some of the agreement observed is contrary to that asserted by Falicov (1962), who echoes Pines regarding the measurement of bandwidths from soft x-ray spectra. Falicov also extrapolates the peak in some manner to get agreement with his results corrected for exchange correlation. When the width of the peak is measured directly, good agreement is obtained for both the peak-width and the bandwidth with Falicov's uncorrected calculations.

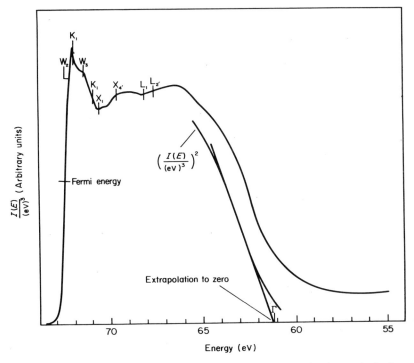

FIG. 9. Aluminium $L_{2,3}$-emission spectrum with the background subtracted, the intensity per unit energy range divided by ν^3 and vertical labelled lines showing the theoretical positions of the van Hove singularities taken from Segall (1961). (Reproduced with permission from Rooke, 1968b.)

Beryllium spectra also show discontinuities that again agree with the theoretical calculations. The results for this case are also presented and discussed in this Volume by Wiech (p. 60) and by Sagawa (p. 32).

In all cases the agreement is good only if the Bohm–Pines corrections for exchange and correlation are ignored or if larger values of β are used. This again supports the idea that soft x-ray emission-spectra can be interpreted in a simple manner.

(iii) *Measurement of Effective Potentials*

The energies of these emission-band discontinuities can be used to measure Fourier components of effective potentials. Using the four orthogonal-plane-wave method (4-OPW), it has been shown that the energies of states at certain symmetry points are related by the Fourier coefficients of the effective potential to the free-electron values (see Ashcroft, 1963). For example, for face-centred cubic (f.c.c.) metals two coefficients are required: V_{111} and V_{200}. At the point X, $(ka/2 = 1,0,0)$ we have

$$E_{X4'} - E_0 = \frac{m}{m^*} \frac{50 \cdot 06}{r_s^2} K_X^2 - V_{200} \qquad (13)$$

and, using equation (8),

$$V_{200} = E_{X4'} - E_0 - (E_F - E_0)K_X^2 = E_{X4'} - E_0(1 - K_X^2) - E_F K_X^2 \quad (14)$$

It can be seen that the larger the values attained by K_X, the less V will depend on E_0, and thus on the experimental error in E_0. Values of the coefficients V_{200} and V_{111} have been obtained for aluminium (Rooke, 1968b) and are shown in Fig. 10. This is a method of simply expressing the agreement shown in Fig. 9, and has the additional benefit of showing the nature of the Fermi surface and, hence, provides a more direct comparison with Fermi-surface techniques (Ashcroft, 1963). The rectangle in Fig. 10 indicates the errors in the measured effective-potentials and it should be noted that these are large compared with the de Hass-van Alphen results and that this type of analysis assumes that the corrections for exchange and correlation can be neglected. It is preferable, therefore, that raw data be published as well as the reduced data in case a more accurate calculation of these corrections may require a recomparison of theory with experiment.

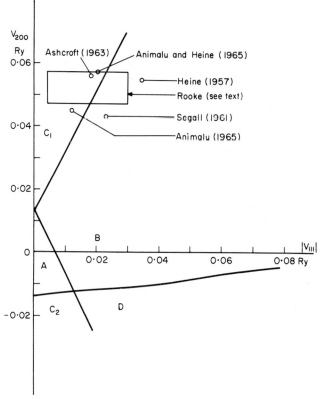

FIG. 10. Fourier coefficients of the effective potentials of aluminium obtained from the $L_{2,3}$-emission spectrum with other estimates of these coefficients. (Reproduced with permission from Rooke, 1968b.)

D. Intensity Distribution

(i) *Transition Theory*

The theory of transition probabilities for soft x-ray spectra has frequently been oversimplified. The formula for the intensity can be written as

$$I(E) \propto v^3 \int_{E = \text{const}} \frac{F(E, \hat{k})}{\nabla_k(E)} \, d^2k \tag{15}$$

where $F(E, \hat{k})$ is the square of the matrix elements of the transition probability

$$F(E, \hat{k}) = \left| \int \psi_k^* \sum_i r_i \, \psi_c \, d\tau \right|^2 \tag{16}$$

ψ_c is the wavefunction of the initial core state and ψ_k is the wavefunction of a state k in the valence band. $F(E, \hat{k})$ is a function of k, but for convenience is written as a function of the energy E and of the direction $\hat{k}$ of k.

Several simplifying assumptions may be made to facilitate evaluation and interpretation of spectral shapes. For example, the intensity relation for the spectrum may be derived by substituting a $1s$ atomic-wavefunction for the core state and using the valence-electron wavefunctions expanded as a series of spherical harmonics. The integral relationships for the spherical harmonics enable this intensity relation to be written

$$\frac{I_K(E)}{v^3} \propto \int \frac{\left| R_{(core, s)} C_p R_p r^3 dr \right|^2}{\nabla_k(E)} \, d^2k \tag{17}$$

where, $R_{(core, s)}$ is the radial part of the core wavefunction, R_p is the radial part of the second harmonic of the valence-band wavefunction, and C_p is a coefficient in the expansion. Equation (17) may be approximately written

$$\frac{I_K(E)}{v^3} \propto \int \frac{\overline{(C_p R_p)}^2}{\nabla_k(E)} \, d^2k \tag{18}$$

where the term $\overline{(C_p R_p)}$ is a weighted average or similar estimate of the terms within the modulus bars of Eq. (17). If we replace the square-of-the-weighted-average in Eq. (18) by an average-of-the-squares, we may write

$$\frac{I_K(E)}{v^3} \propto \int \frac{\int \psi_p^* \psi_p dr}{\nabla_k(E)} \, d^2k$$

$$= N_p(E) \tag{19}$$

where, ψ_p is the component of ψ with p-symmetry, and $N_p(E)$ is the density of p-states.

Similarly, for the L spectra

$$\frac{I_L(E)}{v^3} \propto \int \frac{\left| R_{(core, p)} C_s R_s r^3 dr \right|^2}{\nabla_k(E)} \, d^2k + \frac{2}{5} \int \frac{\left| R_{(core, p)} C_d R_d r^3 dr \right|^2}{\nabla_k(E)} \, d^2k \tag{20}$$

which can similarly be approximately written

$$\frac{I_L(E)}{\nu^3} \propto \int \frac{\overline{(C_s R_s)}^2 + \frac{2}{5}\overline{(C_d R_d)}^2}{\mathbf{V}_k(E)} \mathrm{d}^2 k \tag{21}$$

$$= N_s(E) + \tfrac{2}{5} N_d(E) \tag{22}$$

The factor 2/5 arises indirectly from the selection rule that allows transitions for which $\Delta m = \pm 1$ or 0. On average only 2/5 of the d-states can make a transition to a p-state, although it is not rigorously correct to use an average value and this approximation may therefore predict a distorted shape for the d-bands.

(ii) Measurement of Densities of States

In some cases (see for example Tomboulian, 1957) approximate density-of-states functions have been obtained on separating out the transition probability from Eq. (15) by removing the averaged transition probability from inside the integral; this is an even greater approximation than those described above.

We can quite easily see why this is so. Consider a surface of constant energy whose energy is just sufficient to start the filling of the second Brillouin-zone. For aluminium the states of the second zone are almost completely s-like and the transition probability for L-emission is very nearly a maximum. For the same energy, but for a different direction in k-space, such that the state lies just inside the first zone, the transition probability is almost zero since the state is mostly p-like. An average of these transition probabilities is strongly weighted by the respective areas of the constant energy surface involved, and by the spacing of these surfaces which is proportional to $\mathbf{V}_k(E)$. If the average is to be calculated properly we should use Eq. (17) which will not even separate out the so-called "density of s- and d-states" defined by Eq. (22).

Therefore it is not possible to derive an accurate density-of-states from soft x-ray spectra. Instead, functions defined by Eq. (15) are measured. These functions are in many ways related to the density of states; although where the soft x-ray intensity is increasing with energy, the density of states may decrease, and vice versa.

(iii) Estimation of Transition Probabilities

To calculate the transition probabilities, it is necessary to know the spatial distribution of the lth component of the wavefunction for a state with wave vector k. This can be estimated by using almost any of the known band-structure-calculation techniques. Altmann (this Volume, p. 265) discusses this in more detail and only a few examples are given here.

Segall (1961) used a Green-function technique, and presents some wavefunctions which are suitable only for the region outside the core. He has also shown that these are very similar to the wavefunctions obtained using symmetrized plane-waves. These wavefunctions have been used (Rooke, 1968a) to calculate transition probabilities for the L-emission spectrum from aluminium.

There would be better agreement in the region of the core if these plane-waves were orthogonalized to the core states; i.e. if orthogonal-plane-waves (OPW) were used. Ashcroft (1963) has shown that four OPW's and two Fourier coefficients of the effective potential are sufficient to determine the properties of the Fermi-surface for aluminium. Harrison describes (this Volume, p. 230) how suitable OPW's can be derived from effective potentials.

A simple estimate of the nature of the wavefunctions can be obtained from the following arguments. In an OPW calculation the wavefunction is similar to a plane-wave minus the wavefunctions of the core states. Thus the valence states have wavefunctions which can be expressed as a sum of all the unfilled atomic-state wavefunctions. For states in which we are interested, this will be similar to a sum of the wavefunctions of the first few unfilled atomic states. For example, sodium valence-electron wavefunctions will be $3s$-like with some $3p$-character and $3d$-character and perhaps some $4s$-character. The Fourier transform of an atomic state has the same symmetry as that state. It also has the same number of nodes in k space as in real space. States, $1s$, $2s$ ... ns ... etc., will have large amplitude at $k = 0$ and decrease with k, passing through $n-1$ nodes. All other states np, nd, etc., will have zero amplitude at $k = 0$ and increasing amplitude with k. Hence states with $k = 0$ are entirely s-like; at low k, the p-character increases with k^2, and the d-character with k^4, etc. The plane-wave nature will tend to flatten the k-dependence for small k but its largest effect is to make E a simple function of k and k a good quantum number.

The symmetry of the lattice can be used to simplify the wavefunctions and, in turn, this reduces the number of spherical harmonics required to represent them. As a simple example: draw any one-dimensional function of k which is symmetric about $k = 0$ (this will represent an s-function or a d-function) and repeat at some point $k = G$. Then at a Brillouin-zone boundary where $k = G/2$, the two wavefunctions $\psi(k)$ and $\psi(k+G)$ must be added or subtracted (see Ziman, 1964, p. 71). Hence the s-parts and the d-parts of the wavefunction will be zero when $\psi = \psi(k) - \psi(k+G)$, and large when $\psi = \psi(k) + \psi(k+G)$. On the other hand, antisymmetric functions (p-functions) will be large when $\psi = \psi(k) - \psi(k+G)$ and zero when $\psi = \psi(k) + \psi(k+G)$. Hence the state above the energy-gap at the Brillouin-zone boundary, must have symmetry which is either s-like and d-like, or p-like; the state below the gap will have the opposite symmetry. This illustrates the theory which enabled us, when discussing van Hove discontinuities, to state that one of the states L_1 and $L_{2'}$ will be s-like and d-like and the other p-like. Later in this Volume, Watson, Dimond and

Fabian (p. 53) show how this type of analysis has been used to predict the shape of the magnesium L-emission band and Stott and March (p. 283) show how the symmetrization and shape of the lithium wavefunctions may be responsible for the "anomalous" shape of the lithium emission spectrum.

The success of these calculations is very pleasing and theoreticians engaged in band-structure calculations should be encouraged always to include estimations of transition probabilities or emission-band shapes. We have seen that density-of-states curves cannot be accurately compared with band spectra. We can also conclude that band spectra will not provide the energies of the symmetry points with sufficient accuracy for critical comparison of various band-structure calculations with experiment. It will be interesting to note how sensitive the wavefunctions are to various physical models, and hence to see whether the relative band intensities can be used to confirm the finer details of the theory.

(iv) *Measurement of Transition Probabilities*

Transition probabilities cannot be directly derived from soft x-ray spectra for exactly the same reasons as it was found impossible to estimate a density-of-states curve: the transition probabilities must be unfolded from the $V_k(E)$ term of Eq. (15). Also, since transition probabilities are functions of three-dimensional-space coordinates, it is not possible to estimate them from a one-dimensional curve unless more information is available.

One way of providing more information is to assume that the pseudopotentials are not k-dependent. Then the transition probabilities can be represented by two parameters, the Fourier coefficients of the effective potential, which can both be determined from the one-dimensional curve.

(v) *Measurement of Effective Potentials*

The shape of the emission-band mostly depends on the nature of the wavefunctions and on the density of states, and since both can be derived from pseudopotential theory it should be possible to estimate Fourier coefficients of the effective potential from the relative band-intensities. In principle, it is possible to derive two parameters from two intensity ratios, or from three points on the spectrum; and if the emission intensities are recorded at 150 points, it should be possible to obtain 50 independent estimates of the coefficients.

At present this is not practical because it is impossible to remove the Auger-broadening with sufficient accuracy. Even if widths of the broadening-function were accurately known, we have no deconvolution procedure which can handle a broadening function with varying width. In any case the broadening functions are only approximately Lorentzian; those near the Fermi-edge must be skewed or distorted since there must be no contribution above the Fermi-level. It is clear that the removing of Auger-broadening from spectra requires much more study.

4. INTERPRETATION OF PURE-METAL EMISSION SPECTRA

A. Summary of Features of the Spectra

In order to summarize, we shall briefly discuss the spectra of pure metals and shall see to what extent they are explained by the above theories. This will help also to introduce some of the material to be presented in later chapters of this Volume.

(i) *Lithium*

The anomalous shape of the lithium K-emission spectrum is well known (Crisp and Williams, 1960a): it shows an unexplained hump below the high-energy edge. However, the bottom of the emission band shows the $E^{3/2}$ dependence. To date, no plasmon satellite has been reported.

A lithium K-spectrum is presented in this Volume by Sagawa (p. 34). Two explanations of its shape have been given: one we have already mentioned and is further discussed in Part 3 of this Volume by Stott and March (p. 283); the other attributes the high-energy intensity fall-off to the "pulling down" of energy-states due to the hole in the core state (see Goodings, 1965; Allotey, 1967).

We should expect the latter effect to be particularly large for lithium, because the loss of one electron represents a large portion of the core. Goodings has calculated that the loss of a second electron does not change the spectral shape much further and in the "double-ionization satellite" measurements made by Catterall and Trotter (1958) we should not expect to see the effect of the second "ionization" within the uncertainties of the experiment. In Part 4 of this Volume, Bergerson and Terrell (p. 000) discuss some aspects of the effect of the core-hole on the valence-band.

The Fermi-surface makes no contact with Brillouin-zone boundaries, hence there will be no van Hove singularities observed in the spectrum.

(ii) *Beryllium*

The density of states for beryllium has been calculated by several authors: recently, for example, by Terrell (1966) and Altmann and Bradley (1965). Recent measurements of the Be K-spectrum are reported in this Volume by Sagawa (p. 33), by Wiech (p. 61) and by Watson, Dimond and Fabian (p. 56). The shape of the K-emission-band roughly follows the density-of-states curve, although the hump reported at about 1·6 ev above the bottom of the band does not appear in any of the spectra. There is some doubt regarding the cause of this feature.

The first plasmon satellite ever recorded for beryllium (Watson, Dimond and Fabian, this Volume, p. 56) appears to fit the theories outlined earlier.

(iii) *Sodium*

The sodium $L_{2,3}$-emission spectrum has a parabolic form and reflects the expected density-of-states curve. It cannot be expected to show van Hove discontinuities since the Fermi-surface is spherical and far from making contact with the Brillouin-zone boundaries.

At the Fermi-edge the spectrum shows a small hump which has been attributed by Blokhin and Sachenko (1960) to Auger-broadening. This has been confirmed by the many-body calculations of Glick, Longe and Bose and others, reported on p. 351 of this Volume. It appears that while states at the Fermi-edge are broadened only very little, just below the edge are broadened to such an extent that the total intensity in this region is lowered, leaving a small peak near the edge. Plasmon-satellite data for sodium are reported in the present paper.

(iv) *Magnesium*

Magnesium $L_{2,3}$-emission spectra are discussed later in this Volume by Watson, Dimond and Fabian (p. 49)—and the K-spectrum is presented by Cauchois (p. 77). The observed plasmon satellite is well explained. The L-spectrum is parabolic near the bottom of the band, but has a broad hump further up the band which is due to the electrons in the first two zones being more p-like at higher energies. Overlap into the third and fourth zones involving s- and d-electrons produces the obvious peak at the Fermi-edge as well as a small feature on the high-energy side of the hump. Some van Hove singularities do not show up as discontinuities in the spectra because they are caused by p-like states.

(v) *Aluminium*

Aluminium L-spectra obtained by Wiech (p. 62) and Sagawa (p. 33), and the K-spectra obtained by Cauchois (p. 76) and by Läuger (reported in this Volume by Wiech, p. 62) form an interesting series on aluminium presented in the papers to follow. The plasmon satellite is readily observed and well explained.

There is some doubt regarding the shape of the emission-band and the possibility that an oxide spectrum is present. However, calculations of the shape of both L and K spectra (Rooke, 1968a) have predicted results similar to those reported elsewhere by Ellwood *et al.* (1967) and to the spectra obtained by Läuger (shown here by Wiech, p. 62).

(vi) *Potassium and Calcium*

L and M spectra for potassium have been recorded by Kingston (1951) and by Crisp (1960). Neither has an unexpected shape but no detailed interpretation has been attempted.

Calcium has also been studied by Kingston (1951). Altmann has suggested (at the meeting on which this Volume is based) that this would be a good metal on which to try a full calculation for the shape of the spectrum.

(vii) *Other Metals*

The other spectra that have been recorded are almost entirely from metals for which the d-band overlaps the valence band. The spectra of the transition metals are reviewed later in this Volume by Holliday (p. 105). An interesting calculation of the relative intensities of the nickel spectrum is reported also by Cuthill and co-authors (p. 155) and their calculation shows that the most characteristic feature seen in this spectrum can be attributed to contributions from the d-band. A very extensive analysis reported in the paper here by Liefeld (p. 133) shows that most heavy-metal spectra have overlapping double-ionization satellites, unless the spectra are recorded at threshold potential. Liefeld presents an emission spectrum for zinc, which probably shows for the first time an s,p-band. Since this emission band is very difficult to measure accurately, and the d-band structure will be so fine that it will be lost in the experimental broadening, it is difficult to see how very much information can be obtained from the spectrum. The d-bandwidths are of interest and these are discussed by Holliday (p. 105); there is also great interest in the width of the whole s,p-band, and particularly in the energy-difference between the top of the d-band and the Fermi-edge. However, these cannot be measured until the work of Liefeld has been confirmed and improved to show a definite Fermi-edge.

B. Conclusions Regarding the Interpretation of Spectra

There appears to be general agreement between experiment and theory for many of the light metals. The plasmon satellites are reasonably well explained although the effects of plasmarons have not yet been clearly observed, and more accurate measurements of the satellite intensities are required. The parent bands can probably be interpreted without resorting to the effects of the hole in the core state. These effects cannot alter the measured bandwidths, which now show a good fit with theory despite the effects of exchange and correlation, and leads to the conclusion that the net effect of exchange and correlation is probably negligible.

The band shapes agree semi-quantitatively with theory, but they may be useful only as a general guide to the electronic structure. It is impossible to estimate densities of states from spectra, and it is doubtful whether reliably accurate coefficients of the effective potential can be obtained from them. However, while many small details of the spectra remain unexplained the gross features are now mostly understood, and it should therefore be possible to explain some of the features, in particular the bandwidths, of alloy spectra.

Acknowledgements
The author wishes to thank the U.K.A.E.A. for a Research Fellowship during which the work for this paper was done and also the Science Research Council for a Research Associateship (at the University of Strathclyde) during which this work was completed and prepared for publication. The author is also indebted to Dr. D. J. Fabian and to C. A. W. Marshall for their help in the preparation, and to Dr. L. M. Watson and to R. K. Dimond for valuable discussion.

References

Allotey, F. K. (1967). *Phys. Rev.* **157**, 467.

Altmann, S. L. and Bradley, C. J. (1965). *Proc. Phys. Soc.* **86**, 915.

Animalu, A. O. E. (1965). *Phil. Mag.* (8) **11**, 379.

Animalu, A. O. E. and Heine, V. (1965). *Phil. Mag.* (8) **12**, 1249.

Arakawa, E. T., Hamm, R. N., Hanson, W. F. and Jelinek, T. M. (1965). In "Optical Properties and Electronic Structure of Metals and Alloys", ed. by F. Abelès, p. 374. North Holland Publishing Company, Amsterdam.

Ashcroft, N. W. (1963). *Phil. Mag.* (8) **8**, 2055.

A.S.T.M. (1966). "X-Ray Powder Data", ed. by V. Smith. American Society for Testing Materials, Philadelphia, U.S.A.

Blokhin, M. A. and Sachenko, V. P. (1960). *Izv. Akad. Nauk SSSR Ser. Fiz.* **24**, 397; *Bull. Acad. Sci. U.S.S.S.R., Phys. Ser.* **24**, 410.

Brouers, F. (1964). *Phys. Lett.* **11**, 297.

Brouers, F. (1965). *Physica Stat. Solidi* **11**, K25.

Brouers, F. and Deltour, J. (1964). *Physica Stat. Solidi* **7**, 915.

Catterall, J. A. and Trotter, J. (1958). *Phil. Mag.* (8) **3**, 1424.

Crisp, R. S. (1960). *Phil. Mag.* (8) **5**, 1161.

Crisp, R. S. and Williams, S. E. (1960a). *Phil. Mag.* (8) **5**. 525.

Crisp, R. S. and Williams, S. E. (1960b). *Phil. Mag.* (8) **5**, 1205.

Crisp, R. S. and Williams, S. E. (1961). *Phil. Mag.* (8) **6**, 365.

Ellwood, E. C., Fabian, D. J. and Watson, L. M. (1967). *Metals Mater.* **1**, 333.

Falicov, L. M. (1962). *Phil. Trans.* **A255**, 55.

Ferrell, R. A. (1957). *Phys. Rev.* **107**, 450.

Glick, A. J. and Longe, P. (1965). *Phys. Rev. Lett.* **15**, 589.

Goodings, D. A. (1965). *Proc. phys. Soc.* **86**, 75.

Ham, F. S. (1962). *Phys. Rev.* **128**, 2524.

Hedin, L. (1967). *Solid State Commun.* **5**, 451.

Heine, V. (1957). *Proc. Roy. Soc.* **A240**, 361.

Hunter, W. R. (1964). *J. Opt. Soc. Amer.* **54**, 208.

Jones, H., Mott, N. F. and Skinner, H. W. B. (1934). *Phys. Rev.* **45**, 379.

Kingston, R. H. (1951). *Phys. Rev.* **84**, 944.

Kunz, C. (1965). *Phys. Lett.* **15**, 312.

La Villa, R. E. and Mendlowitz, H. (1965). *Appl. Opt.* **4**, 955.

Landsberg, P. T. (1949). *Proc. Roy. Soc.* **A62**, 806.

Mayer, H. and Hietel, B. (1965). In "Optical Properties and Electronic Structure of Metals and Alloys", ed. by F. Abelès, p. 47. North Holland Publishing Company, Amsterdam.

Pines, D. (1955). *Solid State Phys.* **1**, 367.

Pines, D. (1964). "Elementary Excitations in Solids". Benjamin, New York.

Powell, C. J. and Swan, J. B. (1959a). *Phys. Rev.* **116**, 81.

Powell, C. J. and Swan, J. B. (1959b). *Phys. Rev.* **115**, 869.

Raimes, S. (1954). *Phil. Mag.* (7) **45**, 727.

Raimes, S. (1963). "The Wave Mechanics of Electrons in Metals". North Holland Publishing Company, Amsterdam.

Rooke, G. A. (1963). *Phys. Lett.* **3**, 234.

Rooke, G. A. (1968a). *J. Phys.* C, **1**, 767.

Rooke, G. A. (1968b). *J. Phys.* C, **1**, 776.
Segall, B. (1961). *Phys. Rev.* **124**, 1797.
Skinner, H. W. B. (1940). *Phil. Trans.* **A239**, 95.
Sueoka, O. (1965). *J. Phys. Soc. Jap.* **20**, 2203.
Swan, J. B. (1964). *Phys. Rev.* **135**, A1467.
Terrell, J. H. (1966). *Phys. Rev.* **149**, 526.
Tomboulian, D. H. (1957). *Handb. Physik* **30**, 246.
Watanabe, H. (1956). *J. Phys. Soc. Jap.* **11**, 112.
Weaire, D. (1967). *Proc. Phys. Soc.* **92**, 956.
Wood, R. W. and Lukens, C. (1938). *Phys. Rev.* **54**, 332.
Ziman, J. M. (1964). "Principles of the Theory of Solids". Cambridge University Press.

Soft X-Ray Emission and Absorption Spectra
of Light Metals, Alloys and Alkali Halides

T. Sagawa

Department of Physics, Tohoku University, Japan

ABSTRACT

Soft x-ray emission spectra of the metals Li, Be and Al, and absorption spectra of Be, Al, Al–Mg alloys and of the alkali halide NaCl, are presented. The emission spectra were obtained with a spectrometer employing a recently constructed ultra-high-vacuum x-ray tube, while the absorption spectra were measured with either an orbital radiation synchrotron (at Tokyo) or a quasi-continuum source of the *Vodar* type. Some discussion of the results is given.

1. INTRODUCTION

Soft x-ray spectroscopy is one of the most important fields of physics for providing basic information about the electronic structure of solids. However, its development has been slow due to various experimental difficulties, although recently remarkable progress in experimental techniques has been made.

Photon-counting appreciably enhances sensitivity for the detection of radiation (compared with a photographic plate) and ultra-high-vacuum techniques make possible more stable x-ray sources giving radiation of higher intensity. Further, a new type of radiation source, using an electron synchrotron, has features which are outstanding in respect of radiation intensity, continuity of the spectrum, polarization of radiation, etc.

In this paper some experiments on the emission and the absorption of soft x-radiation by metals, by alloys and by insulators will be described; new techniques were employed to enhance the sensitivity and the accuracy of the measurements. The results obtained are more refined than those reported in earlier investigations, and comparison with the relevant theory appears to be more meaningful.

The results described are emission spectra of lithium, beryllium and aluminium, obtained with an apparatus employing an ultra-high-vacuum x-ray tube operated with low-input power to the target, and absorption spectra of beryllium, aluminium, and aluminium–magnesium alloys,

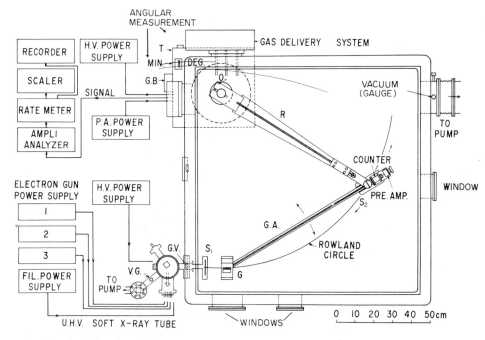

FIG. 1. A plan view of the spectrometer and block diagrams of the various power supplies.

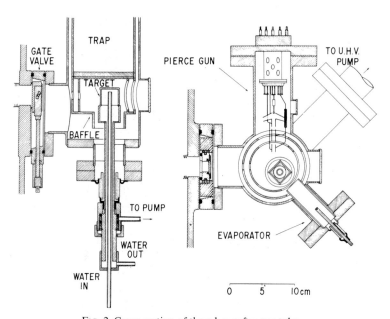

FIG. 2. Cross-section of the u.h.v. soft x-ray tube.

measured with either an orbital radiation synchrotron or a quasi-continuum source of the Vodar type.

Structure in the Be K-emission and Al $L_{2,3}$-emission bands are interpreted in terms of van Hove singularities in the Brillouin zone for the valence band, and sharp absorption bands found in the $L_{2,3}$-absorption spectrum of Na^+ are attributed to spin-orbit splitting of an x-ray exciton.

2. VALENCE-BAND EMISSION SPECTRA OF METALLIC LITHIUM, BERYLLIUM AND ALUMINIUM

A. The Instrument

(i) *Spectrometer Chamber*

A plan view of the spectrometer is shown in Fig. 1 together with a block diagram of the related circuitry. An incidence angle of $87.5°$ is used, with a concave glass diffraction-grating of radius of curvature 2m and 1 152 lines/mm. The holders for the entrance slit S_1 and the grating G are mounted on a flat table, on which the detector carriage moves. The detector scans a spectrum by rotation of the arm R about the axis O; this axis stands on an optical circular table T. The diffracted angle can be read directly on this table to within a relative error of 5 sec of arc. The plane of slit S_2 is kept always at right angles to the diffracted rays by means of a guide-arm GA, which is pivoted just below the grating pole and is driven by a synchronous motor via gearbox GB.

(ii) *X-Ray Tube*

The soft x-ray tube, shown in Fig. 2, is attached to the spectrometer through a thin gate-valve GV (Fig. 1) of 2 cm bore. Flanges are sealed with gold O-rings. The target is water-cooled and is surrounded by a copper baffle cooled by a liquid nitrogen trap just above the target. Opposite the electron gun, which is of the Pierce-type, is installed a small furnace for evaporating target material. The pressure is continually measured by a Beyard-Alpert

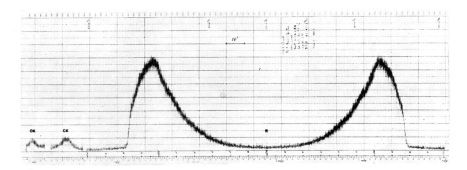

FIG. 3. An original chart record for the Be K-emission band.

gauge, and the tube attains a vacuum of 3×10^{-9} torr after baking; under operating conditions the vacuum is of the order of 10^{-8} torr.

(iii) *Detector*

Most of the work reported here was performed with a gas-flow proportional counter of the type used by Holliday (1967). A 50:50 argon–methane mixture (P-10 gas) or an argon–ethanol mixture (33:67) was employed in the counter.

B. Experimental Results

Typical chart records of the beryllium K-emission spectrum are shown in Fig. 3. The scanning is from right to left and R indicates a reversal of the traverse direction of the detector; tube current was 2·3 mA and the vacuum was $\sim 4\cdot5$–$8\cdot0 \times 10^{-8}$ torr. Count-rate at the peak was ~ 680/sec.

To check the extent of oxidation and carbon deposition on the target surface, the K-emission bands for oxygen and carbon were recorded after a return scan-time of about 50 min required for a Be K-emission band-spectrum.

Interpretation of this emission band as a density-of-states curve, is made from the original chart record using the following relationships:

$$I(E)/v^2 \propto n(\theta)/v^5 \propto \overline{|Mik|^2}\, N(E) \tag{1}$$

where $I(E)$ is the radiation intensity per unit photon-energy at energy E, $n(\theta)$ is the count-rate when the diffraction angle is 90–θ, and Mik is the matrix element for the transition i–k. $I(E)$ curves, derived from a mean curve averaged over many recordings, are shown in Fig. 4 for the Li K spectrum, in Fig. 5 for the Be K spectrum and in Fig. 6 for the Al $L_{2,3}$ spectrum.

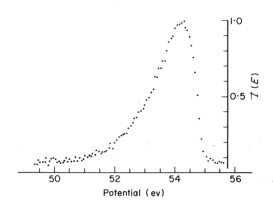

FIG. 4. $I(E)$ curve for K-emission band of metallic lithium.

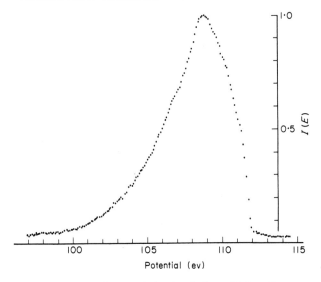

FIG. 5. $I(E)$ curve for K-emission band of metallic beryllium.

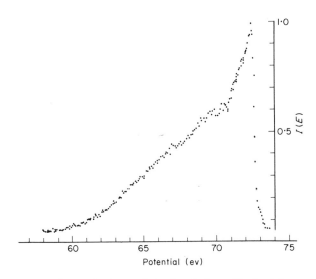

FIG. 6. $I(E)$ curve for $L_{2,3}$-emission band of metallic aluminium.

C. Discussion

(i) K-*Emission Spectrum of Metallic Lithium*

Our result does not show any important differences from those obtained under normal vacuum conditions. The unusual profile near to the Fermi-edge is still observed (Fig. 7), and compares well with the theoretical $N(E)$ curve of Ham (1962), calculated using the Green-function method. In order to interpret this profile, it is necessary to examine certain collective effects of the electrons such as, for example, electron–hole scattering resonance proposed recently by Allotey (1967), and plasmons or excitons. The curve calculated by Allotey is shown in Fig. 7.

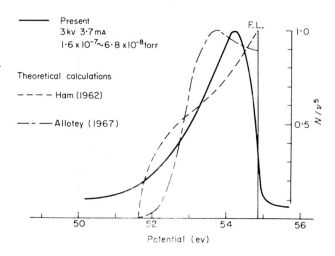

FIG. 7. Comparison of the N/v^5 curve for the K-emission for metallic lithium with the theoretical $N(E)$ curves of Ham (1962) and Allotey (1967).

(ii) K-Emission Spectrum of Metallic Beryllium

Figure 8 shows a variation in emission-band profile obtained with different tube operating conditions; the solid line is the spectrum obtained under ultra-high-vacuum (u.h.v.), and low target-input ($\sim 10w$) and the dotted line is the spectrum obtained under vacuum of 10^{-6} torr and target input of $\sim 200w$. Differences between the profiles are noticeable in respect of the height of the emission-edge, the bandwidth, and the tailing at both low and high energies.

The $N(E)$ histograms for beryllium, calculated in detail by Loucks and Cutler (1964) using the orthogonalized-plane-wave method (OPW), and by Terrell (1966) using an augmented-plane-wave calculation (APW), are shown in Fig. 9 and are compared with our $n(0)/v^5$ curve. General agreement between theory and experiment is good, considering the fact that the K-emission corresponds only to that part of the density of states of the valence band which has p-symmetry. In particular, the positions of the edge and the peak agree well with the OPW curve. The intensity on the low-energy side is lower than predicted in the theoretical curve, and this is clearly due to the s-like character of the valence band in this energy range; this can be seen from the $E(k)$ curves obtained by Loucks and Cutler (Fig. 10).

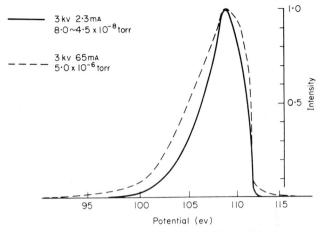

FIG. 8. Variation in *K*-emission spectrum of metallic beryllium, on changing the tube operating-conditions.

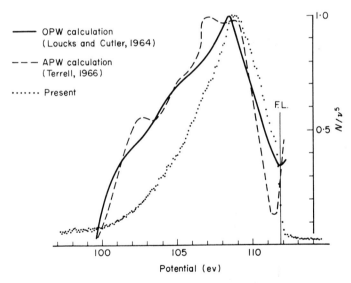

FIG. 9. Comparison between N/v^5 curve and the theoretical $N(E)$ curves (Loucks and Cutler, 1964; Terrell, 1966) for metallic beryllium.

A small kink at G (Fig. 10) is also indicated in the experimental curve and suggests an overlap of two $E^{3/2}$-curves starting from the point Γ_1^+ and A_1. Discontinuities appearing in the emission band are indicated by capital letters in alphabetical order. Some may be assigned to van Hove singularities as shown in Table I.

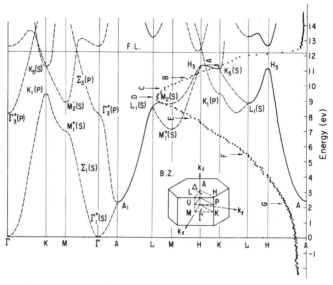

FIG. 10. Comparison between N/v^5 curve and the $E(k)$ diagram (Loucks and Cutler, 1964) for metallic beryllium.

TABLE I. Comparison of structures in the Be K-emission spectrum with points of singularity in the theoretical band structure

Characteristic point in the spectrum	Band-state	Type of van Hove singularity	Experimental energy-difference from the Fermi level (ev)
A	H_3	M_3	0·9
D	K_1	M_2	2·9
E	Γ_3^+	M_1	4·4

(iii) $L_{2,3}$-*Emission Spectrum of Metallic Aluminium*

Remarkable differences are found in the results reported by various workers (Crisp and Williams, 1960; Wiech, 1966), as shown in Fig. 11. From our experience it seems that the lower the vacuum, the higher the hump at 67 ev. The rather large hump in the profile obtained by Crisp and Williams might be ascribed to the effect of oxidation, but it is difficult to account for the marked difference observed in the result obtained by Wiech also using u.h.v. conditions. There is a possibility, however, in the instrument used by Wiech, of a reflection anomaly at the surface of the grating which is coated with aluminium.

The bandwidth of 10·5 ev, determined by a best-fit parabola, corresponds well to the theoretical value of 11·4 ev predicted by Segall (1961) using a

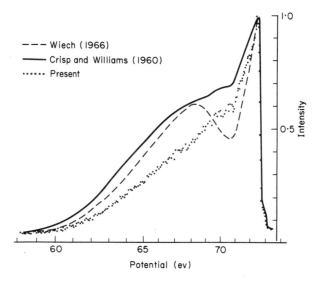

FIG. 11. Comparison between Al $L_{2,3}$-emission spectra reported by various workers.

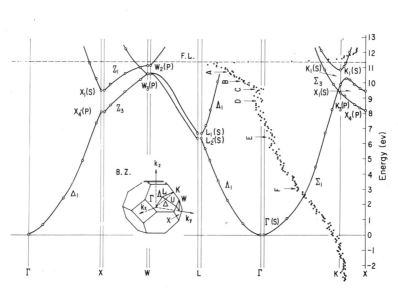

FIG. 12. Comparison between N/v^5 curve and the $E(k)$ diagram (Segall, 1961) for metallic aluminium.

Green-function calculation. In Fig. 12, our $n(\theta)/v^5$ curve is compared with the $E(k)$ curves obtained by Segall. As with beryllium, the fine-structure in the emission band is indicated by capital letters and respective assignments to van Hove singularities are examined in Table II.

TABLE II. Comparison of structures in the Al $L_{2,3}$-emission spectrum with points of singularity in the theoretical band structure

Characteristic point in the spectrum	Band-state	Type of van Hove singularity	Experimental energy-difference from the Fermi level (ev)
A	K_1 (3rd zone)	M_1	0·8
B	The peak between K and X	M_1	1·4
C	X_1	M_0	1·9
E	L_1	M_0	5·3

Details of the results described above are reported by Sagawa and Aita (1968).

3. SOFT X-RAY ABSORPTION SPECTRA OF METALS, ALLOYS AND ALKALI HALIDES

Most of the absorption measurements reported previously were made using line-spectra emitted from either a vacuum spark or a gas discharge. These sources are inadequate for high resolution spectroscopy due to the limited number of spectral lines they provide, and suffer also from the draw-back that slits and the absorbing specimen are easily contaminated by sputtered dusts. A synchrotron radiation-source is entirely free from these disadvantages and moreover has the following advantages: (a) the radiation is the strongest among soft x-ray sources now available; (b) its continuum has a perfectly smooth profile; (c) the profile can be predicted exactly by classical electrodynamics (Schwinger, 1949); (d) the radiation is highly polarized in the orbital plane; (e) the radiation constitutes very nearly parallel rays. These features make it very promising as a light-source for solid-state spectroscopy; e.g. for measurements of absorption, reflection, fluorescence-emission, photoelectric effects and optical constants.

In this report some absorption data obtained with the Tokyo synchrotron as a source (Sagawa et al., 1966a) are presented, together with absorption spectra of sodium halides obtained with a Vodar-type source (Damany et al., 1966). The absorption measurements were made with grazing-incidence spectrographs.

A. Soft X-Ray Absorption Spectra of Metallic Beryllium and Aluminium

(i) Beryllium K-Absorption

This is shown in Fig. 13, where it is compared with the data reported by Johnston and Tomboulian (1954) and with the theoretical $N(E)$ curve obtained by Loucks and Cutler (1964). The marked difference of the present result from that of Johnston and Tomboulian is clearly due to an insufficient number of lines per unit wavelength emitted from the x-ray source employed by these investigators. The general behaviour of our absorption profile is very near to the theoretical $N(E)$ curve.

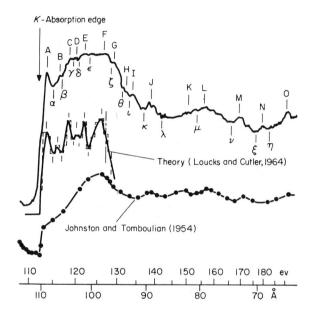

FIG. 13. *K*-Absorption spectrum of metallic beryllium (reproduced with permission from Sagawa *et al.*, 1966a).

(ii) *Aluminium* $L_{2,3}$-*Absorption*

The $L_{2,3}$ spectrum of metallic aluminium is shown in Fig. 14 where the K spectrum is shown also for comparison; the anomalies due to reflection by the glass grating show up well. It is interesting to note, that as far as energy distances from edges are concerned, absorption maxima correspond well with one another. The most remarkable difference of the $L_{2,3}$ from the K spectrum is the large and broad maximum at 96 ev, denoted by F. A similar feature is observed in the $L_{2,3}$-absorption of the oxides and alloys of aluminium and also in the energy-loss spectrum of high-energy electrons. It may be caused by a delayed onset of atomic transitions of a $2p \rightarrow nd$ type.

B. $L_{2,3}$-Absorption for Aluminium–Magnesium Alloys

The aluminium and magnesium $L_{2,3}$ spectra obtained for Al–Mg alloys are shown in Fig. 15; the absorber was prepared as layers, such as Al(300 Å)–Mg(500 Å–Al(500 Å). It should be noted that the discontinuous jump at 170·56 Å, expected in the region of the Al $L_{2,3}$-absorption edge, is very poorly defined. Samples containing lower concentration of magnesium than the sample whose spectrum is illustrated, show similar absorption

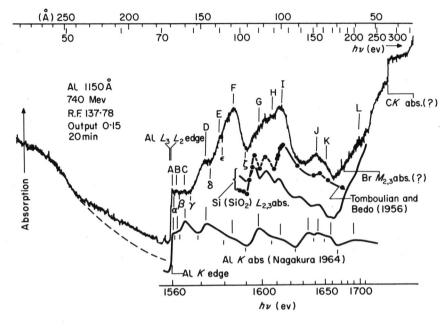

FIG. 14. $L_{2,3}$-Absorption spectrum of metallic aluminium (reproduced with permission from Sagawa *et al.*, 1966b).

curves though the Al $L_{2,3}$ absorption-edge gradually recovers with increasing content of aluminium. The results suggest that alloying is in progress in these composite films and that variation of electronic structure takes place. Similar effects have been noted by Crisp and Williams (1960) in the L-emission spectrum of aluminium evaporated on magnesium. The broad maximum at 96 ev (130 Å), noted previously, is still observed (Fig. 15); on the other hand, the $L_{2,3}$-absorption edge for magnesium is seen sharply.

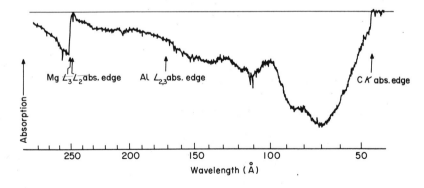

FIG. 15. $L_{2,3}$-Absorption spectra for both elements in Al–Mg alloy.

These results are qualitatively consistent with the $L_{2,3}$-emission spectra for more concentrated Al–Mg alloys, reported by Appleton and Curry (1965).

C. Soft **X**-Ray Absorption Spectra of Alkali Halides

The chloride ion (Cl^-) $L_{2,3}$-absorption spectra for various alkali chlorides, LiCl, NaCl, KCl, RbCl and CsCl, have been measured with the use of synchrotron radiation (at Tokyo) at both room temperature and liquid nitrogen temperature. In addition the Na^+ $L_{2,3}$-absorption spectra of sodium halides, NaCl and NaBr, have been obtained using a quasi-

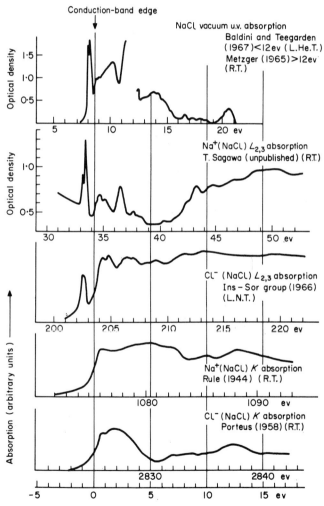

FIG. 16. Comparison between various soft x-ray and vacuum-u.v. absorption spectra of NaCl. (L.He.T. = liquid helium temperature; R.T. = room temperature; L.N.T. = liquid nitrogen temperature.)

continuum source of the *Vodar*-type (at Sendai). Details of these results are reported elsewhere (Nakai and Sagawa, 1968) and are not presented here except for the case of NaCl which is taken as a typical example.

Various soft x-ray and vacuum ultraviolet (vacuum-u.v.) absorption spectra for NaCl are compared in Fig. 16, where the Cl$^-$ K-absorption and $L_{2,3}$-absorption spectra (respectively Parratt, 1959; Sagawa *et al.*, 1966a), the Na$^+$ K-absorption and $L_{2,3}$-absorption spectra (respectively Rule, 1944; the present work), and the vacuum-u.v. absorption spectra (Metzger, 1965; Baldini and Teegarden, 1967), are compared on a normalized energy scale. The life-time broadening plus instrumental resolution are about 0·01 ev for the Na $L_{2,3}$, 0·1 ev for the Cl$^-$ $L_{2,3}$, 0·6 ev for the Na$^+$ K, and 0·7 ev for Cl$^-$ K spectra; so that the Na$^+$ $L_{2,3}$-absorption data should provide the most detail among these spectra.

There are some similarities between the vacuum-u.v. and the $L_{2,3}$ spectra; especially for Na$^+$. The peaks A and B in the Na$^+$ $L_{2,3}$, and the peak A in the Cl$^-$ $L_{2,3}$ spectra, are due to x-ray excitons. The observed intensity-ratio of the peaks for the Na$^+$ $L_{2,3}$ spectrum is about 1:5 (A:B) and their separation is 0·27 ev. These values are surprising since a splitting due to a spin-orbit interaction in the Na $L_{2,3}$ states for the free atom, should result in an intensity-ratio of 2:1 and a separation of 0·20 ev. This discrepancy can be reasonably explained by exchange interaction between the excited electron and the inner positive-hole. In practice, the theory of Onodera and Toyozawa (1967) for the excitons of alkali halides gives, for the ratio of the oscillator strength for the peaks A and B, the relation

$$I_B/I_A = \cot^2{(\gamma - \phi)} \tag{2}$$

and for their energy separation

$$E_B - E_A = [(\Delta - \lambda/3)^2 + 8\lambda^2/9]^{1/2} \tag{3}$$

which gives

$$[(E_B - E_A)^2 - 8\lambda^2/9]^{1/2} + \lambda/3. \tag{4}$$

Δ is the exchange energy; λ is the spin-orbit splitting in the free atom; $\gamma = \arctan{\sqrt{2}}$; and ϕ is given by the relation, $\tan{2\phi} = 2\sqrt{2}\Delta/(3\lambda - \Delta)$.

In our case with $\lambda = 0·20$ ev and $E_B - E_A = 0·27$ ev, the values obtained for Δ and for I_B/I_A are respectively 0·26 ev and 6·0 ev, which agree well with the observed results.

In the K-absorption spectrum no appreciable exciton peaks are found. The system may be represented qualitatively, by a simple hydrogen-like model. An excited electron moving in an effective Coulomb field, due to the inner hole, has an infinite series of bound states converging to a state of zero binding energy. If the excited electron has p-symmetry as for the cases of vacuum-u.v. and $L_{2,3}$-absorption, the only allowed transitions are to states with s-symmetry ($1s$, $2s$ ..., ns, etc.). Then the transition to a $1s$-state gives rise to the first exciton peak, isolated from others, while the transitions to the ns states ($n \geqslant 2$) are smeared out by various broadening effects, such as electron–phonon interactions. Moreover, the absorption band due to transition to ns states overlaps with the $E^{1/2}$ absorption profile, starting from the bottom of the conduction band, and therefore is observed only as a modifying structure. In the case of K-absorption spectra, only the transitions to np states ($n \geqslant 2$) are allowed and are smeared out by various broadening effects. The absorption process for transitions to the conduction band results in an $E^{3/2}$ profile, as is well known, and only a gradual increase in the absorption profile is expected for the K spectra, which is consistent with experimental results.

References

Allotey, F. K. (1967). *Phys. Rev.* **157**, 467.

Appleton, A. and Curry, C. (1965). *Phil. Mag.* **12**, 245.

Baldini, G. and Teegarden, K. (1967). *Phys. Rev.* **155**, 896.

Crisp, R. S. and Williams, S. E. (1960). *Phil. Mag.* **5**, 1205.

Damany, H., Roncin, J. Y. and Damany-Astoin, N. D. (1966). *Appl. Opt.* **5**, 297.

Ham, F. S. (1962). *Phys. Rev.* **128**, 82, 2524.

Holliday, J. E. (1967). *In* "Handbook of X-Rays", ed. by E. F. Kaelble, chap. 8. McGraw-Hill, New York.

Johnston, R. W. and Tomboulian, D. H. (1954). *Phys. Rev.* **94**, 1585.

Loucks, T. L. and Cutler, P. H. (1964). *Phys. Rev.* **A133**, 819.

Metzger, P. H. (1965). *J. Phys. Chem. Solids* **26**, 1879.

Nagakura, I. (1964). *Sci. Rep. Tohoku Univ.* (I) **48**, 37.

Onodera, Y. and Toyozawa, Y. (1967). *J. Phys. Soc. Japan* **22**, 833.

Parratt, L. G. (1959). *Rev. Mod. Phys.* **31**, 616.

Rule, K. C. (1944). *Phys. Rev.* **66**, 199.

Sagawa, T. and Aita, O. (1968). *J. Phys. Soc. Japan,* in press.

Sagawa, T., Iguchie, Y., Sasanuma, M. and others (1966a). *J. phys. Soc. Japan* **21**, 2602.

Sagawa, T., Iguchie, Y., Sasanuma, M. and others (1966b). *J. phys. Soc. Japan* **21**, 2587.

Segall, B. (1961). *Phys. Rev.* **124**, 1797.

Schwinger, J. (1949). *Phys. Rev.* **75**, 1912.

Terrell, J. H. (1966). *Phys. Rev.* **149**, 526.

Tomboulian, D. H. and Bedo, D. E. (1956). *Phys. Rev.* **104**, 590.

Wiech, G. (1966). *Z. Physik* **195**, 490.

Soft X-Ray Emission Spectra of Magnesium and Beryllium

L. M. Watson, R. K. Dimond and D. J. Fabian

Department of Metallurgy, University of Strathclyde, Glasgow, Scotland

ABSTRACT

Mg $L_{2,3}$-emission and Be K-emission spectra have been measured using a one-metre grating spectrometer. Accurate data were collected using a moiré-fringe measuring system with which the spectrometer is servo-operated to provide stepwise scanning, with summing of the spectra. The Mg spectrum shows fine-structure, and the processed data are compared with density-of-states distributions estimated from a nearly-free-electron model, using firstly the energy-gaps calculated by Falicov, and secondly those estimated by Ketterson and Stark. Good agreement is found with the model using the smaller energy-gaps obtained by Ketterson and Stark. Plasmon satellites for both metals are presented and discussed.

1. INTRODUCTION

In this paper accurate intensity measurements of the magnesium $L_{2,3}$-emission and the beryllium K-emission spectra are discussed. Both the parent emission-bands and the low-energy satellites have been measured and a number of interesting features have emerged. Before presenting these results we describe the experimental techniques employed in sufficient detail to enable the reader to judge the reliability of the results, and of any unexplained feature observed in the spectra.

2. EXPERIMENTAL

A. The Spectrometer

The development of the present instrument has been described in detail elsewhere (Campbell *et al.*, 1967; Watson *et al.*, 1967). Its general arrangement is illustrated in Fig. 1; for the work reported here, the moiré-fringe system for angular measurement has been mounted outside the vacuum chamber, with the reading head attached to the detector-carriage arm through a one-inch diameter shaft that also forms the pivot for adjustments of diffraction grating and source slit.

The x-ray tube is made of glass, and the target-holder assembly of stainless steel. The metal specimen to be examined is bonded to the water-cooled target holder using a conducting Araldite cement, and forms the cathode. The target is maintained at 3 kv positive with respect to the electron filament, and a bombarding current of 4–6 mA is normally employed. The electron gun consists of a nickel filament, coated with oxides of strontium and barium, and of a focusing grid that produces a line focus of ~1 mm × 15 mm on the target. Both the target holder and electron gun can be rotated and also adjusted laterally with the instrument under vacuum and operating, to assist with finding the best take-off angle of the x-rays and optimum filament position for maximum emission-intensity. Rotation of the target holder also allows a bellows-mounted scraper (see Fig. 1) to be used to clean the specimen surface under vacuum. X-Rays emitted from the target pass through a vacuum isolation-valve to the spectrometer chamber.

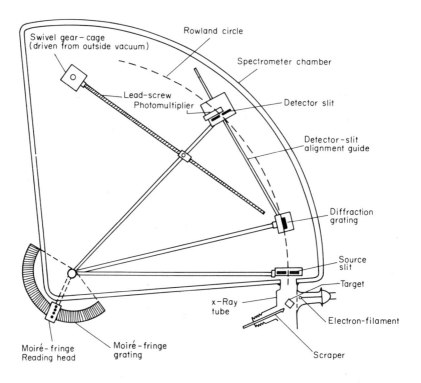

Fig. 1. Spectrometer layout, showing drive mechanism and mounting of moiré-fringe grating and reading head.

The instrument employs a concave grating and the source slit, diffraction grating and detector slit are all mounted on a Rowland circle. For the work described here the diffraction grating used was a Bausch and Lomb, platinum-coated blazed grating with 600 lines/mm, a radius of curvature of 99·88 cm and a blaze angle of 2° 4′.

The spectrometer chamber and x-ray tube are evacuated independently, using mercury-vapour diffusion pumps and liquid nitrogen cold-traps; the vacuum under operation is $\sim 2 \times 10^{-6}$ torr. The soft x-ray detector is a *Bendix M.*306 photo-electronmultiplier using crossed electric and magnetic fields, and has a tungsten photocathode. It is mounted on a carriage which locates the detector slit, and is driven around the Rowland circle by means of a lead-screw. Both the source and detector slit-widths are set at 50 μm, giving a theoretical resolution of 1Å (calculated from geometrical considerations; see Campbell *et al.*, 1967). The grating equation may be expressed as

$$g\lambda = e \left(\cos \frac{\alpha}{2} - \cos \frac{\phi}{2} \right) \tag{1}$$

where α and ϕ are the angles subtended at the centre of the Rowland circle by the chords joining, respectively, the source-slit to the grating and the grating to the detector-slit; λ is the wavelength of the diffracted radiation, g is the spectrum-order and e is the line-spacing of the grating. In this instrument α is set at 11° (corresponding to a grazing angle of 5° 30′) and the angle ϕ is measured to an accuracy of 0·002° with the moiré-fringe measuring system, as described elsewhere (Watson *et al.*, 1967).

For a chosen wavelength region, spectra are recorded in the following manner: the corresponding portion of the Rowland circle is divided into a large number of steps, the width of each being less than the resolving power of the instrument; the detector, controlled by the moiré-fringe measuring and servo system, is instructed to traverse these steps one at a time and to count the photon intensity after each for a preselected time; the angle (ϕ) and the photon-count at each position are punched into paper-tape.

This allows spectra of low intensity to be scanned in a time short enough to keep specimen contamination at an acceptable level, the large statistical counting-error being reduced by recording several spectra and summing them; the contamination effects are prevented from accumulating by cleaning the target under vacuum using the scraper, between scans. A correction is made for the small amount of contamination during each scan by measuring the fall-off in intensity with time. This is done by recording the count rate at point ϕ_1, chosen in a reasonably intense part of the spectrum, before and after each scan. The fall-off is due only to contamination since all power supplies are stabilized to better than 0·1% with additional emission-stabilization of the anode filament current.

B. Data Processing

The accumulated spectra, on punched paper-tape, are processed by computer which performs the following operations.

(1) The photon counts at corresponding angular positions for all scans are summed

$$\sum_s n_\phi^s = N_\phi$$

where n_ϕ^s is the number of photon counts in the fixed time interval at angular position ϕ, for the scan denoted by s; and N_ϕ is the total count-intensity recorded at the point ϕ in the spectrum.

(2) The statistical counting error is calculated for each summed count-intensity

$$\Delta N_\phi = \pm(N_\phi)^{1/2}$$

(3) The angular position measurements are converted to an energy-scale using the relation

$$E = \frac{gch}{e\left(\cos\dfrac{\alpha}{2} - \cos\dfrac{\phi}{2}\right)}$$

where c and h are respectively the velocity of light and Planck's constant in appropriate units.

(4) N_ϕ and ΔN_ϕ are divided by v^5; that is, by v^2 to convert intensity at wavelength λ to intensity at frequency v (or, $I(\lambda)d\lambda$ to $I(v)dv$), and by v^3 to take account of the v^3-dependence of the transition probabilities.

$$P(v) \propto v^3 |\langle f | \sum_i r_i | k \rangle|^2$$

where $|k\rangle$ and $|f\rangle$ are respectively the initial-state and final-state wave-functions.

(5) The background count (chiefly the bremsstrahlung continuum) is subtracted after averaging the first ten and last ten count-intentities recorded in the background regions to the low-energy and high-energy sides of each of the summed spectra. These readings are taken at points well clear of the spectral tailing, and it is assumed that the background intensity varies linearly between the low-energy and high-energy points of the spectrum.

(6) A correction is made for the intensity-loss due to target contamination as already indicated, using photon counts taken at a suitable point ϕ_1 before and after each scan. Assuming that the intensity fall-off with time occurs linearly (this has been measured and found approximately to hold) each corrected count-intensity N_ϕ^c is obtained using the relation

$$N_\phi^c = N_\phi \left[\frac{t}{t + \left(\dfrac{A}{B} - 1\right)z} \right]$$

where t is the total number of steps in the scan, z is the number of steps up to the point of the spectrum with count-intensity N_ϕ, A is the average of all photon counts at ϕ_1 taken after the scan and B is the average of all photon counts at ϕ_1 taken before the scan. The most common causes of target contamination are carbon deposition, due to cracking of organic materials in the electron beam, and surface oxidation due to residual oxygen in the system. In our instrument the first-order carbon K-emission due to contaminants is very weak and the fifth and sixth orders, which would occur in the region under examination, are negligible. On the other hand, oxidation of the surface is a considerable problem at pressures of 10^{-6} torr, and the spectrum may suffer distortion due to oxide absorption and emission. The extent of this effect is at present unknown.

3. RESULTS AND DISCUSSION

A. The Magnesium $L_{2,3}$-Emission Band

In Fig. 2(a) we show the results from one series of 30 separate scans of the magnesium $L_{2,3}$ spectrum, summed and processed in the manner outlined above. In the parent emission-band, discontinuity features considered to be real and due to the effects of valence-band structure, are indicated at A and B. A third prominent discontinuity has been indicated at C, but was not consistently observed in all series of summed spectra. Other small discontinuities appear but were never reproducible.

Five series of summed magnesium spectra were taken with the present instrument; these were respectively, three series of 20 separate scans, one of

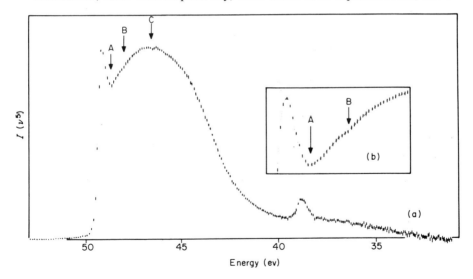

FIG. 2. Magnesium $L_{2,3}$-emission spectrum: (a) The summed and processed results (see text) of 30 scans, and inset (b) portion of the summed and processed results (without background correction) for 50 scans taken with half the step-size, giving twice the number of points for a given wavelength region of the spectrum.

30 scans and one of 50 scans, each series extending to a different wavelength of the spectrum and taken under different instrumental conditions. The only reproducible discontinuities, clearly identifiable in each series of emission spectra, were A and B; the feature at C, when observed, came at consistent energies but is probably not a band-structure effect since no correlation could be obtained with known features of the valence band. For A and B, on the other hand, such correlations are obtained as discussed below. Figure 2(b) (inset) shows a portion of the results (without background correction) for the series of 50 scans, taken over the parent emission-band with the step-size selected to give twice the number of points for a given wavelength region as recorded in the spectrum of Fig. 2(a).

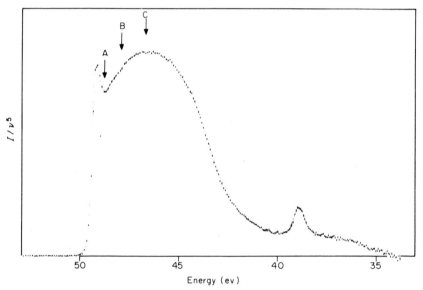

FIG. 3. Magnesium $L_{2,3}$-emission spectrum obtained with an instrument in Western Australia similar to that used in the present work (R. K. Dimond, unpublished).

Figure 3 shows a magnesium $L_{2,3}$ spectrum obtained with a similar instrument, in Western Australia by one of the present authors (Dimond), using an equivalent multi-channel step-scanning and summing technique. The discontinuity features A, B and C are prominent here also, and occur at energies that agree very closely with the present work.

Table I lists the energies, measured from the bottom of the emission-band, at which these features occur in spectra from the two instruments, and compares the measured energy-values with those estimated from the discussion which follows in the next section. In each case the low-energy limit of the emission bands (corresponding to the bottom of the valence bands) was determined by plotting the square of intensity as a function of energy, and extrapolating the linear portion of this curve to zero. The bandwidths are measured from the bottom of the emission band to the mid-point of the high energy edge.

TABLE I. Measured features of the Mg $L_{2,3}$-emission band

Feature	Strathclyde	W. Australia	Calculated
bandwidth	6.86 ± 0.03	6.81 ± 0.03	—
A	6.1 ± 0.1	6.1 ± 0.1	6.2
B	5.4 ± 0.1	5.4 ± 0.1	5.4
C	4.0	4.1	—

Values (ev) are compared for the energies at which consistent discontinuities have been observed in spectra obtained with the present instrument and with a similar instrument in Western Australia; the calculated values for A and B are those estimated from a nearly-free-electron model using the energy-gaps obtained by Ketterson and Stark.

It is valuable to compare in detail such spectra taken independently with two different spectrometers. For example, an additional point of interest is the difference in the intensity-ratio of the high-energy peak to the hump in the middle of the spectrum. The ratio is smaller for the spectrum shown in Fig. 3 than for that in Fig. 2(a). This is probably an effect due to differing wavelength-dependence of the photo-electron efficiencies of the two photo-cathodes used. For the spectrum obtained in Western Australia, a copper–beryllium cathode was used in the detector and for the spectrum of Fig. 2, obtained with the present instrument at Strathclyde, a tungsten photo-cathode was employed. When copper–beryllium photocathodes have been used in the present instrument, we have observed a similar peak-to-hump ratio as that obtained with the instrument in Western Australia.

B. Density of States for Magnesium

An attempt has been made to calculate a density-of-states curve for magnesium and to explain the spectrum in a manner similar to that used by Rooke (1968) for aluminium. The first, second and third Brillouin-zone surfaces for a hexagonal close-packed crystal, in the extended zone scheme, are shown in Fig. 4. Each zone can accommodate two electrons per unit cell; that is, one electron per atom. In a free-electron model the constant-energy surfaces in k-space are spheres, and for a divalent metal the Fermi-surface will occupy regions of the first, second, third and fourth zones. The density of occupied-states for completely free electrons in each of the first to fourth zones was calculated, and is shown in Fig. 5. To obtain a more realistic density-of-states, energy-gaps were introduced and the curves

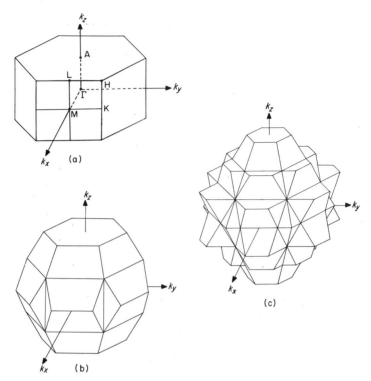

FIG. 4. Brillouin-zone surfaces for an h.c.p. crystal: (a) first Brillouin-zone, (b) second Brillouin-zone, and (c) third Brillouin-zone.

have been distorted to accommodate these gaps, maintaining unchanged the area under the occupied region of the curve. Two density-of-states curves were obtained in this manner; the first using energy-gaps computed by Falicov (1962), and the second using energy-gaps estimated by Ketterson and Stark (1967).

Falicov lists the band-energies at symmetry points, both with and without accounting for correlation and exchange effects. Since his computed Fermi-energy corrected for correlation and exchange is ~9·4 ev, and uncorrected is ~7·1 ev, we have used the uncorrected results. The energy-gaps in this case are quite large resulting in considerable distortion from the free-electron density of states. The amount of distortion introduced is somewhat arbitrary and many overall shapes of the final curve could result; however, these would not differ greatly from each other, and the energies at which van Hove singularities occur are accurate to within the accuracy of the energy-values calculated by Falicov. The large energy-gaps produce gross features in the density-of-states curve (Fig. 6(a)) and these would be expected to have a marked effect on the spectrum. This does not appear to be the case; the only gross feature in the spectrum is at A in Fig. 2.

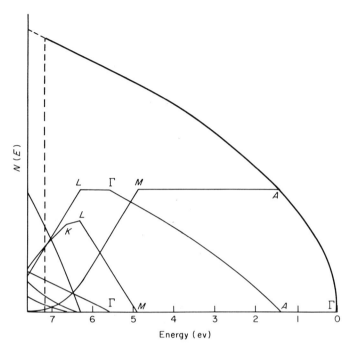

FIG. 5. Free-electron density of states for magnesium, showing contributions from the 1st, 2nd, 3rd and 4th zones.

The second density-of-states curve using the results of Ketterson and Stark, differs considerably from the first and is shown in Fig. 6(b). Here the energy-gaps are much smaller and the resulting deviation from the free-electron parabola is greatly reduced. The energy gap at M was taken as negligibly small.

Since the L_2 and L_3 core-levels have p-like symmetry, and because the selection rules prohibit p state to p-state dipole-transitions, any state with p-like symmetry in the band will not contribute to the emission spectrum. A crude estimation of the effect of the transition probabilities has been introduced. For this the probability of transitions from s- and d-states is assumed to be 1, and from p-states to be zero. An estimate of the distribution the different states throughout the band is then obtained as follows. (1) The ratio of the number of "s- and d-" to the number of p-states is assumed to decrease linearly with energy from 1:0 at the bottom of the band to 0:1 at the top of the band. (2) Where group theory suggests differently (Altmann and Bradley, 1965) changes are made as follows: (a) a discontinuity occurring where the state is entirely p-like is smoothed out. This occurs at Γ_3^+; (b) where the state is entirely s- and d-like the transition probability is assumed to rise to 1 for that point of the band. This is the case at Γ_4^-; (c) where the state is s-, p- and d-like the transition probability is assumed

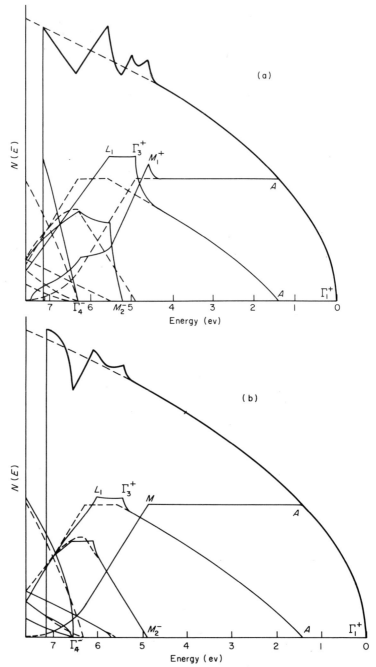

FIG. 6. Estimated density-of-states distribution for magnesium, after introducing the energy-gaps at Brillouin-zone surfaces: (a) using the energy-gaps calculated by Falicov (1962), and (b) using the energy-gaps calculated by Ketterson and Stark (1967).

to be $\frac{1}{2}$, and discontinuities arising where this is the case are somewhat less sharp, as at the point L. (3) Auger-broadening effects were "sketched in" to produce changes similar to those calculated for aluminium by Rooke (1968).

The estimated curves obtained are shown in Fig. 7. It is clearly seen from these curves that the energy-gaps estimated from Fourier-coefficients of potential by Ketterson and Stark, lead to better agreement with the experimental results that the energy-gaps obtained by Falicov. The discontinuities due to the overlap into the third and fourth zones at L and into the third zone at Γ_4^-, come respectively at 6·53 and 5·72 ev; when these are divided by the ratio of the experimental bandwidth to the free-electron bandwidth, they are reduced respectively to 6·2 and 5·4 ev, and agree—to within the accuracy of the calculation—with the energies of the points A and B on the experimental curves. The point C remains unexplained and occurs at an energy at least 0·5 ev lower than any discontinuity caused by an energy gap at the symmetry point M. It may be due to an overlapping oxide-band; this remains to be investigated.

We conclude that our results support those of Ketterson and Stark, and that magnesium is very nearly free-electron-like.

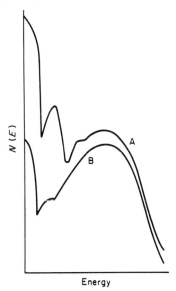

FIG. 7. Approximate effect of transition probabilities and Auger-broadening on calculated density-of-states distribution: A using curve obtained in Fig. 6(a), and B using curve obtained in Fig. 6(b).

C. Beryllium K-Emission Band

Figure 8 shows the results of 30 scans of the first-order Be K-spectrum, and Fig. 9 the results of several scans of the third-order spectrum. The parent emission-band in Fig. 8 is quite featureless due to the poor resolution (equivalent to 1 ev) in this wavelength region of the first-order spectrum. However, the summed third-order spectrum (Fig. 9) shows the existence of the high-energy edge although the error bars are large due to the low overall intensity. The third-order spectrum also exhibits a shape which is in general agreement with the band-structure calculations of Terrell (1966).

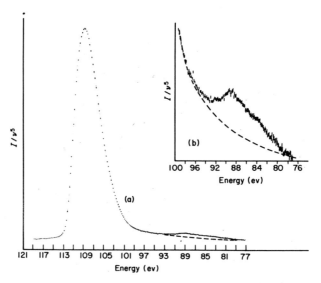

FIG. 8. First-order Be K-emission spectrum: (a) the summed and processed results (see text) of 30 scans, and inset (b) plasmon satellite, on an enlarged intensity-scale.

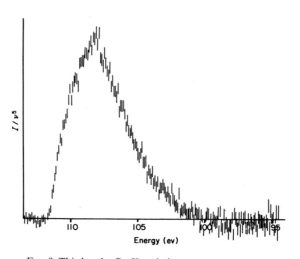

FIG. 9. Third-order Be K-emission spectrum (see text).

D. Plasmon Satellites

In Fig. 8 a plasmon satellite, not previously measured, is clearly detected in the low-energy tail and is shown inset on an expanded intensity scale. The theory of the emission and shape of such satellites is fully described in Part 4 of this Volume by Brouers and Longe (p. 329), by Glick, Longe and Bose (p. 319), and by Hedin (p. 337). In Fig. 10 we show the beryllium plasmon satellite obtained from the Be K-emission spectrum of Fig. 8 by subtracting the extrapolated low-energy tail of the parent emission-band. At the high-energy end of the satellite the intensity rises slowly and then more rapidly to the maximum. If the more rapidly rising portion is extrapolated to zero intensity, and is taken to be the start of the plasmon satellite, then the high-energy end of the satellite occurs at 18 ± 0.5 ev below the high-energy edge of the parent band. This value agrees well with the characteristic energy-loss experiments which give plasmon energies of 19 ev for beryllium (Powell, 1960).

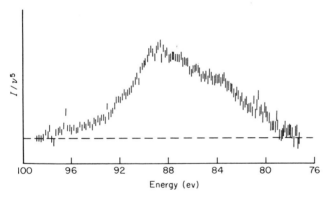

FIG. 10. Beryllium plasmon satellite, obtained from first-order spectrum (Fig. 8) by subtracting the extrapolated low-energy tail of the parent emission-band.

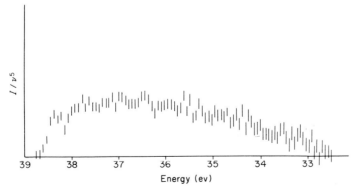

FIG. 11. Magnesium plasmon satellite, obtained from first-order spectrum (Fig. 2(a)) by subtracting the extrapolated low-energy tail and L_1–L_3 line.

Figure 11 shows the plasmon satellite of magnesium similarly obtained by subtracting both the extrapolated low-energy tail of the parent emission-band and also in this case, the L_1–L_3 line; the shape, or broadening, of the latter is assumed to be symmetrical about a vertical line through its peak. The satellite profile obtained is in good agreement with that published by Rooke (1963); and the high-energy end of the satellite comes at 10·6 ev below the high-energy edge of the parent band, which agrees extremely well with the value of 10·6 ev predicted by characteristic energy-loss experiments (Powell, 1960).

Acknowledgements

The authors wish to thank the Science Research Council for a grant in support of this work and for the award of Research Fellowships to two of us (L.M.W. and R.K.D.). We are also indebted to G. A. Rooke and C. A. W. Marshall for their help in the interpretation of data, to G. M. Lindsay and A. F. McDougall for assistance with the experiment, and to Professor E. C. Ellwood for his support and encouragement.

References

Altmann, S. L. and Bradley, C. J. (1965). *Rev. Mod. Phys.* **37**, 33.
Campbell, J. K. R., Watson, L. M. and Fabian, D. J. (1967). *J. Sci. Instrum.* **44**, 21.
Falicov, L. M. (1962). *Phil. Trans.* **A255**, 55.
Ketterson, J. B. and Stark, R. W. (1967). *Phys. Rev.* **156**, 748.
Powell, C. J. (1960). *Proc. Phys. Soc.* **76**, 593.
Rooke, G. A. (1963). *Phys. Lett.* **3**, 234.
Rooke, G. A. (1968). *J. Phys. C*, **1**, 767.
Terrell, J. H. (1966). *Phys. Rev.* **149**, 526.
Watson, L. M., Dimond, R. K. and Fabian, D. J. (1967). *J. Sci. Instrum.* **44**, 506.

Soft X-Ray Emission Spectra and the Valence-band Structure of Beryllium, Aluminium, Silicon and some Silicon Compounds

G. Wiech

Sektion Physik der Universität München, Germany

ABSTRACT

The Be K-emission band of beryllium, the Al $L_{2,3}$-emission band of aluminium, and the Si $L_{2,3}$-emission bands of silicon, SiC and SiO_2 are examined. A two-metre concave-grating spectrometer was used, with x-ray tube pressures of $\sim 10^{-8}$ torr to keep target contamination to a minimum. The Si $L_{2,3}$-emission from Si and from SiC shows a quite different intensity-distribution from that from SiO_2. For comparison with density-of-states calculations the K-emission and L-emission bands are superimposed, using assumptions regarding the transition probabilities, the Auger effect and the intensity-ratio $K_\beta-L_{2,3}$; the results agree well with the density-of-states calculations of Kane. The agreement between experimental and calculated bandwidths and density-of-states distributions for Si and SiC, for Al, and for Be is discussed.

1. INTRODUCTION

In the last few years a great deal of work has been done on the investigation of x-ray emission bands, especially the light elements, and in calculating the electronic band structure and the density of states of numerous elements and compounds. If, for a given substance, the intensity distribution of the emission bands and the theoretical data are available, the problem remaining is to correlate x-ray emission with band-structure calculations.

The most suitable elements for such studies are those of the third-period. Their complete x-ray emission spectrum (Fig. 1) consists—if we neglect the satellites—of the K_β-emission band, the $L_{2,3}$-emission band and the $K_{\alpha_{12}}$ doublet. The measurement of the K_α lines make possible the energy-correlation of the K-emission and L-emission bands.

As a consequence of selection rules the K-emission bands give information about the p electrons, and the L-emission bands about the s- and d-electrons of the valence band. We obtain full information only if we measure the complete spectrum. This has been done for aluminium and elemental silicon, and for silicon in silicon carbide and silicon dioxide. The K-emission of the second-period element beryllium has also been investigated, but here the

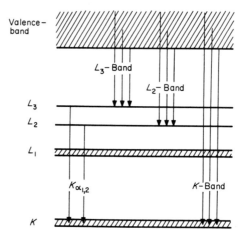

FIG. 1. The occupied energy-states and x-ray transitions for a third-period element.

x-ray band-spectra data are incomplete, since the Be K-emission band gives information about only the p-electrons of the valence band, and no information about the s-electrons.

2. EXPERIMENTAL

For measurement of the L-emission from Al and Si, and the K-emission from Be, a grazing-incidence concave grating spectometer was used (Wiech, 1966). The blazed grating (blaze angle $1°\,31'$) has a radius of curvature of 2m, has 600 lines/mm, and is aluminized. The detector is an open magnetic-electronmultiplier with a tungsten photocathode. In order to avoid target contamination, or at least to keep its influence small, the vacuum in the x-ray tube was $\sim 10^{-8}$ torr.

3. RESULTS

A. Beryllium

Figure 2 shows a step-by-step measurement of the Be K-emission band which has a maximum at A $(113\cdot83\,\text{Å})$ and two discontinuities B and C at respectively $112\cdot4\,\text{Å}$ and $\sim 111\cdot3\,\text{Å}$. The target input was 3 kv and 0·5 mA. In the wavelength region from $111\cdot2\,\text{Å}$ to $117\,\text{Å}$ the statistical counting error is $0\cdot7\%$; in other parts of the spectrum it rises to 3%. The results are in good agreement with the K-emission band measured by other workers (Holliday, 1967a; Lukirskii and Brytov, 1964). By comparing the Be K-emission band with the density of states for beryllium, calculated by Loucks and Cutler (1964) (Fig. 3), we find satisfactory agreement on the high-energy side; but as already mentioned, the x-ray emission information is incomplete.

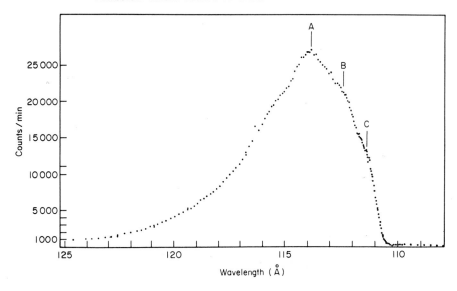

FIG. 2. Be K-emission band from metallic beryllium.

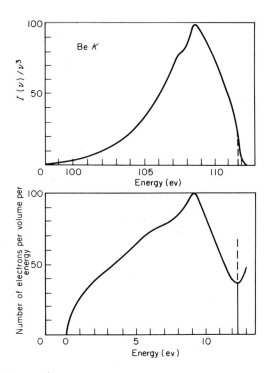

FIG. 3. $I(v)/v^3$-curve and density of states (Loucks and Cutler, 1964) for beryllium.

The discontinuity B (Fig. 2) is absent in the theoretical curve. In the $I(v)/v^3$ curve no corrections have been made for instrumental distortion, or for the width of the K level.

B. Aluminium

The Al K-emission band (K. Läuger, unpublished), the Al $L_{2,3}$-emission band (Wiech, 1966) and the density-of-states curve for aluminium (Ashcroft,

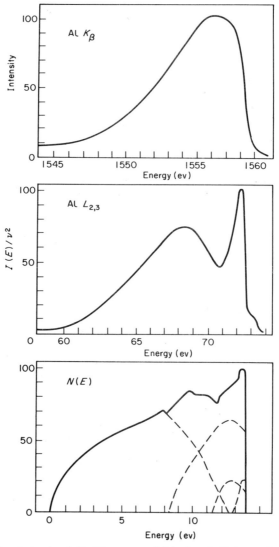

Fig. 4. Al K-emission band (K. Läuger, unpublished) and Al $L_{2,3}$-emission band (Wiech, 1966) for metallic aluminium; and the density of states of the valence band for aluminium, calculated by Ashcroft (1966). The Fermi energy is assumed to be 13·5 ev.

1966) are shown in Fig. 4. The energy-difference between the energy scales for the $K\beta$-emission and the L_3-emission bands can be obtained from the K_1 line (see Fig. 1) which has an energy of 1486·66ev (K. Läuger, unpublished).

The minimum in the intensity distribution for the L-emission, near 71 ev, is more pronounced here than in the measurements by other workers (Rooke, 1966; Fomichev, 1967). It is difficult to interpret this discrepancy, since the wavelength dependence of the reflecting power of the grating and the quantum-efficiency of the photocathode of the multiplier, for different experimental instruments, are mostly unknown. Nevertheless, on the whole all the L-emission bands appear similar to their corresponding density-of-states curves and indeed often the L-emission alone is compared with the density of states. On the other hand, the K-emission band differs markedly from the density-of-states curve, and one would not compare these two. Possibly the similarity of the L-emission band to the density-of-states curve is due to the fact that the K-emission shows no strongly marked feature that would seriously modify the "sum" of the two; however, transition probabilities may have an important influence.

C. Silicon

In the case of silicon, where both the K-emission and the L-emission bands have pronounced structure, it becomes quite obvious that both have to be considered to get satisfactory agreement with the calculated density-of-states. The experimental results for the Si K-emission and the Si $L_{2,3}$-emission from elemental silicon, from silicon carbide and from silicon dioxide are represented in Figs. 5 to 7. For correlation of the two energy-scales the energy of the $K\alpha_1$ line was chosen, because the L_3-emission is more intensive than the L_2-emission, and no separation of their respective emission bands was observed.

The K spectra and the $K\alpha_1$ lines shown in Figs. 5–7 were measured with a bent-crystal spectograph, and with photographic-photometric recording, by Kern (1960) and Krämer (1960). By comparing the K and L spectra of the element (Fig. 5), we find three characteristics which appear to have a more general meaning in so far as they are also observed in the spectra of SiC and SiO_2. (a) The high-energy edges of the K-emission and L-emission bands coincide. (b) The bandwidth for the K and L spectra are nearly identical; although there appears to be a tendency for the width of the K-emission band to be smaller than that of the corresponding L-emission band. From this we conclude that s- and p-electrons are present in the whole energy-range of the valence band: i.e. there is no region with only s- or only p-electrons. The K and L spectra of magnesium and aluminium show a similar behaviour, but this is not so for phosphorus and sulphur where the L-emission bands are much broader than the K. (c) The intensity distribution of the K and L spectra are complementary in the following

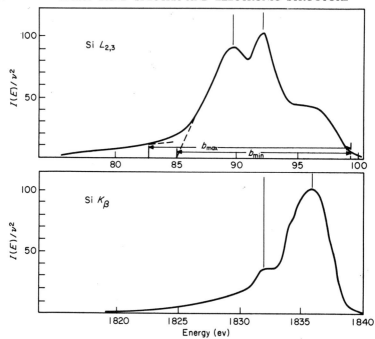

FIG. 5. Si $L_{2,3}$-emission and Si K_β-emission bands from silicon.

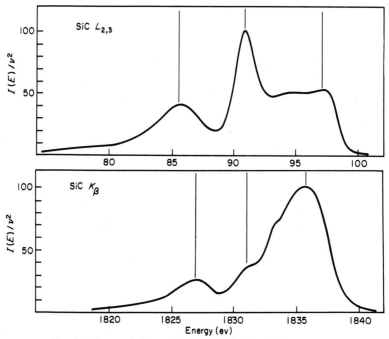

FIG. 6. Si $L_{2,3}$-emission and Si K_β-emission bands from silicon carbide.

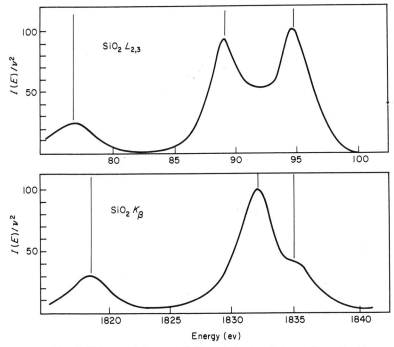

FIG. 7. Si $L_{2,3}$-emission and Si K_β-emission bands from silicon dioxide.

sense; generally a region of high intensity of the K-emission corresponds to a region of small intensity of the L-emission, and vice versa.

From the experimental results we can see that in silicon, p-electrons predominate at the top of the valence band, and s- and d-electrons in the middle and lower parts of the valence band. The same is found for SiC (Fig. 6). The similar shape of the emission bands of Si and SiC is due to the crystal structures of these materials which have respectively, diamond-type and zinc-blende-type lattices.

It may be mentioned that the intensity distribution of the spectrum of diamond (Holliday, 1967) agrees in detail with that of silicon, and that the intensity distribution of the $L_{2,3}$-emission from phosphorus in some group III–V compounds is very similar to that of the Si $L_{2,3}$-emission from silicon and from silicon carbide (G. Wiech, to be published).

The silicon K-emission and L-emission from SiO_2 (Fig. 7) are quite different from those from Si and SiC. The shape of the L-emission band and the energy-value of the maximum agree well with the measurements of Ershof and Lukirskii (1967). The p-electrons mainly occupy the middle of the valence band, whereas s- and d-electrons form peaks near the top and the bottom of the valence band. Due to the large energy-gap (about 8 ev) in SiO_2, there is no tailing to the low-energy side of the spectrum as for Si and SiC.

4. COMPARISON WITH THEORETICAL CALCULATIONS

A. Bandwidths

A comparison of valence bandwidths for Si and SiC, calculated by different authors, with the emission bandwidth derived from x-ray data is made in Table I. The experimental b_{min}-values of the L band, as defined in Fig. 5, agree satisfactorily with the values for the K-emission band given by Kern (1960).

TABLE I. Comparison of the calculated width of the valence bands with the experimental width of the emission bands

	Silicon	Silicon carbide	Reference
Calculated width of the valence band (ev)	10·3 and 20·3	—	Kleinman and Phillips (1960)
	13	—	Cardona and Pollak (1966)
	12·25	—	Kane (1966)
	11	16 α-form	Bassani and Yoshimine (1963)
	—	20·4 β-form	Kobayasi (1958)
K-Emission bandwidth (ev)	13·5	16·5	Kern (1960)
$L_{2,3}$-Emission bandwidth (ev)	b_{min} 14·4	17·5	
	b_{max} 16·5	19·2	Present work (see Fig. 5)

For the case of silicon the theoretical results are (with one exception) in the 10–13 ev range. These values are somewhat smaller than the bandwidths derived from experiment. For the carbide the width of the valence band has been calculated for both the α- and β-forms. The calculated widths for these two valence bands differ by more than 4 ev. This contradicts the experimental results; the L-emission bands of the α- and β-forms have been measured,

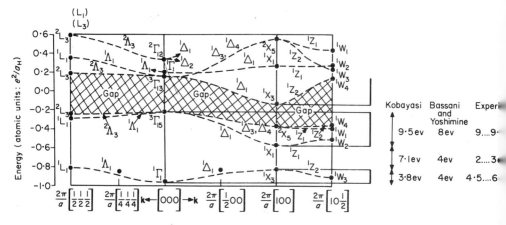

FIG. 8. Band structure for the β-form of SiC, calculated by Kobayasi (1958). The energy-values were taken from the calculations of Kobayasi (1958), Bassani and Yoshimine (1963), and from the x-ray data of the K-emission and L-emission bands.

and are nearly identical, which means that the calculation made by Kobayasi (1958) yields a bandwidth which appears to be too large.

The band structure of the carbide shows a peculiarity: the valence band is split into two bands, separated by an energy-gap. This is shown in more detail in Fig. 8. For the occupied bands the theoretical and the experimental values are in satisfactory agreement. In the K-emission and L-emission bands of the carbide the only indication of the energy-gap is an intensity minimum. The intensity does not fall off to zero because there is a low-energy tail to the upper band, caused by Auger-transitions. Although it is difficult to determine the width of the energy-gap from the experimental results, these show unambiguously that the value obtained by Kobayasi for the energy-gap, and consequently for the total bandwidth, is too large.

For SiO_2, no band calculations are so far available.

B. Density of States

A density-of-states distribution for silicon has recently been calculated by Kane (1966) using the pseudopotential method, and an attempt has been made to compare it with a curve resulting from the superposition of the x-ray K-emission and L-emission bands (see Fig. 9).

The density of states $N(E)$ is connected with the x-ray intensity $I(E)$ by

$$I(E) \simeq v^2 \int_s \frac{P(E,k)}{|\nabla_k E|} \, dS \tag{1}$$

where $P(E,k)$ is the matrix element of the transition probability, S is an isoenergetic surface in the reciprocal lattice, k is the wave-vector, v is the frequency and E the energy. Since this integral has not to date been calculated, we assume P to be independent of k. Thus we obtain the well-known formula

$$I(E) \simeq v^2 \, P(E) \int_s \frac{dS}{|\nabla_k E|} = v^2 \, P(E) \, N(E) \tag{2}$$

Separating $N(E)$ into two parts, one representing the s-electrons, the other the p-electrons, we have

$$N(E) = N_s(E) + N_p(E) \tag{3}$$

We obtain information about N_s from the L-emission band, and information about N_p from the K-emission band. Applying Eqs. (2) and (3) to these bands, we obtain

$$N(E) \simeq \frac{R I_p(E)}{v_K^2 P_p(E)} + \frac{I_s(E)}{v_L^2 P_s(E)} \tag{4}$$

R being the intensity-ratio K_β-$L_{2,3}$ of the normalized K-emission to L-emission, taking account of the fact that no absolute intensities were measured.

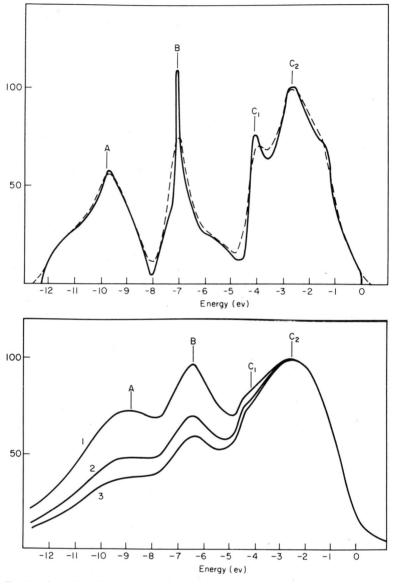

FIG. 9. *Above:* Density of states for silicon (Kane, 1966: full line); density of states curve folded with a Guassian of half-width 0·5 ev (dashed line). *Below:* The K-emission and $L_{2,3}$-emission bands of silicon (from Fig. 5) superimposed. The intensity-ratio R $(K-L_{2,3})$ is 1:1 for curve 1, 2:1 for curve 2 and 3:1 for curve 3. The curves were normalized after superimposing; the zero of energy is arbitrary.

Since the transition probabilities are undetermined, $N(E)$ cannot be derived from Eq. (4). Further difficulties arise from the Auger effect the influence of which is also unknown.

The simplest assumption we can introduce for the transition probabilities is that $P_s \simeq P_p \simeq$ const. If we superimpose the K-emission and L-emission bands for silicon, and take respectively the values of 1, 2 and 3 for the parameter R, we obtain the curves 1, 2 and 3 in Fig. 9. In order to take account of the spectral "window" of the spectrometer, and the width of the K-level, the theoretical curve was folded with a Gaussian of a half-width of 0·5 ev. The result is the dashed line in Fig. 9. The features of the experimental curves and the calculated curve agree in the main. The peaks A and B of the superimposed curves are due to the L band, and the features C_1 and C_2 to the K band.

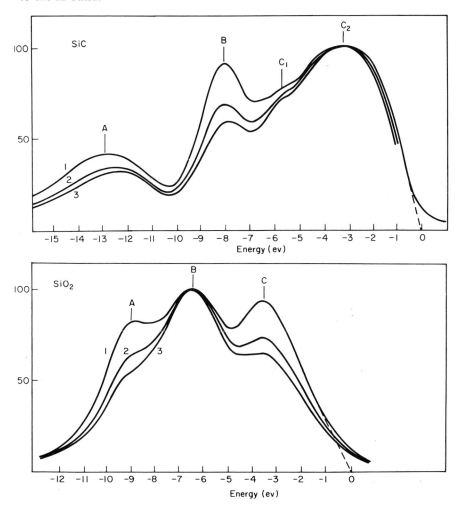

FIG. 10. The K-emission and $L_{2,3}$-emission bands superimposed for SiC (from Fig. 6), and for SiO$_2$ (from Fig. 7). The intensity-ratio $R(K-L_{2,3})$ is 1:1 for curve 1, 2:1 for curve 2 and 3:1 for curve 3. The curves were normalized after superimposing.

If we assume the transition probability P to be a linear function of E, and then vary the values for P_s and P_p, the superimposed curves always show either three maxima, with varying intensity-ratios, or two maxima plus a more or less pronounced hump to the low-energy side; the energy-value of the maxima of these curves is only little affected by the choice of value for the parameters $P(E)$ and R.

Applying the same procedure of superimposing the K-emission and L-emission bands for SiC and SiO_2, we obtain the curves shown in Fig. 10. Since in the case of SiC and SiO_2 we are dealing with compounds, the results are tentative with regard to the intensity-ratios of the peaks, but the energy-values of the maxima are most probably reliable. These may be useful in calculations of electronic bandstructure and the density of states.

References

Ashcroft, N. W. (1966). *In* "Optical Properties and Electronic Structure of Metals and Alloys", ed. by F. Abelès, p. 336 (in paper by Wooten *et al.*). North Holland Publishing Company, Amsterdam.

Bassani, F. and Yoshimine, M. (1963). *Phys. Rev.* **130**, 20.

Cardona, M. and Pollak, F. H. (1966). *Phys. Rev.* **142**, 530.

Ershov, O. A. and Lukirskii, A. P. (1967). *Soviet Phys.—Solid State* **8**, 1699.

Fomichev, V. A. (1967). *Soviet Phys.—Solid State* **8**, 2312.

Holliday, J. E. (1967a). *In* "Handbook of X-Rays", ed. by E. F. Kaelbe, chap. 8. McGraw-Hill, New York.

Holliday, J. E. (1967b). *J. Appl. Phys.* **38**, 4720.

Kane, E. O. (1966). *Phys. Rev.* **146**, 558.

Kern, B. (1960). *Z. Physik* **159**, 178.

Kleinman, L. and Phillips, J. C. (1960). *Phys. Rev.* **118**, 1153.

Kobayasi, S. (1958). *J. Phys. Soc. Japan* **13**, 261.

Krämer, H. (1960). Dissertation, Univ. München.

Loucks, T. L. and Cutler, P. H. (1964). *Phys. Rev.* **133**, A819.

Lukirskii, A. P. and Brytov, I. A. (1964). *Sov. Phys.—Solid State* **6**, 32.

Rooke, G. A. (1966). *In* "Optical Properties and Electronic Structure of Metals and Alloys", ed. by F. Abelès, p. 336 (in paper by Wooten *et al.*). North Holland Publishing Company, Amsterdam.

Wiech, G. (1966). *Z. Physik* **193**, 490.

Sur Les Spectres X des Metaux—Quelques Commentaires et Exemples

Y. CAUCHOIS

Laboratoire de Chime Physique de la Faculté des Sciences de Paris, France

ABSTRACT

This paper presents the simple theoretical models and formulae that are used in the author's laboratory for relating experimental results, obtained from the soft x-ray spectra of metals, with the density of electron-states. Some examples are given of studies carried out on Al and Mg *K*-emission bands (for the metals and oxides), observed using both direct and fluorescence excitation.

New results obtained by a Franco-Italian research group[†] are presented on ultra-soft x-ray absorption spectra of heavy metals using continuous radiation from the Frascati synchrotron. Unexpected variations of the photoelectric absorption-coefficients were observed, and are explained by means of a non-hydrogenic atomic model, with summing of partial photo-ionization cross-sections related to various outer-shells.

1. INTRODUCTION

Vers 1939, les mémoires et livres de Mott, Skinner, Brillouin ou Farineau, en particulier, ont attiré l'attention sur l'intérêt direct des spectres x pour atteindre les distributions des états d'énergie électroniques occupés ou non dans les solides (Cauchois, 1948). Depuis lors la question est périodiquement reposée de savoir si les bandes d'émission x et les spectres x d'absorption permettent effectivement de connaître ces distributions. Les expérimentateurs s'activent ou se détournent des réalisations et études correspondantes et ne bénéficient que sporadiquement de l'appui des théoriciens du solide. Notre Laboratoire est l'un des rares qui ait poursuivi de manière ininterrompue depuis cette époque, des recherches dans ce domaine ; il s'est attaché à améliorer les moyens d'analyse spectroscopique dans toute la gamme des rayons x mous et ultra-mous et à préciser les conditions de travail.

La conférence sur laquelle est fondé ce livre a marqué une période d'enthousiasme d'écoles théoriques envers les problèmes du spectroscopiste et l'on devait s'en réjouir à priori, si toutefois leurs conclusions ne s'avéraient pas trop décourageantes et si leur champ d'application ne se montrait pas trop particulier. Le texte rédigé par Rooke (ce Livre, p. 3) m'amène à penser que des schémas et remarques très simples pourraient peut-être encore avoir

71

† Istituto Superiore di Sanita, Rome.

quelque intérêt, afin de rappeler sur quelles bases nous faisons reposer les dépouillements de nos données brutes d'expérience et les précautions prises pour définir les conditions mêmes de celles-ci.

2. SCHEMA THEORIQUE

Nous nous plaçons ici dans un schéma théorique où l'on puisse admettre que le nombre n de photons (observable par détecteurs électroniques), émis par transitions des électrons de la bande de valence vers une lacune électronique formée dans un niveau atomique profond, est proportionnel au produit d'une probabilité de transition par une densité d'états. Passant à l'intensité (observable dans le cas du récepteur photographique) nous écrivons :

$$I(v) \propto v\, p(v)\, N(\varepsilon) \tag{1}$$

où, dans le cas d'un métal (cf. Fig. 1),

$$hv = E_x - (\varepsilon_F + e\phi) + \varepsilon = E_{ox} + \varepsilon$$

ε_F = énergie de la surface de Fermi et ϕ = potentiel de Richardson. L'équation (1) peut être récrite en tenant compte de $p(v) \propto v|M|^2$

$$I(v) \propto v^2 |M|^2\, N(\varepsilon) \tag{2}$$

où l'élément de matrice M se réfère à des sommes d'expressions du type

$$\int \psi_f^* \operatorname{grad} \psi_i \, d\tau$$

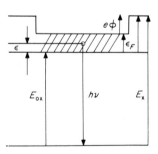

FIG. 1. Schéma des niveaux d'énergie d'un métal.

comportant l'opérateur gradient et que nous écrirons symboliquement $G(\varepsilon)$.

Les notations i et f se rapportent aux états initiaux et finaux, c'est-à-dire, d'une part à la lacune x et aux électrons de valence et d'autre part aux différentes lacunes dans la bande de valence en présence du cortège électronique complet de l'ion métallique. En fait, on suppose généralement que les calculs portent sur l'approximation à 1-électron sans interactions. Dans des cas spéciaux, impliquant les approximations dipolaires et hydrogénoïde ou à champ central pour l'atome, monoélectronique avec des conditions à la limite, pour les états du solide, on peut retenir pour M des expressions de la forme

$$M \propto v \sum \int \psi_f^* \, r \, \psi_i \, d\tau$$

r étant l'opérateur coordonnées.

Nous écrirons symboliquement $R(\varepsilon)$ pour les fonctions de probabilité :

$$\sum \int \psi_f^* \, r \, \psi_i \, d\tau$$

liées aux forces d'oscillateur en jeu.

Dans ces conditions, on écrira

$$p(v) \propto v^3 |R(\varepsilon)|^2$$

et

$$I(v) \propto v^4 |R(\varepsilon)|^2 \, N(\varepsilon) \tag{3}$$

ou

$$n(v) \propto v^3 |R(\varepsilon)|^2 \, N(\varepsilon)$$

Certains auteurs confondent les expressions de I et de n. Certains autres introduisent plus ou moins explicitement dans l'expression de l'intensité, un facteur supplémentaire v^{-2}, ce qui risque d'entrainer des ambiguités pour l'expression de $I(v)$. Il s'agit d'un facteur dont on doit éventuellement tenir compte pour passer d'un tracé expérimental $I(\lambda)$ à la courbe $I(v)$, conformément à $I(v)dv = I(\lambda)d\lambda$ avec $d\lambda = -\dfrac{c}{v^2}dv$. Remarquons aussi que la déformation des courbes $n(v)$ ou $I(v)$, par rapport à $N(\varepsilon)$ du fait de facteurs v^2, v^3 ou v^4, reste généralement négligeable dans les limites d'une bande x d'émission du domaine de la spectroscopie cristalline ; il n'en est pas de même, en général, dans le domaine de la spectroscopie par réseaux tangents.

Dans des notations et approximations analogues à celles que j'ai indiquées pour l'émission, le coefficient d'absorption photoélectrique μ, et la section efficace de photoionisation de la couche interne σ, s'écrivent :

$$\mu(v) \propto \sigma(v) \propto 1/v^2 \, p(v) \, N(\varepsilon)$$

ou

$$\propto 1/v \, |G(\varepsilon)|^2 \, N(\varepsilon)$$

ou

$$\propto v \, |R(\varepsilon)|^2 \, N(\varepsilon) \tag{4}$$

les fonctions de probabilité G et R se référant alors aux états initiaux : ion sans lacune, états de conductibilité vides, et aux états finaux ; ion avec lacune interne, états de conductibilité dont 1 est occupé.

Les calculs effectués dans le cas de l'atome libre mais dans différentes approximations pour les fonctions d'onde monélectroniques, conduisent à des résultats fort différents, comme nous le verrons ci-dessous. Un certain nombre de remarques s'impose encore, par exemple dans le cas de bandes x d'émission. Abstraction faite des fonctions instrumentales, la distribution d'intensité $I_x(v)$ déduite de l'expérience dans le cas d'une lacune dans le niveau profond x $(= K, L_1, L_2, L_3, \ldots)$ apparaît comme un produit de convolution de $I(v)$ avec la distribution $\mathscr{L}_x(E)$ de ce niveau profond; celle-ci, liée à sa durée de vie et due aux transitions radiatives et Auger, peut être exprimée par une fonction de Lorentz, tout au moms pour les niveaux suffisamment internes, dont cependant l'énergie se trouve modifiée avec la structure électronique et particulièrement avec la liaison chimique (Sureau et Berthier, 1963). Dans le cas d'une transition atomique pure entre x et x' nous écririons:

$$I_x(v) \propto \mathscr{L}_x * \mathscr{L}_{x'}$$

Dans le cas d'une bande d'émission x, nous écrivons par exemple schématiquement:

$$I_x(v) \propto I(v) * \mathscr{L}_x \quad \text{et} \quad I_x(v) \propto v^4 |R(\varepsilon)|^2 N(\varepsilon) * \mathscr{L}_{x'}$$

sous les approximations indiquées. Mais l'observation introduit les caractères spectraux propres à l'appareillage utilisé, entre autres ceux relatifs au spectrographe lui-même et à son détecteur. S'il s'agit de mesures dans le domaine de la spectroscopie cristalline, la fonction de diffraction (diffraction pattern) du cristal établie dans un modèle plus ou moins idéalisé, ses imperfections à l'état plan ou courbé, par rapport à ce modèle, les défauts de construction ou éventuellement de focalisation etc. ... interviennent pour déterminer les courbes de distribution d'intensité observées, d'une manière que nous schématisons comme suit:

$$I_{\text{obs}}(v) \propto I_x(v) * I_{\text{instr}}(v)$$

I_{instr} représentant la fonction instrumentale.

Ainsi, pour remonter de la courbe de distribution d'intensité observée à la courbe "vraie" d'intensité $I(v)$, il est nécessaire de connaître les fonctions $\mathscr{L}_x$ et I_{instr}; et d'effectuer les déconvolutions correspondantes. Ce travail généralement effectué à partir de formes schématiques pour celles-ci, est toujours difficile, et comporte le risque de faire apparaître dans la courbe $I(v)$ qu'il détermine, des structures fallacieuses. Il est plus facile et plus sûr de se limiter à des mesures de largeurs à mihauteur spécialement intéressantes pour des bandes assez étroites et symétriques, comme celles dues à des transitions atomiques ou avec des états d; ou encore d'effectuer des corrections sur des positions relatives ou largeur totale de bandes x.

Revenons à l'expression (1). On dit souvent que, l'intensité au début d'une bande K varie comme $\varepsilon^{3/2}$, et d'une bande $L_{2,3}$ comme $\varepsilon^{1/2}$. Il paraît banalde rappeler que ceci résulte d'approximations simples pour les probabilités de transition et de l'approximation des électrons libres pour $N(\varepsilon)$; les densités d'états étant représentées de manière plus acceptable par cette approximation au fond de la distribution de valence que vers la surface de Fermi par exemple. On trouve déjà dans Mott et Skinner une présentation où $N(\varepsilon)$ s'exprime en tenant compte du fait que les états quasi-libres dans le solide dérivent d'états atomiques à caractère s, p ou d:

$$N(\varepsilon) = N_s(\varepsilon) + N_p(\varepsilon) + \dots$$

une fonction de probabilité de transition partielle convenable devant être appliquée à chaque terme de la somme. On admet en effet que la fonction applicable à une transition vers un état interne à caractère p, c'est-à-dire à $N_s(\varepsilon)$, est une constante, tandis que celle applicable à un niveau interne s, c'est à dire à $N_p(\varepsilon)$, est proportionnelle à v c'est-à-dire à ε.

La fonction de probabilité à appliquer à N_s étant alors à peu près une constante, et celle à appliquer à N_p à peu près proportionnelle à v et les densités d'états étant dans certaines limites à peu près proportionnelles à $\varepsilon^{1/2}$, les variations indiquées en $\varepsilon^{3/2}$ et $\varepsilon^{1/2}$ pour les bandes K et $L_{2,3}$ se trouvent justifiées (compte non tenu des facteurs v^3). Des remarques analogues joueraient pour μ ou σ.

3. DES RESULTATS ET DE LEUR DISCUSSION

La vocation même de notre laboratoire nous amène à considérer en outre, d'aussi près que possible, à quel état physicochimique correspond en fait le spectre observé. On doit se poser la question de savoir, en particulier, comment définir et contrôler la nature des échantillons à l'étude. Nos expériences comportent, entre autres, un contrôle par diffractions x et électronique avant et après la prise du spectre. Mais quelles que soient les précautions prises, il est évident que la couche émissive à la surface d'un échantillon plan métallique par exemple, a une épaisseur très différente et peut comporter des modifications physiques et chimiques d'importance relative plus ou moins grande suivant le domaine spectral analysé (il en est ainsi dans le cas des émissions K ou $L_{2,3}$ d'un même métal), et en fonction des conditions expérimentales. Le mode d'excitation—direct par électrons (Bonnelle, 1966) ou indirect par photons (C. Bonnelle et C. Sénémaud, à paraître) d'énergie plus ou moins élevée—intervient aussi très fortement. La contribution du fond continu provenant entre autres du rayonnement de freinage, doit être soustraite au spectre d'excitation primaire, ce que nous effectuons à l'aide d'émetteurs auxiliaires de numéro atomique voisin.

A. Etude des Bandes d'Emission de l'Aluminium et du Magnésium

A titre d'exemple, nous signalons deux cas étudiés dans notre laboratoire: celui des bandes K de l'aluminium et du magnésium.

Dans l'un et l'autre cas, des résultats apparemment différents ont été obtenus suivant le mode d'excitation, la haute tension appliquée au tube, la nature initiale de l'échantillon, le degré de vide dans l'enceinte où il est placé, etc.… Il est manifeste que l'excitation par fluorescence dans nos conditions d'expériences, évite au moins une partie importante de l'oxydation ; ce qui se traduit par des changements de forme de la bande principale et un affaiblissement considérable de la bande $K\beta$ en particulier.

La bande $K\beta'$ observée dans l'oxyde (Sénémaud, 1968) serait due à des transitions d'électrons $2s$ de l'oxygène vers les lacunes $1s$ des ions du métal, alors que la bande principale $K\beta$ comporterait des transitions d'électrons $2p$ de l'oxygène vers $1s$ du métal (Bonnelle, 1968). Les valeurs théoriques approchées dont nous disposons ($\varepsilon_{2s}-\varepsilon_{2p}$ de la notation usuelle) pour les écarts d'énergie $2s2p$ dans les ions O^{2-} et O^-, ainsi que les séparations des niveaux mesurées optiquement pour les ions libres, sont en accord acceptable avec l'interprétation proposée pour $K\beta'$, dans le cas de Al_2O_3 et de MgO.

La Figure 2 montre l'une des courbes d'intensité obtenues pour l'aluminium par Sénémaud, à partir d'un spectre photographié dans un spectrographe à cristal courbé par excitation indirecte. L'échantillon, un monocristal préparé par fusion de zone, était placé extérieurement au tube à rayons x primaires. On remarquera la forme relativement "creuse" du flanc de

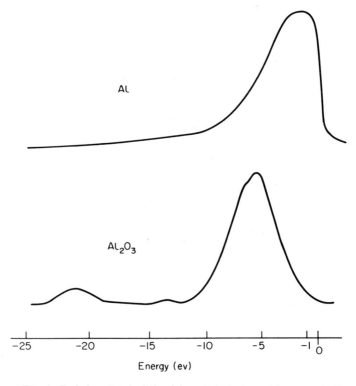

FIG. 2. Emission $K\beta$ de l'aluminium (métal et oxyde) par excitation "indirecte".

basse fréquence de la bande principale dans une région habituellement "gonflée" par superposition de l'émission parasite due à l'oxydation, ainsi que la difficulté d'en définir la limite de basse fréquence. Il faut noter aussi que de tels spectres ne montrent plus l'émission $K\beta'$, dont l'existence a souvent été indiquée pour le métal à environ 16 ev de $K\beta$. Or c'est à cette position que se situerait $K\beta'$ de Al_2O_3. On sait de plus que la valeur de la perte unitaire d'énergie caractéristique des électrons la plus intense dans l'aluminium métallique est de 15 ev; elle est attribuée aux oscillations de plasmas dans le volume du métal. La présence de contaminations oxygénées dans la couche émissive d'un échantillon de ce métal pourrait donc faire croire à la manifestation plus ou moins intense des plasmons dans son spectre K; or, si elle existe, elle met en jeu des intensités relatives très faibles. N'y aurait-il pas lieu de s'affranchir avec certitude de telles contaminations, toujours plus probables pour le rayonnement x très mou, avant de faire reposer d'importants efforts théoriques sur les caractères d'une bande L large, extrêmement faible, signalée elle aussi à une distance d'environ 15 ev de la bande $L_{2,3}$ de l'aluminium mais qu'il est difficile de préciser sur un "fond continu" de distribution incertaine.

La Figure 3 reproduit des courbes inédites obtenues pour le magnésium au cours d'une étude critique comparée effectuée par Sénémaud pour Mg

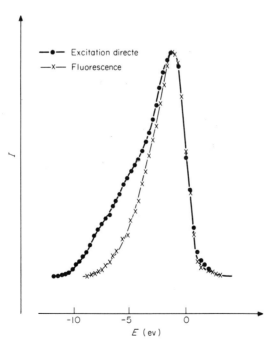

FIG. 3. Emission $K\beta$ du magnésium métallique par excitations "directe" et "indirecte".

et MgO, par excitation secondaire. Ici encore la bande correspondant au métal présente une forme plus pure que celles précédemment publiées. Cependant, un travail effectué dans notre laboratoire par excitation électronique directe (Callon, 1959), avait déjà montré que la forme globale observable résultait de la composition d'une émission due à l'oxyde avec une bande plus étroite et à un seul maximum due au métal. Dans ce mode d'excitation, l'allure de la courbe globale qui est modifiée par la profondeur du cheminement efficace des électrons, dépend de la tension appliquée au tube (Bonnelle, 1966).

Dans l'excitation par fluorescence, l'émission $K\beta'$ de l'oxyde a pratiquement disparu. Compte tenu de la largeur du niveau K et de l'effet instrumental, la largeur de la distribution $N(\varepsilon)$ de Mg "métallique" de la relation (3), serait ici d'environ 7 ev, ce qui reste en bon accord avec les observations de Rooke (1963) pour la bande $L_{2,3}$ et les calculs de Falicov (1962) sans correction de Bohm et Pines. Pour Mg la valeur du plasmon de volume (10·0 à 10·5 ev) ne coïncide pas avec la séparation $K\beta$ métal–$K\beta'$ oxyde, ce qui facilite la recherche d'une éventuelle émission due au plasmon.

B.   Absorption des Rayons X Ultra-Mous par les Métaux Lourds

Pour terminer, revenons à l'absorption. Nous rappellerons que l'utilisation du rayonnement orbital du synchrotron à électrons de Frascati, rend possible à un groupe de chercheurs de notre laboratoire en collaboration avec un groupe de l'Istituto di Sanita de Rome, d'effectuer, entre autres, des mesures continues en fréquence des coefficients d'absorption de couches métalliques minces dans le domaine x ultra-mou. Une série de métaux lourds à déjà été étudiée : Au, Bi, Pb, Pt, Ta, etc.... à l'aide d'un spectrographe, à réseau de Siegbahn, construit dans notre laboratoire.

Dans les conditions où des discontinuités d'absorption N ou O à attendre ne sont que faiblement ou pas observables, des variations importantes de μ, alors tout à fait inattendues, ont été mises en évidence, par Jaegle et al. (1966 et 1967), comme le montre la courbe de la Fig. 4 pour l'or entre 120 et 26 Å, c'est-à-dire pour des énergies de 100 à 450 ev. Les mesures faites sur le bismuth puis d'autres métaux, confirment le phénomène qui se manifeste sous l'apparence d'une région de transparence relativement large, plus ou moins accentuée, avec quelques structures secondaires peu prononcées.

L'interprétation de cet effet a été donnée par Combet-Farnoux (1967) et par Combet-Farnoux et Heno (1967). Le calcul de $\sigma(\nu)$ se fait, non en considérant le rôle d'une seule couche électronique, mais en composant les sections efficaces partielles des différentes couches les plus extérieures. Pour chacune, le calcul est analogue à celui effectué par Cooper (1962, 1964) pour des atomes légers. On part d'une relation du type (4) où ψ_i est une fonction d'onde atomique, ψ_f une fonction d'onde relative à l'ion positif avec l'électron dans son potentiel central, ψ_i et ψ_f étant des produits

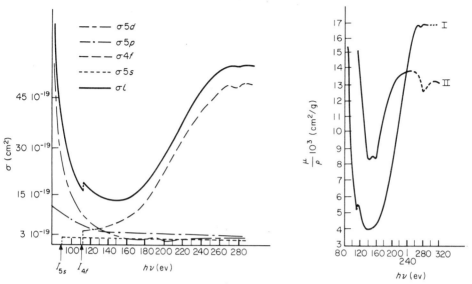

FIG. 4. Absorption de l'or entre 120 et 26 Å. Courbe I, les prévisions théoriques; courbe II, les résultats d'expérience.

antisymétrisés de fonctions d'onde monoélectroniques; on admet d'ailleurs le théorème de Koopmans. Le calcul qui ne prend en considération pour les états finaux que le transfert dans le continuum et se réfère à l'atome libre, suffit, du fait de l'introduction du champ central, pour rendre compte des courbes expérimentales obtenues avec des métaux; alors que l'approximation hydrogénoide ne laisse attendre que la variation en "dents de scie" avec discontinuités.

Bibliographie

Bonnelle, C. (1966). *Ann. Phys.* **1**, 439.

Bonnelle, C. (1968). "Physical Methods of Inorganic Chemistry", Chap. 2. Wiley, New York.

Callon, P. (1959). *C.R. Acad. Sci., Paris* **248**, 1985.

Cauchois, Y. (1948). "Les Spectres de Rayons X et la Structure Électronique de la Matière". Gauthier Villars, Paris.

Combet-Farnoux, F. et Heno, Y. (1967). *C.R. Acad. Sci., Paris* **264**, B138.

Combet-Farnoux, F. (1967). *C.R. Acad. Sci., Paris* **264**, B1728.

Cooper, J. W. (1962). *Phys. Rev.* **128**, 681.

Cooper, J. W. (1964). *Phys. Rev. Lett.* **13**, 762.

Falicov, L. M. (1962). *Phil. Trans.* **255**, 55.

Jaegle, P. et Missoni, G. (1966). *C.R. Acad. Sci., Paris* **262**, B71.

Jaegle, P., Missoni, G. et Dhez, P. (1967). *Phys. Rev. Lett.* **18**, 887.

Rooke, G. A. (1963). *Phys. Lett.* **3**, 234.

Sénémaud, C. (1968). Thèse, Paris.

Sureau, A. et Berthier, G. (1963). *J. Physique* **24**, 672.

K-Absorption of Boron in Boron Trioxide

L. Jacob,[†] R. Noble, H. Yee and B. Fraenkel[‡]

Department of Physics, University of Liverpool, England

ABSTRACT

Foils of B_2O_3 were supported on a specially designed frame so that they could be moved behind the main slit of a one-metrre concave grating spectrograph. In this way it was possible to avoid electron bombardment, specimen contamination and surface charges, and to keep the film intact for the long exposures required to obtain K-absorption spectra.

The absorption curves for B_2O_3 show considerable structure. There is a small absorption-edge at 69·4 Å and a sharp absorption line at 65 Å, followed by several small edges to a maximum at about 62 Å; this is considered to be the main absorption-edge, corresponding to a transition from the K level to the bottom of the empty conduction band. To the high-energy side of the main absorption-edge the structure is associated with the density-of-states of unoccupied levels, and structure to the low-energy side is ascribed to the existence of "exciton" levels.

1. INTRODUCTION

Much interest in recent years has centered on the group of solid non-metals in the periodic table that begins with boron, and whose members have high refractive indexes. In fact, the latter criterion has been linked to the property of photoconductivity; at the same time it has been shown that many of these non-metals possess semiconducting properties. A general account of the form of the band-structure, for both metals and non-metals, has been given by Skinner and O'Bryan (1940) in several surveys of the field. In the case of boron trioxide Skinner found band-doubling in the emission spectrum of the boron ion; it was therefore of interest to examine the absorption spectrum of the oxide, to elucidate its band-structure and to look for any special features of the absorption process.

Absorption studies of this kind, in the soft x-ray region from 100 to 1000 Å, have been made on selenium by Givens and Siegmund (1952); on iron and iron oxide by Givens and Carter (1953); and on tellurium by Woodruff and Givens (1955), and by Goffe *et al.* (1955). Our objective was to make a study of the fine structure of the absorption-edge of boron, on the same lines as

† Present address: Department of Natural Philosophy, University of Strathclyde, Scotland.
‡ Present address: Hebrew University of Jerusalem, Israel.

that already carried out by Skinner and Johnston (1937) for the metals Li, Mg, Ni and Cu.

Differences in the two cases arise, apart from considerations of shell structure, from differences in the character of the absorption-edge in metals and non-metals.

Owing to the technical difficulties associated with the production of pure-boron films of the required thickness, it was decided to use films of B_2O_3 as absorbers; though an account has been given by Moss (1952) of a method of producing thin films of boron by evaporation from a graphite crucible at about 2500°C. Comparison was made with the emission spectrum of B_2O_3 in order to throw light on the effective gap-width in the oxide.

2. THE VACUUM SPECTROGRAPH

We used the same spectrograph as described by Skinner (1940), but modified for absorption work. It had a concave glass grating with 3×10^4 lines per inch, of one-metre radius of curvature. Thin *Ilford-Q* plates were bent to the curvature of the Rowland circle, and the main slit (of 0·03 mm width) had two auxiliary defining slits placed between it and the grating. The square-sectioned water-cooled target was coated on its three sides respectively with aquadag, B_2O_3, and boron; a 22-carat gold strip (of 0·3 mm thickness) was gold-soldered to the remaining face to provide the background of *N*-radiation used for the absorption work.

This target remained clean throughout the long exposure required, was unaffected by opening the chamber to air, and gave reproducible direct-background radiation for the calibration of the plate emulsion.

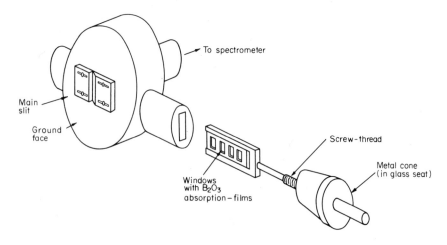

FIG. 1. The absorption specimen-chamber.

A specially designed absorption-chamber, shown in Fig. 1, was placed between the flange of the x-ray tube and the flange supporting the main slit. It held a framework of six windows each 10 mm × 2 mm, separated a distance of 7 mm between centres. Usually, three adjacent windows were covered by the absorbing film, the remainder left open. A ground metal cone, which rotated in a vacuum-greased glass socket, moved the windows behind the main slit; either introducing the absorbing film into the path of the x-ray beam, or allowing it to pass unimpeded. The framework was calibrated by focusing a microscope at the centre of the windows in turn. Exposure times for the absorption were in the region of 10 h using 3kv, 50mA; emission spectra of carbon and of the oxide were obtained in 0·5–2 h.

3. PREPARATION OF B_2O_3 ABSORBING FILMS

There are severe difficulties involved in making foils for absorption in the x-ray region between 50 and 500 Å, since very thin films are required and it is necessary to control their homogeneity. These problems apply particularly to foils or screens involving the use of chemical compounds as absorbers: used, for example, in seeking information on the effect of chemical combination on the position, width, slope, or structure of the absorption-edge.

Further, it is of the greatest importance that an absorber should transmit sufficient intensity to the photographic plate or recording device, to ensure that the response-time will not be too long. We have therefore been faced with the problem of providing an absorbing foil of B_2O_3, and it is useful to summarize our experience with the methods employed to this end.

(a) Deposition of finely powdered B_2O_3 on to a film of formvar in chloroform which was picked-up and held on a window: films were found to have minute gaps in the powdered surface which rendered them unreliable.

(b) Painting of a suspension of finely powdered B_2O_3 in amyl acetate on to a *formvar* base: gave a coherent film, but of thickness not controllable and usually too thick for the wavelength-range used.

(c) Mixing of aqueous solutions of various strengths of B_2O_3 with the *formvar* solution and casting on to a water surface: films were patchy and poorly coherent.

(d) Painting of aqueous solutions of B_2O_3 of various strengths on to *formvar* base: after drying, gave good coverage but poor cohesion and excessive thickness $(5–10\mu)$.

(e) Dipping of an electron microscope grid (6–8 mesh/mm) into molten B_2O_3 gave excellent adhesion with a glass-like film, but excessive thickness.

The successful technique consisted of blowing bubbles from the end of a Pyrex or brass tube, 3 mm internal diameter, after heating in a flame the powdered B_2O_3 collected on the end of the tube; the melt forms an almost clear liquid, filling a 3–5 mm length of the tube. The tube was withdrawn from the flame and allowed to cool a little, then blown out into an elongated bubble of irregular form. When first blown the bubbles are transparent and exhibit thin-film interference colours in one or more places on the surface; it is possible to pick off a suitable region, of thickness ~ 500 Å, on to a hot metal frame soon after their formation. The range of thickness, determined by both Michelson and Rayleigh interferometers lay between 200 and 500 Å; no claim is made to absolute values.

The films when allowed to stand in air for some time became cloudy due to weathering by the atmosphere, and the initial compressional strain pattern became irregular and more pronounced with time; newly made films were placed in a desiccator and used within an hour of formation.

4. PLATE CALIBRATION

There are two separate problems involved in calibrating the photographic plate: first, the dispersion of the spectrograph over the wavelength-range used; second, the response of the emulsion over the range of intensities encountered during the period of the exposure.

The dispersion was obtained by photographing the various orders of the carbon spectrum (first order ~ 44.8 Å) on the same plate as that used for the absorption; the necessary adjustments to the shutter of the spectrograph were made to secure correct positioning on the plate. We were primarily concerned with the 50–80 Å region of the boron absorption-band; this lies between the first and second orders of the carbon K-emission band, and since these were clearly delineated in the microphotometer traces, we regard a linear interpolation of the spectral features as sufficiently accurate to provide a reliable wavelength-scale over the range of the absorption-band.

The second problem involving the response of the emulsion was much more difficult. We proceeded in the same manner as Skinner, taking a direct photograph of the background radiation with no absorber present, and arranging the exposure to be such that the average intensity over the wavelength range was close to that obtained in the much longer exposure through the absorbing foil. Thus, by trial and error, we obtained a scale of background-radiation intensity within limits of intensity of the order of that transmitted through the absorber.

Application of the Bunsen-Roscoe reciprocity law then enables the illumination on the plate to be determined in terms of the photographic density and the time of exposure; that is, providing it is assumed that the response of the emulsion does not depend on the wavelength over the range recorded. This assumption is reasonable and was adopted in determining the response curve of the plate.

However, it was necessary to make an experimental check on the reciprocity law, because its validity has never been properly established in the soft x-ray region. This was done using the $Q1$ type of emulsion in a higher dispersion spectrograph; that is, over a range of about 20–50 Å. The procedure was to make two exposures for periods of time t_1 and t_2 in the ratio 2:1; there was, of course, a region of overlap where the exposure-time was $(t_1 + t_2)$ and use was made of intensities in this region to supplement the intensity-scale derived from the other exposures on the plate.

Deflections of the microphotometer trace for the first exposure were plotted against those for the second. For any initial arbitrary deflection in the first exposure, it was possible to read from this curve, values of deflection for twice, four times, eight times, etc., the initial exposures; that is, a curve of relative exposure against deflection could be drawn to represent the response curve of the emulsion. The whole process was repeated using various initial values for the arbitrary deflection, and an average response-curve was obtained which could be regarded as a reasonably reliable calibration of the plate, within the range of densities likely to be encountered. Repeating the exposures with other plates of the same batch gave uniform results.

The microphotometer readings from which the intensities are plotted affect the form of the response curve, and therefore the following considerations are made. If we assume that an exposure E versus density D curve (or "H and D" curve) for the emulsion in the soft x-ray region is like that in the optical region, then within the latitude (or sensitivity-range) of the plate.

$$D \propto \log E, \quad \text{that is:} \quad \log \frac{1}{T} \propto \log E$$

where T is the ratio of light-transmission of exposed and unexposed film (measured with the microphotometer). Thus, $1/T \propto E \propto I$, where $I =$ intensity of incident x-radiation.

With galvanometer recording,

$$\frac{\theta_l - \theta_0}{\theta_p - \theta_0} \propto I$$

where θ_l is the photometer deflection for unblackened plate, θ_0 is the deflection for zero light-intensity and θ_p is the deflection for light transmitted by exposed portion of plate.

Since $\theta_l - \theta_0$ is a characteristic of the plate and microphotometer arrangement adopted, and for any fixed geometry is a constant of the system, the intensity I of the diffracted radiation reaching the plate is proportional to $1/(\theta_p - \theta_0)$. The relationship between $1/(\theta_p - \theta_0)$ and relative exposure should thus be linear over that range of exposures permitted by the latitude of the plate. For the carbon background there is reasonable linearity up to relative-exposure ratios of 50:1. Due to variable factors during an exposure, the linear response extends over a smaller range, 30:1.

It is well known that during times of exposure, of the order of hours, conditions in the x-ray tube mainly affecting the surface state of the target do not remain constant. This means that the blackening on the photographic plate represents an integrated effect of an intensity which is varying with time in some uncontrollable manner. Experimentally, it is easy to gauge the spread of intensity-ratios on the plate; for a fixed ratio of exposure-times, for example, it can vary 50% either way when this ratio is 2:1.

To obtain the effective intensity, use is made of the region of overlap in the following manner. A relative-exposure curve is plotted by reading-off pairs of values from exposures in the ratio $R:1$. This represents the response curve for both exposures; hence, it includes intensities ranging from the arbitrary unit chosen on one curve, to R^n on the other. We examine the region of overlap, and at wavelengths corresponding to those already taken to provide the response curve, we read-off the corresponding deflections for the intensities $(1+R)R$, $(1+R)R^2$, etc., and plot them. If R (the ratio of exposure-times, i.e. the ratio of intensities) fits the reciprocity law, these two curves should coincide over the limit of intensities permitted by the plate.

In practice they are never superimposed because of the variable factors in the source; it is therefore necessary to know the true value of R in order to obtain the correct intensity-ratios from the microphotometer traces. This is achieved by assuming a series of values of R close to the actual ratio used and replotting the response curves for each. The correct value of R yields a curve for the region of overlap which coincides with that on either side; in our case the approximate ratio was 1·6:1 when the actual ratio used was 2:1. To illustrate this a series of curves for a gold target is shown in Fig. 2.

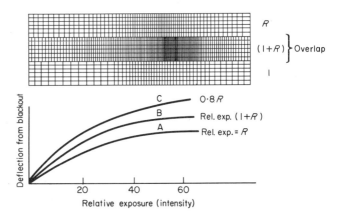

Fig. 2. Response curve for the photoplate emulsion.

Curves A and B respectively represent the response curves for relative exposures of R and $(1+R)$; they do not coincide. Curve C represents that derived from A and B separately, when the value of R was reduced to nR, with $n = 0·8$. This variation of about 20% in R is thus associated with the variation in x-ray output under the conditions used.

We conclude from our experiments that it is likely that the reciprocity law would hold under conditions where the intensity was rigorously controlled, even over long periods of time—at least up to the limit of intensity defined by the latitude of the emulsion. It was significant that the linearity of response, up to intensity ratios of 20:1, covers all the variations in intensity over the absorption band; it was only by using the kind of technique described that an effective exposure-time scale was obtained enabling interpretation and correction of intensity variation, from which the average absorption curve of the foil was determined.

5. THE ABSORPTION BANDS

Using the response curve, the microphotometer readings for both the direct and absorbed beams were respectively interpreted to give reliable values of the intensities I_d and I_a, and to provide the absorption curve.

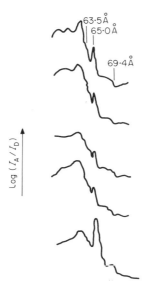

FIG. 3. *K*-Absorption curves for B_2O_3.

A series of such curves is shown in Fig. 3, where considerable structure can be seen and several features are clear. There is a small absorption-edge at about 69·4 Å and a fairly sharp absorption line at 65·0 Å. Beyond this, the absorption continues with several small edges to a maximum near 62 Å, which is believed to be the main absorption-edge and corresponds to a transition from the *K*-level to the bottom of the conduction band. The structure beyond this edge is most probably connected with the form of the density-of-states curve for the normally unoccupied levels.

We must postulate a special origin for the long-wavelength and short-wavelength peaks. The small absorption-edge at 69·4Å may have a similar origin to the band-doubling observed by Skinner and O'Bryan (1940) in emission experiments on oxides; that is, after emission the lattice may not be left in its lowest state. The mechanism involved in producing this discrete excited-state, involves the transfer of an electron from an oxygen ion to a neighbouring positive ion. The emission band at the same wavelength as the pure-element emission, corresponds in the case of B_2O_3 to emission from positive ions; that at lower energies, to emission from neutral atoms. In the present results, the absorption-edge at 69·4 Å may perhaps correspond to absorption by the few neutral atoms present in the lattice; the more normal absorption takes place in the positive ions, some 16 ev below the main absorption.

The thickness of the majority of films used in the absorption work was estimated from the displacement of fringes in an interferometer, but no accurate values can be claimed, and we set a limit of 500 Å for the maximum permissible thickness. The spread in thickness is reflected in the absorption curves, which resemble each other in general features, both in form and position. One difference, arising from variations in thickness of film, is the variation in scattered intensity to be expected; this, if serious, might considerably affect the absorbed intensity. To check this, and also to see if there existed reasonable proportionality between the values of μt given by the $\log (I/I_d)$ curves, μt is plotted for different wavelengths for three of the films (Fig. 4). The lines are reasonably parallel; this indicates a uniform variation in absorption at the different wavelengths.

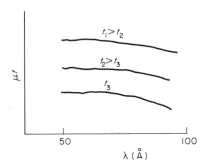

Fig. 4. Variation of μt with wavelength for B_2O_3 films. Range of thickness $\sim 2:1$ (thinnest ~ 250 Å).

6. EXCITON-LEVELS IN ABSORPTION BANDS

An average curve for the absorption, obtained from an examination of a large number of absorption bands, noting the common features, is presented in Fig. 5 together with the B_2O_3 emission band for comparison. The main absorption-edge occurs some 7 ev from the emission-edge and gives a measure of the gap-width, i.e. the energy required to take an electron from the top of the valence band in B_2O_3 and place it at the bottom of the conduction band. The fine-structure to the low-energy side of the main absorption band is attributed to the formation of exciton-levels, produced below the conduction band by the additional core-change of the ionized x-ray level.

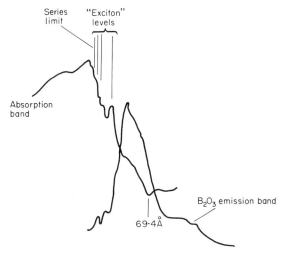

FIG. 5. Average absorption-curve, with the B_2O_3 emission band superimposed.

These are not of the usually accepted type of exciton which is an association of a hole in the valence band with a corresponding electron in the conduction band. Combinations of the latter kind can have separations of several atomic diameters and, on moving through the crystal, can transfer energy to say, an F-centre, liberating a photoelectron. The bound quantum states of such a system are hydrogen-like in character, and can be predicted with the aid of a modified Bohr-formula, involving the dielectric constant of the material. The kind of exciton postulated (Fig. 6) cannot move through the crystal, since the hole occurs in an inner atomic level, and the hole–electron separation is of the order of an atomic radius; the energy levels are no longer hydrogen-like but are similar to those of a free atom.

The core of the atom behaves rather like that of an atom with the atomic number of the next element in the periodic table, i.e. carbon, and the exciton levels will hence correspond to transitions from the $1s$-level to the $2p$-level of carbon. The strong absorption line, which is a feature of all the spectra, is taken to represent the transition to the first excited p-state of carbon, and other excited p-states are found to be in reasonable agreement with the absorption-edges shown in the main absorption band.

At least the first three are resolved, which is shown schematically in Fig. 7 where the overlapping of the levels is comparable with, or exceeds, their separation.

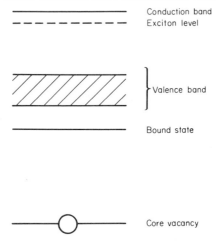

FIG. 6. Formation of exciton: vacancy in the core.

The exciton-levels converge to a series limit when the 1s-electron is taken into the 3p-levels of carbon, and the separation expected from the energy-level diagram of carbon agrees with that shown by the experimental curves, i.e. about 7 ev.

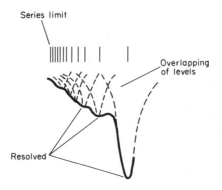

FIG. 7. Overlapping of exciton-levels affecting the resolution.

7. SUMMARY AND CONCLUSIONS

The gap between valence and conduction bands in B_2O_3 is about 7 ev. A series of exciton levels is proposed to account for the absorption edges found in the main absorption band. Because of the missing 1s-electron in the core, these are associated with the atomic levels of carbon, the next element in the periodic table. The weak absorption on the low-energy side is attributed to a small number of neutral atoms of boron present in the lattice; and the stronger absorption to the positive ions. The transfer of an electron from the oxygen atom in the compound to a neighbouring positive-ion shifts the energy levels, and the discrete excited-state is left some 15 ev above the ground state.

Acknowledgements

We wish to record our gratitude to the late Professor H. W. B. Skinner for suggesting the problem, for placing his spectrographs at our disposal and for encouragement given during the course of the work.

References

Skinner, H. W. B. (1939). *Phys. Soc. Reports* **5**, 257.
Skinner, H. W. B. (1940). *Phil. Trans.* **239**, 95.
Givens, M. P. and Carter, D. E. (1953). *Phys. Rev.* **90**, 379.
Givens, M. P. and Siegmund, W. P. (1952). *Phys. Rev.* **85**, 313.
Woodruff, R. W. and Givens, M. P. (1955). *Phys. Rev.* **97**, 52.
Goffe, W. L., Koester, C. J. and Givens, M. P. (1955). *Phys. Rev.* **100**, 1112.
Skinner, H. W. B. and Johnston, J. E. (1937). *Proc. Roy. Soc.* **161**, 420.
Moss, T. S. (1952). "Photoconductivity", p. 78. Butterworths, London.
Skinner, H. W. B. and O'Bryan, H. M. (1940). *Proc. Roy. Soc.* **176**, 229.

A Survey of Experimental Factors
and Studies of some *K*-Emission Spectra
Using Fluorescence Excitation

A. FAESSLER

Sektion Physik des Universität München, Germany

ABSTRACT

Some of the problems of obtaining accurate emission-band data are surveyed, with particular regard to the method of excitation for the x-ray emission. With electron-bombardment excitation only metallic materials and a few compounds with high melting-point such as oxides and carbides can be investigated. Fluorescence excitation permits the examination of a large variety of compounds of all possible bonding types and would have considerable advantages for all target materials. The results for a number of systems, using fluorescence excitation, are discussed.

1. INTRODUCTION

A main objective of this book is to discuss the results of soft x-ray spectroscopy in the light of theoretical calculations of the energy band-structure of solids, chiefly of metals and alloys. Detailed features in the intensity distribution of emission bands have been compared with the results of theoretical work. However, we should ask ourselves whether it is not too early to engage in extended discussion of details which have yet to be shown, beyond doubt, to be reproducible.

It is clear in soft x-ray spectroscopy that it is often not easy to estimate correctly the reliability of the observed intensity distribution for an emission or absorption spectrum. The true intensity distribution may be affected by numerous experimental factors, some of which have been mentioned in the foregoing papers.

2. EXPERIMENTAL FACTORS THAT AFFECT THE SOFT X-RAY EMISSION

A. Excitation

(i) *The Problem*

In the ultra-soft x-ray region, all emission spectra so far have been measured with electron-bombardment (or primary) excitation—with the

exception of fluorescence analysis, which is an application of x-ray spectroscopy not considered here, and for which apparatus of very small resolution
is generally used. Fluorescence (or secondary) excitation has been employed
for the study of emission bands only in the region below 10 Å. The question
that arises is whether the structure of an emission band depends upon the
method of excitation.

At Munich, before we began our studies of valence-electron transitions,
we extensively investigated the influence of chemical bond on the wavelength
of the K_α doublet of the elements silicon, phosphorus and sulphur (see for
example Faessler, 1963). In order to study a great variety of compounds,
fluorescence excitation was used. Special demountable secondary x-ray
tubes, originally developed for what is now called fluorescence analysis
(Alexander and Faessler, 1931), give good intensities and were found to be
very suitable for this type of investigation.

(ii) The K_α Doublets of $_{14}Si$, $_{15}P$ and $_{16}S$

The application of fluorescence excitation to the study of the wavelength
dependence of the K_α doublets of the elements silicon, phosphorus and
sulphur, immediately gave consistent results; while primary excitation gave
results which were very often obscured by chemical disintegration effects on
the target. It was found that the shifts of the K_α lines are quite considerable:
for example, in the case of sulphur a maximum shift of 3 Å was observed,
corresponding to 1·3 ev, which is a surprisingly high value considering the
fact that the bond energies themselves amount to a few electron-volts only.
In particular, the study of a large number of sulphur compounds revealed
the following rules for the shifts of the K_α doublet: firstly the lines of the
atom in a compound are shifted towards shorter wavelengths or higher
energies—relative to the lines of the free element—if the emitting atom has
a positive charge, and toward longer wavelengths or lower energies, if it
has a negative charge; secondly, with increasing valency (or oxidation
number) and increasing electronegativity of the partner element in the
compound, the K_α lines are shifted to shorter and shorter wavelengths. Thus
the values observed for the wavelengths of the K_α lines, give information
on the charge of the atom in the substance. Knowledge of the charge of the
atom allows, for example, conclusions with regard to the type of chemical
bond in the compound under investigation. This approach to the problem
of bond character has led, among other things, to interesting results in a
study of silicon–metal compounds, of which some are metallic phases, and
others are compounds with ionic-type bonds.

(iii) Is There a Special Cathode-ray Effect?

Returning to the problem of the two types of excitation, we occasionally
used primary excitation in our investigations in order to save time; the
exposure-time with fluorescence excitation is intolerably high for compounds
containing one or more elements of high atomic-number. We found that

certain groups of compounds—such as oxides, carbides and a few salts of oxygen acids with high melting-points—when supported on a noble-metal target and bombarded with electrons, were not decomposed or affected.

Therefore, when we began valence-electron transition studies (over 10 years ago) with an investigation of the S $K\beta$-emission band of rhombic sulphur and a few sulphides and sulphates (Faessler and Schmid, 1954), we used secondary as well as primary excitation, because the avoidance of very long exposures was a still more difficult problem in this case than with the stronger $K\alpha$ lines. Some of the compounds were chemically unaffected by electron bombardment, when platinum targets were used and when the anode current-density at the focal spot was kept low. This could be shown by observing the $K\alpha$ doublets, for which the wavelengths were known from earlier measurements, or by taking *Debye-Scherrer* photographs of the substances removed from the target after exposure to the electron beam.

In some cases we found small differences between the spectrum of a substance obtained with fluorescence excitation and that obtained with electron-impact excitation. Since it was not possible to explain these differences by chemical changes, we concluded that the two kinds of excitation might not give entirely identical spectra, and suggested the possibility of a special cathode-ray effect, to be studied in further investigations.

Later we repeatedly came back to this question, but in spite of much effort it was not possible to confirm the above observations. The most careful study made was of the complete Si K-emission spectra of elementary silicon, of SiO_2 (quartz) and of SiC (carborundum). The first investigation was carried out by Kern (1960) who used electron-impact excitation of the K-emission bands after making sure that there were no chemical changes of the material on the target. Later these measurements were repeated by Krämer (1960) who used the same spectrograph, but with fluorescence excitation. The agreement of the two sets of spectra was as good as could possibly be expected. In fact, an example was never found of the two types of excitation giving different spectra for the same stable substance; only when some chemical decomposition, or change of the substance, is caused by the electron-bombardment do we find differences between the spectra excited by electron impact and those excited in fluorescence.

B. Limitations for the Application of Primary Excitation

(i) *Observed Effects in the* K *and* L *Spectra of Light-metal Oxides*
With the results of Kern and Krämer in mind, we were very surprised, in the course of a study of the Si L-spectra of the same three substances, to make the observation (Wiech, 1967) that SiO_2 slowly decomposes under electron bombardment. If a few runs are made with the same specimen on the target, the characteristic intensity-distribution for SiO_2 changes slowly in such a way that one finally observes the band structure of elementary

silicon. Similar effects have been found by Wiech (1966) in the case of the Al L-emission band of Al_2O_3.

The explanation of these apparently contradictory observations is less difficult than may be expected. It is possible that a slow disintegration always occurs when SiO_2 is bombarded with electrons: if the disintegration proceeds slowly from the surface into the interior of the crystal grains, then the effect of the disintegration may not be noticeable with K-radiation emerging from deeper layers, but it may be observable in studying ultra-soft L-radiation emitted from the outermost layers which have been decomposed in the course of the electron bombardment. An explanation which seems to be more likely, is that the differences found in the spectrum of SiO_2 were a result of the rather different vacuum in the two sets of experiments: the $K\beta$ spectra were taken with a vacuum of 10^{-5}–10^{-6} torr, and the L spectra with a much higher vacuum of the order of 3×10^{-8} torr. It is probable that with the lower vacuum a re-oxidation occurs of the silicon or lower silicon oxides formed by the electron bombardment of the SiO_2. With higher vacuum the oxygen produced by the decomposition of SiO_2 is pumped away and increasing quantities of lower oxides, and finally elementary silicon, are formed during the exposure of the specimen.

Whatever the explanation for the decomposition of SiO_2, the above example indicates the caution necessary in studies of spectra of bound elements when primary excitation is used. To date, for reasons of poor intensity, only primary excitation has been used in soft x-ray spectroscopy, and many of the earlier observations with photographic recording, which frequently required very long exposures and comparatively high anode-currents at the focal spot, will have been distorted by chemical changes of the substance under electron bombardment.

The example shows that even with very much reduced tube current and short exposure-times, one has to consider such effects. Our studies of the influence of chemical bond on x-ray spectra, over many years, have shown that the number of compounds that may be exposed to electron bombardment on the target of an x-ray tube without decomposition is rather limited. These observations indicate that the effect may be greater in soft x-ray spectroscopy where the current density at the focal spot is usually small, but where the radiation is emitted from a thin surface layer only.

(ii) *Remarks Regarding Pure Metals and Alloys*

One would expect conditions to be generally more favourable in the study of metallic specimens, for which the cooling of the target is more effective so that its temperature in the focal-spot region may be lower than for powdered material of low thermal-conductivity. However, even with alloys, one should make sure that the specimen composition does not alter. The constituents of an alloy, if it is composed of very different metals, may have different rates of evaporation. Also, under the action of the bombarding electrons, it is possible that diffusion processes may occur, with the result that the percentages of the constituents in the emitting area are different from those of the bulk material.

It follows that it is most desirable to use fluorescence excitation in the emission spectroscopy of soft x-rays; not only would we then practically always avoid chemical changes of the substances under investigation, but this kind of excitation would allow the study of a much greater variety of chemical substances.

However, at present and for some time to come we are compelled for reasons of intensity to use primary excitation not only in soft x-ray spectroscopy, but frequently in more general x-ray spectroscopy. Therefore it is worthwhile making systematic investigations of the physical and chemical processes in the focal-spot region of an x-ray tube target. Very little work has been done in this direction since the early days of the application of x-ray spectra to the qualitative and quantitative determination of chemical elements (see for example Coster and Nishina, 1925; Glocker and Schreiber, 1928; Hevesy and Faessler, 1934), for the reason that the excitation has been usually by primary x-radiation from a sealed x-ray tube.

C. Self-absorption

There are additional experimental factors that affect the structure of a spectrum in such a way that the true intensity distribution may be rather different from the observed. One, of principal importance, is the absorption of the radiation in the specimen itself. Liefeld, in Part 2 of this Volume (p. 133), draws our attention to this point which has been overlooked by many workers. Liefeld shows some striking examples of the effect; but it may be the case that self-absorption does not always have as much effect as he observes for the *L* spectra of the first transition series elements, where both emission-edges and absorption-edges are rather broad, and in addition overlap is considerable. In cases where the emission and absorption edges are steeper and have less overlap, there may only be a small self-absorption effect which alters very little the position and the slope of the emission-edge; in favourable cases there may be no effect at all. It is necessary to study every case carefully and often systematic experiments will be necessary to account for the effects of self-absorption.

D. Other Factors

Additional experimental effects must be considered that may influence the band spectrum and cause the observed intensity-distribution to require several corrections. Wiech (1966), in reporting the Al *L*-emission band studied with a new concave-grating spectrograph, has discussed all such factors; target contamination (mainly by carbon), the wavelength dependence of reflectivity of the grating, the energy dependence of the quantum yield of the detector, background effects and finally the resolution and the "window effect" of the slits of the spectrograph. We do not know the full influence of these factors (see Wiech, 1966) and it will require considerable effort before we are in a position to account fully for them all.

References

Alexander, E. and Faessler, A. (1931). *Z. Physik* **68**, 260.
Coster, D. and Nishina, J. (1925). *Chem. News* **130**, 149.
Faessler, A. and Schmid, E. D. (1954). *Z. Physik* **138**, 71.
Faessler, A. (1963). *In* Proc. 10th Colloq. Spectroscopicum Internationale, p. 307.
Glocker, R. and Schreiber, H. (1928). *Ann. d. Physik* **85**, 1089.
Hevesy, G. v. and Faessler, A. (1934). *Z. Physik* **88**, 336.
Kern, B. (1960). *Z. Physik* **159**, 178.
Krämer, H. (1960). Dissertation, Munich.
Wiech, G. (1966). *Z. Physik* **193**, 490.
Wiech, G. (1967). *Z. Physik* **207**, 428.

Part 2

HEAVY-METAL AND ALLOY SPECTRA AND COMPARISON WITH OTHER OPTICAL METHODS

Soft X-Ray Emission Bands and Bonding for Transition Metals, Solutions and Compounds

J. E. HOLLIDAY

Edgar C. Bain Laboratories, U.S. Steel Corporation, Monroeville, Pennsylvania, U.S.A.

ABSTRACT

A number of emission bands from transition metals and from their compounds and solutions, have been compared with theoretical band-structure calculations and specific-heat measurements. Some qualitative agreement was obtained with band theory after correcting the emission bands for instrumental and inner-level broadening, for overlapping bands and for background (though not for self-absorption which does not appear to affect the shape of the band for transition metals as much as has been anticipated), and with special care taken to avoid surface contamination.

Observed changes in intensity distribution and peak wavelength for the emission bands from alloys and compounds, relative to those for the pure metal, were not readily correlated with the band picture. However, good qualitative agreement was obtained with electro-negativity and bonding concepts; and the results suggest that more theoretical work should be done to relate soft x-ray emission bands to electro-negativity and bonding.

PREFACE: SOME INTRODUCTORY REMARKS TO PART 2

A Role for Soft X-Ray Spectroscopy

One purpose of this conference was to narrow the gap between theory and experiment. This gap has been attributed to a number of instrumental and secondary electron-transition effects that mask the true shape of the emission band, and some of these effects have been outlined by Faessler in his concluding remarks to Part I. Since the original classic measurements made by Skinner, there has been considerable effort to narrow the gap by paying careful attention to instrumental effects. This is to be seen in the papers presented in this Volume; however, the discussion in these papers does not reflect complete agreement on the relative importance of such effects as self-absorption, surface contamination, excitation voltage and methods of excitation.

At the excitation voltages normally used (~ 4 kv) surface contamination can strongly influence the shape of the emission-band, and careful attention must be given to surface cleanliness. In this regard either ultra-high-vacuum experiments (at pressures $\sim 10^{-10}$ torr) or self-absorption experiments (which do not involve the target surface) are needed to evaluate the relative importance of surface contamination and self-absorption in producing the observed dependence of the emission-band on excitation voltage. To determine the effect of target excitation on the shape of the emission band, and the importance of using near-threshold excitation voltages, extensive comparisons are required of emission bands excited by x-rays with those excited by electrons at various voltages. However, even after the careful attention already paid to instrumental and secondary electron-transition effects, and regardless of which experimental factor is found to influence most the shape of the emission band, the gap between theory and experiment has not been significantly reduced. This is especially true of emission spectra from alloys and metal compounds where recent results have in some cases even widened the gap.

It might be possible to devise new or more refined corrections to experimental emission bands that would bring theory and experiment closer together; past progress along this line does not appear to justify further large-scale effort, although the work of Rooke, Sagawa, Watson, Dimond and Fabian, and Wiech described in Part 1, and by Cuthill in Part 2, may prove a useful approach for pure metals.

The large changes in intensity distribution and in the peak wavelengths produced by alloying, or by chemical change, clearly show that important information concerning the electronic nature of solids is to be gained from soft x-ray spectroscopy and since these parameters are not readily correlated with the band model, it would be of interest to determine whether they can be better correlated with other concepts, without the need for numerous corrections. For example, one of the most sensitive experimental parameters is peak-wavelength shift, and it has been shown that these shifts correlate well with changes in electron distribution; while the changes in intensity distribution of the emission band have been correlated with bonding. Ziman (1967) has pointed out that the bond picture is complementary to the band model and he has urged that the mathematical relation between these two models be explored.

Thus one way to reduce the gap between theory and experiment may be for soft x-ray spectroscopists to correlate their experimental measurements with the bond picture, leaving the theoreticians to relate the bond picture to the band model. If this approach were taken it would appear that considerable information concerning electronegativity, bonding, bond-energy and the rôle of resonance in bonding could be obtained from soft x-ray spectroscopy without the need for numerous corrections. Since these properties are of considerable interest in metallurgy and solid state physics, soft x-ray spectroscopy might make a greater contribution to these fields.

1. INTRODUCTION TO THE PRESENT INVESTIGATION

In the study of the x-ray emission bands from transition metals, considerable advances have been made since the early measurements by Skinner *et al.* in 1954. Correlations of experimentally observed transition-metal emission bands with the results of theoretical band-calculations have not been as good as those for the so-called ideal metals, namely lithium, sodium, potassium, magnesium and aluminium. Various explanations have been proposed for the lack of good correlation, such as the effects of instrumental broadening, inner-level width, self-absorption, satellites and surface contamination. In the study of transition-metal spectra which is described in the present paper, careful attention was given to the above factors in order to make a more meaningful comparison with theoretical band-structure. However, the results show that although certain anomalies, such as the lack of sharp emission-edges in transition-metal spectra, can be explained and some correlation can be made with theoretical bandwidths, there are still significant differences between the theoretical band-structures and the soft x-ray emission bands. This is especially evident from the study of changes in the emission bands when the transition metal is combined with other elements in a solution or a compound. In view of these results, we particularly consider the above question of whether the band model really provides the best guide or whether the concepts of electronegativity and bonding provide a better model for relating to soft x-ray emission of alloys and compounds.

2. EXPERIMENTAL

In the present investigation many instrumental refinements were incorporated into the spectrometer to minimize the experimental uncertainties already mentioned. To keep carbon contamination to a minimum, both the x-ray tube and spectrometer chambers are fitted with mercury-vapour diffusion pumps and the x-ray tube is operated at pressures in the region of 10^{-8} torr. An arrangement is provided in the x-ray chamber for cleaning the target by ion-bombardment after the initial electron-bombardment. It has been shown (Holliday, 1967a,b) that, even in a relatively clean vacuum, carbon contamination continues as long as the target is degassing; the carbon combines with a metal target to form a carbide, resulting in an extraneous contribution to the observed intensity distribution for the emission band from the metal. It is important to clean all targets after the initial degassing, in order to avoid distortion of the emission bands by chemical combination of the target with surface contaminants.

The spectrometer is a conventional grating instrument with a Rowland-circle radius of 50 cm (Fig. 1). To obtain high resolving-power, a 3600 groove/mm grating, with platinum surface and a $1°$ blaze-angle, is used with 25μ slit-widths. When there is sufficient intensity the emission bands are measured

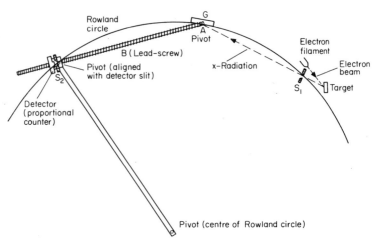

FIG. 1. Spectrometer arrangement, showing positioning of the diffraction grating (G) and the source and detector slits (S_1 and S_2) on the Rowland circle. The arm B is a lead screw (36 threads to the inch) driven from the grating pivot A, and acts also as an alignment guide for the detector slit S_2.

in the second order. Constant motion is provided by a lead-screw (Fig. 1) which is driven from a point directly beneath the diffraction grating. A travelling microscope is used to align the slits and the grating in the horizontal plane, and provides greater precision than obtained earlier using a telescope arrangement.

The detector is a gas-flow proportional counter with a Formvar window (0·03 mg cm^{-2}), using P-10 gas (90%–10% argon–methane mixture) at a pressure of ~ 10 torr. The data are plotted automatically using the step-by-step method with sufficient photon-counts taken at each angle setting to give a statistical deviation of 1·5% or smaller. Several scans are made of each spectrum to confirm the fine structure. The electron-beam current is usually between 1 and 1·4 mA, and the excitation voltage ~ 4 kv.

The lead-screw drive and improved alignment have increased the resolution, in the 10–30 Å region, over that obtained with the arrangement previously employed (Holliday, 1967a). Poor alignment may be the reason why gratings have not previously given as good a resolution as crystals in this wavelength region. It appears that the resolution of the present grating-spectrometer is now comparable with those employing crystals, because the double peak in the Fe L_3-emission band (transitions $4s + 3d \rightarrow 2p$), which has only recently been seen with a crystal analyzer (by Cauchois and Bonnelle, reported in this Volume, p. 165, and by Fischer and Baun, 1967a) can now be observed with our instrument (see Fig. 2). The width at half-maximum-intensity ($W_{1/2}$) for the Ti L_3-emission band, with the improved spectrometer, was 2·8 ev, which is the same as reported by Lukirskii and Brytov (1964); the $W_{1/2}$ for their instrumental-error curve was 0·02 Å, or

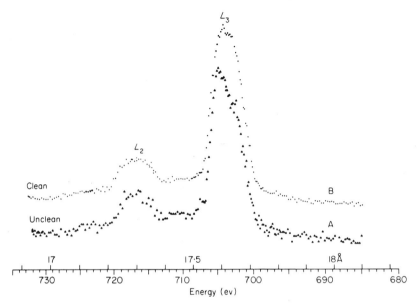

Fig. 2. The Fe $L_{2,3}$-emission band (2nd order) for iron: A, before cleaning and B, after cleaning by ion-bombardment. Note the presence of a double L_3 peak.

0·35 ev. Our value gives a resolving power of 1300 at 27·5 Å, in the first order, for the present improved instrument. For the second-order Fe L_3-emission band, this would give a $W_{1/2}$ of the instrumental-error curve of approximately 0·01 Å, or 0·3 ev, which represents a resolving power of 2400 at 17·5 Å in the second order.

Cauchois and Bonnelle (this Volume) and Bonnelle (1966), using a gypsum crystal, reported for their instrumental-error curve a $W_{1/2}$-value of 0·3 ev at 17·5 Å. The $W_{1/2}$ for the Fe L_3-emission band with our improved spectrometer is 4·7 cv, while Bonnelle reported a $W_{1/2}$ of 5·0 ev when the two peak intensities were above the $W_{1/2}$ point on the Fe L_3 band. These results show that grating spectrometers can be made with comparable resolution to crystal spectrometers, at about 17 Å; they will become superior at longer wavelengths because the energy resolution of a grating spectrometer increases with increasing wavelength. Not all the spectra presented in this paper were measured with the improved spectrometer, because there has been insufficient time to remeasure them all.

3. TRANSITION-METAL SPECTRA

A. Corrections to the Emission Bands

When comparing emission bands with theoretical band calculations it is important that the emission bands be corrected for instrumental broadening and for inner-level width. The method we have used to correct for these effects has been described in detail elsewhere (Holliday, 1967a). The necessary expression is the convolution integral

$$P(E') = \int_{-\infty}^{+\infty} T(E)\, S(E' - E)\, dE \qquad (1)$$

where $P(E')$ represents the emission-band intensity distribution obtained after correction for overlapping spectra, $T(E')$ is the required intensity distribution for the emission band before broadening, and $S(E' - E)$ is the smearing function which contains the broadening effects of the inner-level width and instrumental error. The known functions are $S(E' - E)$ and $P(E)$, and Eq. (1) was solved graphically using an analog computer; Tomboulian (1957) and Shaw and Jossem (1959) also discuss the solution of this equation by graphical methods.

Another parameter of considerable importance in the correction of emission bands, is self-absorption. Chopra and Liefeld (1964) showed that in the Ni L_3-emission band, the edge-width and peak position change with excitation voltage (between 2 and 15 kv) which they attribute to self-absorption effects in the target. However, there are other factors that could contribute to changes in emission edges with excitation voltage; for example, the contribution to the metal emission band from a surface oxide, or from a

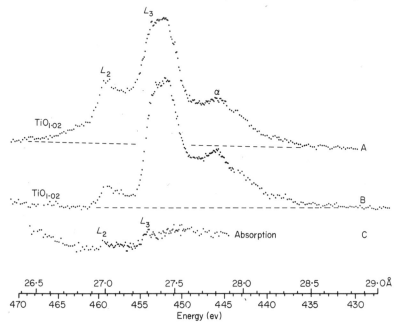

FIG. 3. Ti $L_{2,3}$-emission bands for titanium metal contaminated by oxygen: A, with no absorbing film between detector and diffraction grating and B, with a 1000 Å film of titanium metal between detector and grating. Curve C is the absorption spectrum for the titanium film.

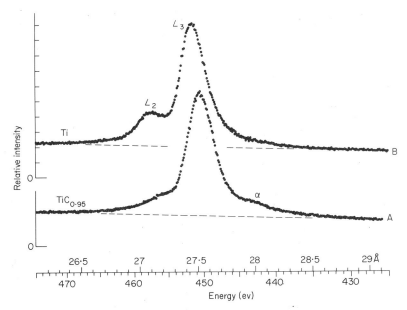

FIG. 4. The Ti $L_{2,3}$-emission band: A, for $TiC_{0.95}$ and B, for titanium metal. For $TiC_{0.95}$ the L_3 peak has shifted to lower energy.

carbide due to carbon contamination, will decrease with increasing excitation voltage. Recent experiments at this laboratory have shown that surface contamination can contribute to the metal spectra, even at an excitation voltage of 4 kv, if the target is not thoroughly degassed and cleaned. In curve A (Fig. 3) and curve B (Fig. 4) we show the Ti $L_{2,3}$-emission bands from titanium measured under two different surface conditions. Curve A (Fig. 3) was obtained after the titanium was subjected to electron bombardment in poor vacuum which resulted in a large amount of surface oxidation. Curve B (Fig. 4) was obtained after the titanium was thoroughly degassed and cleaned by ion bombardment.

Comparison of the two Ti L_3-emission bands shows that surface contamination can change the slope and shape of the emission edge, and shift the position of the peak, from the true shape and position for the pure metal (the $L_{2,3}$ peak was shifted 1·1 ev relative to that for pure titanium). The Ti L_3-emission from titanium in Fig. 4 shows an additional peak on the low-energy side of the L_3 band which has not been seen before. This peak, which has been labelled peak α has become more prominent for the L_3-emission band from contaminated titanium. Peak α is also more prominent for the Ti L_3-emission band from TiC. The emission band from TiO (see Fig. 21) is very similar to that from contaminated titanium (A, Fig. 3), and indicates that the oxygen on the surface has combined with titanium forming a titanium oxide.

In order to show that changes in emission bands due to surface contamination are not unique to titanium (because of its strong chemical reaction with oxygen and carbon), the Fe $L_{2,3}$-emission band of pure iron was measured before and after cleaning and is shown in Fig. 2. Before cleaning, the L_3-emission edge rises steadily to the peak and descends quite rapidly to ~ 17.65 Å, where a change in slope occurs. The band has the appearance of a double peak, with the high-energy peak having a greater intensity. After cleaning (curve B, Fig. 2) the evidence for the double peak is clearer, and the intensity of the high-energy peak has been depressed relative to that of the low-energy peak. Also, at a point (~ 17.58 Å) about $\frac{3}{4}$ of the way up to the peak, the L_3-emission edge bends over more than that for the contaminated iron. The increased sharpness of the edge near the top, for the contaminated iron spectrum was also observed for the Fe L_2-emission band. It was found that the shape of the Fe $L_{2,3}$-emission band from 50%–50% Fe–Fe$_3$C (Fig. 5) has nearly the same shape as curve A, Fig. 2. The surface carbon has reacted with the iron forming an iron carbide.

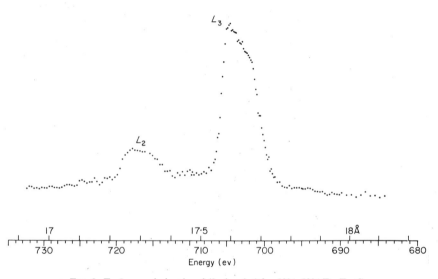

FIG. 5. Fe $L_{2,3}$-emission band (2nd order) for 50%–50% Fe–Fe$_3$C.

It might be argued that removing surface contamination allows the emitted x-rays to come from deeper within the metal, with consequently greater self-absorption by the pure metal. The argument usually given for the effect of self-absorption in a metal is that it sharpens the emission edge, but this directly contrasts with the observed greater slope of the emission edge, which was noted and discussed above.

In order to observe possible effects of self-absorption without the compli-cating effects of surface contamination, a 1000-Å film of titanium metal was arranged so that it could be moved into or out of the radiation path between counter and grating, without disturbing the vacuum. The wavelength positions of the Ti L_2 and L_3 absorption edges are shown by the curve C in Fig. 3. The absorption edge of the Ti film occurred at 27·27 Å (Bearden, 1964, gives a value of 27·28 Å) and the L_3-emission edge of Ti begins to rise at 27·21 Å.

The self-absorption measurements showed that there was no change in the L_3-emission edge when the titanium film was placed in the path of the titanium $L_{2,3}$-radiation. To answer the objection that there might be insufficient overlap between the emission and absorption edges, because the oxygen absorbed by the titanium may have shifted the absorption edge slightly, the titanium film was placed in the path of the titanium $L_{2,3}$-radiation from a TiO target (B, Fig. 3) for which the $L_{2,3}$-emission edge begins to rise at 27·14 Å, with the top of the edge coming at 27·29 Å. There should be complete overlap between the Ti L_3-absorption edge of the film and the Ti L_3-emission edge from TiO. The results of these measurements show that for the Ti L_3-emission band from TiO, the edge and shape were not affected by self-absorption. Although the base of the Ti L_3-emission edge is depressed in curve B relative to curve A (Fig. 3), the L_3 edge begins to rise at the same wavelength for both curves. The base and peak were normalized, and the Ti L_3 edges from both curves coincided. Thus, when the absorption and emission edges overlap the absorption edge does not cut off the emission edge as has been indicated. However, it may be seen from comparing curves A and B that self-absorption does reduce the L_2 to L_3 intensity-ratio. Thus, reported changes of the L_2 to L_3 ratio with excitation voltage, for transition metals, appear to be due to self-absorption; and the reported differences in L_2 to L_3 ratio for the same excitation voltage, may be due to different x-ray take-off angles which result in varying amounts of self-absorption. Skinner et al. (1954) reported for the transition metals an L_2 to L_3 ratio of 0·1, while in our laboratories a value of 0·25 was obtained (Holliday, 1962); for the Fe $L_{2,3}$-emission, self-absorption did not affect the L_2 to L_3 ratio. Another effect of self-absorption is the complete suppression of the high-energy tail of the L_2 emission band; this tail is probably a satellite. In curve B (Fig. 3) the background remains flat until the bottom of the L_2-emission edge.

Although self-absorption does effect the L_2 to L_3 intensity-ratio it does not appear to have any effect on the L_3-emission edge or on the fine structure beyond the edge. It is therefore concluded that observed changes in the L_3-emission bands before and after cleaning, Figs. 2, 3A and 4B, are due to surface contamination and not self-absorption. Also, reported changes in the L_3-emission band with voltage appear to be largely due to surface contamination. However, this does not mean that self-absorption could never effect the shape of emission bands. For example, Cauchois and Bonnelle (in this Volume), and Fischer and Baun (1967b), report results for rare-earth elements which indicate that the empty $4f$ states cause a self-absorption which strongly affects the shape of the emission band.

Considerable tailing has been observed in the emission bands of metals. This tailing is more pronounced to the low-energy side and has made the determination of bandwidths extremely difficult. Skinner (1940) attributed the broadening to Auger transitions within the band, and Rooke (1968) has performed calculations to correct for Auger-broadening. However, for the Ti $L_{2,3}$-emission bands from titanium (Fig. 21) the improved resolution has revealed the tail to be made up of a series of peaks which are probably satellites. There are at least two prominent peaks, and there may be more. In the case of iron (Fig. 2) there is much less tailing to the bottom of the band. Chopra and Liefeld (1964) have attributed these satellites to target voltages that are greatly in excess of the excitation potential of the energy-state involved. They found a reduction in tailing effects for the Ni L_3-emission band from nickel when the voltage was reduced from 2 kv to within 10% of the excitation potential of the L_3-band (~ 850 v).

It is interesting to note in the measurements made by Bonnelle et al. (1966), of the L_3-emission from Ni and from NiO, that less tailing is shown to the high-energy side of the L_3-emission edge for NiO than for Ni. It has been argued that there is no oxide on the surface of nickel when the metal is heated at about 800°C. MacRae (1964) observed that in some instances it is possible to remove absorbed oxygen from the surface of a nickel crystal by heating to ~ 800°C for 30 sec in ultra-high vacuum (2×10^{-10} torr). MacRae pointed out that this was due to diffusion of oxygen into the bulk of the crystal and not to the evaporation of absorbed oxygen; moreover, he stated that after several 30-second cycles of such treatment it was no longer possible to remove the oxygen by heating alone. It does not appear that this technique can be applied to soft x-ray spectroscopy to obtain a L_3-emission band from nickel free from NiO, with excitation voltages in the 1–2 kv region. Using lower voltages does not always reduce satellites. The Ti $L_{2,3}$-emission band from TiO was measured at various excitation voltages, between 5 kv and 1·2 kv (Holliday, 1967b), and it was found that the satellite peak, α in Fig. 21, was not reduced in intensity relative to the L_3 peak but appeared to increase with decreasing voltage.

B. Corrected Emission Bands

Since there appears to be more disadvantage than advantage to the exciting of spectra with voltages that are less than 1·1 times the excitation voltage, the emission bands of the transition metals were measured at 4 kv. The satellites in the tails were corrected by linear extrapolation. The edge-width and base-width are listed in Table I, corrected for inner-level width, for the small instrumental error, for satellites and for background slope. The L_3-level widths are those given by Parratt (1959). The base-width includes the edge-width, which from the above results was shown to be real and not altered by self-absorption. The base-width is seen to increase going from

TABLE I. Corrected transition-metal emission bands

Metal	Edge-width[†] (ev)	Base-width[‡] (ev)	$W_{1/2}$ (ev)	Calculated bandwidth[§] (ev)
	1st series L_3-emission band $(3d + 4s \rightarrow 2p)$			8·9
Ti	1·1	3·8[‡]	2·05	6·5
				2·7
V	1·6	6·6	3·0	
Mn	2·2	7·6	3·4	
Fe	2·2	7·4	4·1	
	2nd series M_5-emission band $(5p \rightarrow 3d)$			
Y	0·18[‡]	~2·18	1·6	3·3
Zr	0·75	4·95	2·95	3·9
Nb	0·0[‡]	4·2	2·75	
Mo	1·5	5·7	3·5	

[†] Extrapolated and includes edge-width.   [§] Altmann (this Volume, p. 277).
[‡] Corrected for temperature.

titanium to manganese as expected. However, there is no significant change in base-width from manganese to iron. Also shown in Table I are bandwidths calculated by Altmann (this Volume, p. 277) using three different potential-fields. Altmann states that the width of 2·7 ev was calculated with an unrealistic crystal-potential and that the smallest bandwidth likely for titanium is 4 ev. The experimental width of the Ti L_3-band depends on whether the emission edge of the L_3-emission band is a satellite or part of the band. The base-width of the Ti L_3-emission band including the emission edge is 3·8 ev, and without it is 2·7 ev. Since the value of 3·8 ev is in better agreement with the value calculated by Altmann, it appears that the edge-width is part of the band.

Even after making all the above corrections the emission edges for V, Mn and Fe do not have zero width. This does not mean that soft x-ray spectra cannot give information concerning the density of states, since the emission edge may be part of the density-of-states curve (as for Ti) rather than being the Fermi-edge. This is supported by a comparison with the specific-heat data obtained by Cheng et al. (1962) and shown in Fig. 6. The total density of states reaches a maximum between Cr and Mn, and is nearly at a minimum between Mn and Fe. The emission edge of the Fe and Mn L_3-spectra probably corresponds to the sloping part of the density-of-states curve, and the Fermi-edge is probably too small at Fe and Mn to be observed above the background. From the total density of states, indicated in Fig. 6, we should expect vanadium to show a sharp emission-edge; although it does show a sharper edge than Fe or Mn, the indication is that it should be sharper than 1·6 ev.

With regard to shape, the specific-heat measurements show that the Fe band has a double peak. As indicated above, the Fe L_3-emission band

(Fig. 2) also shows two peaks. Since the emission-edges were shown to be part of the density of states rather than the Fermi-edge, temperature corrections were not made to the edge-width.

The M_5-emission bands (transitions $5p \rightarrow 3d$) for Y, Zr, Nb, and Mo have

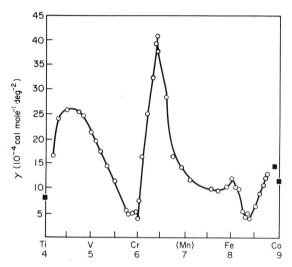

FIG. 6. Electronic specific-heat coefficients for body-centred cubic alloys of the first-series transition elements. (Reproduced with permission from Cheng *et al.*, 1962.)

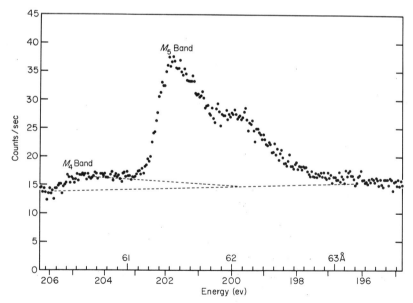

FIG. 7. The Nb $M_{4,5}$-emission band ($5p \rightarrow 3d$ transition) for niobium metal, obtained using a blazed aluminium grating (2160 lines/mm). (Reproduced with permission from Holliday, 1966.)

also been measured. These emission bands occur between 60 Å and 100 Å. Since the $5p$-level is unfilled in the free atom, the $5p$-band must overlap with the $4d$-band resulting in a considerable amount of mixing of the wavefunction. The fact that the $5p$-band does not have a free-electron-like shape but does give reasonable correlations in shape with the d-band calculated from specific-heat data, indicates that there is a large number of electrons with d-type wavefunctions in the $5p$-band.

The uncorrected and overlapping $M_{4,5}$-emission band of niobium is shown in Fig. 7. Even without corrections the Nb M_5-emission appears to have a sharp edge and there is evidence for a double peak which is characteristic of the density-of-states distribution for many of the transition metals. Wiech (1964) has also measured the M_5-emission band of Nb; his results indicated a second peak but it is not as well resolved as that shown in Fig. 7. This is probably because Wiech used a 2-metre grating with 600 lines/mm while the present results were obtained using a 1-metre grating with 2160 lines/mm. The rise in the background with increasing wavelength is due to the intense Nb $M_5 N_3$-line at 64·2 Å.

The M_5-emission bands for Y, Zr, Nb and Mo were corrected for inner-level width, instrumental error, M_4 overlap and background. The corrected emission bands are shown in Figs. 8–10. All of these M_5 bands show a

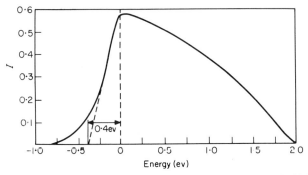

Fig. 8. Y M_5-emission band for yttrium metal, corrected for inner-level width, instrumental error, and background.

second peak except that for yttrium. A third peak is observed for zirconium but is probably a satellite. The M_5-emission bands of Mo and Nb are interesting; beyond the peak both have identical shapes and widths within experimental error. However, the emission-edge width for niobium is much less than that for molybdenum. The emission-edge widths obtained by linear extrapolation are shown in Table I. The temperature correction was 0·32 ev, and after correcting for target temperature the width of the emission edge for niobium was zero, but the emission for molybdenum still had an edge-width of 1·18 ev. Mo does not have zero edge-width for the same reasons as

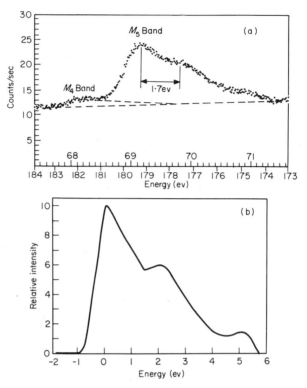

FIG. 9. Zr M_5-emission band for Zr metal ($5p \rightarrow 3d$ transition): (a) the $M_{4,5}$ band as obtained using a blazed platinum grating (2160 lines/mm) and (b) the M_5-emission band corrected for instrumental error, inner-level width, M_4 overlap, and background.

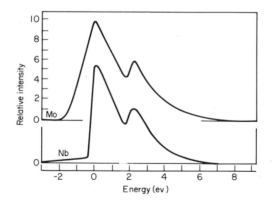

FIG. 10. The M_5-emission bands ($5p \rightarrow 3d$ transition) for Mo metal, and for Nb metal (Fig. 7), corrected for instrumental error, inner-level width, M_4 overlap and background. The intensity distributions of the two M_5 bands are identical beyond the peaks (which have been normalized).

given above for the metals of the first transition series. This can be seen from the total density-of-states curve obtained from specific-heat measurements by Van Ostenburg *et al.* (1963) and shown in Fig. 11. The density-of-states curve is a maximum at niobium but is nearly a minimum at molybdenum. The apparent edge-width for molybdenum is actually evidence of the rise in the density of states between 1 and 0 ev. It appears that the Fermi-edge of

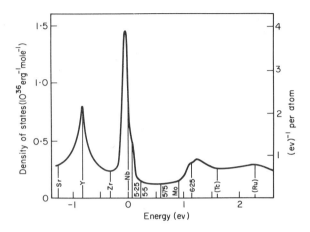

FIG. 11. Density of states obtained from the low-temperature specific-heat data for the $4d$ transition-metal series and their alloys. The Fermi-levels of these metals and alloys are shown. (Alloys are specified by the number of electrons per atom, e/a, where e/a = 5 for Nb metal and 6 for Mo metal.) (Reproduced with permission from Van Ostenburg *et al.*, 1963.)

molybdenum is lost in the background as is apparently the case for the Mn and Fe L_3-emission bands. Applying this type of argument (and the curve in Fig. 11) to the remaining metals of the second transition series below molybdenum, niobium will have an emission edge of nearly zero width, that for zirconium will be broad and yttrium will have a zero edge-width. The corrected emission-edges of Nb, Zr and Y closely follow this pattern (Table I). Only Nb and Y were corrected for temperature (0·32 ev) since these were the only metals for which the emission edge corresponded to the Fermi-edge.

Although there is qualitative agreement between the corrected emission bands of Y, Zr, Nb and Mo and the total density-of-states curve of Fig. 11, there is little quantitative agreement. If the Mo M_5-emission band edge corresponds to the sloping part of the density-of-states curve then the edge-width for molybdenum should be 0·9 ev according to the curve of Fig. 11, but from Table I its value is 1·5 ev. Furthermore, from this curve the second peak of the Nb and Mo M_5-emission band should occur at approximately 0·8 ev from the first, but from Fig. 10 the separation is 2·3 ev. In addition Zr shows a second pronounced peak which would mean, according to the total density-of-states curve, that Nb and Mo should have three peaks; but the emission bands for Nb and Mo in Fig. 10 show no evidence for a third peak.

Altmann (this Volume, p. 277) has calculated the bandwidth for Zr and Y and his results are listed in Table I. The experimental bandwidth for zirconium, of 5·3 ev, is larger than the calculated value of 3·9 ev, while the experimental value of 2 ev for yttrium is rather less than that calculated by Altmann. However, due to the overlap of the Y M_2M_4-line there is some uncertainty in the exact location of the bottom of the Y M_5-emission band. Another anomaly in the bandwidths listed in Table I, is that the bandwidth of Zr is larger than Nb whereas it might be expected to be smaller.

Although there is some agreement with density-of-states curves obtained from specific-heat measurements, and with theoretical band-calculations, there is still considerable room for improvement in correlating soft x-ray spectra with density-of-state curves. In general, the large corrections that are required to soft x-ray spectra, before comparison can be made with band calculations, tend to obscure important changes in the emission bands that would be useful in solid-state physics and metallurgy. The second part of this paper will discuss the changes that occur to metal emission-bands when the metal is alloyed or combined chemically, and the significance of these changes with regard to the understanding of bonding and electron distribution.

4. BONDING

The above results show that it is possible to make some correlation between pure-metal emission bands and theoretical band-calculations. The band theory has not been very useful in predicting or explaining the changes in emission bands resulting from alloying and chemical combination. For example, measurements of soft x-ray emission-bands of alloys made by Curry (in this Volume, see p. 177) indicate that the alloys do not have common valence-bands, and that there is no appreciable change in the energy of the emission edge. In addition, the results obtained by Curry for alloys, and the results obtained in our laboratories on compounds (Holliday, 1967c) show that emission bands are not sensitive to crystal structure. In the case of transition-metal carbides, where the rigid-band model might be expected to hold, large changes with chemical combination have been observed in the bottom of the carbon K-emission band. It is obvious that these results are contrary to those predicted by the band theory.

Since some of the reported changes in emission bands (see Mueller *et al.* 1966; Holliday, 1967c) have been correlated with changes in electronegativity and bonding, a more extensive study of the correlation would be fruitful. The effects of alloying and of chemical combination on the emission bands can be put into three broad classes.

1. In the first class, the only significant changes are in the intensity and

relate to the percentage change in composition. The emission bands from the elements involved in the alloy or compound do not change in shape or wavelength. Alloys are the most common type of solid in this category. In general there is only a small difference in electronegativity between the elements, and the bonding is nearly 100% metallic.

2. The second class is one in which changes are observed in the intensity distribution of the band, but there is no significant shift in wavelength of the top or bottom of the band; although there may be small wavelength shifts in the peaks. In general, the electronegativity difference of these solids is greater than for those of Class 1. The solids that fit into this class usually have a combination of metallic-ionic or covalent-ionic bonds with an ionic character of less than ~50%. Solids in this category are either alloys or compounds.

3. The third class is characterized by a shift in wavelength of the top and bottom of the emission band. The shift is accompanied by large changes in intensity distribution of the band. The elements involved usually have large differences in electronegativity, with >50% ionic character in the bond. Most of the solids in this class are compounds. The wavelength shift can be correlated with the number of electrons transferred from one element to the other. Faessler (1962) has pioneered the work of correlating wavelength shift with valence.

A detailed discussion is given here of emission-band changes in Classes

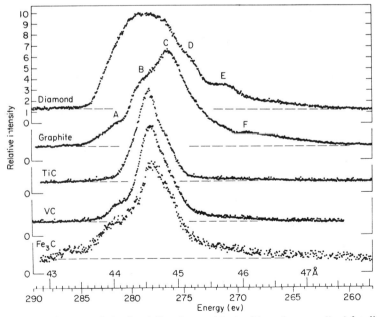

FIG. 12. Carbon K-emission band ($2p \rightarrow 1s$ transition), with peaks normalized, for diamond, graphite, TiC, VC and Fe_3C. (Reproduced with permission from Holliday, 1967.)

2 and 3; no discussion of the Class 1 is given, and the reader is referred to the work of Clift *et al.* (1963) on Cu–Ni, and to that of Curry on Cu–Zn (this Volume, p. 175).

(i) *Emission-band Changes of Class 2*

One of the best examples for this Class is the carbon *K*-emission band from transition metal carbides. Figure 12 shows the C *K*-emission band from TiC, VC and Fe₃C compared to the C *K*-emission bands of graphite and diamond. The peak wavelengths of these C *K*-emission bands are listed in Table II. The results obtained by a number of other workers for the C *K*-emission bands from graphite and diamond have been compared elsewhere with the results in Fig. 12 (Holliday, 1967c).

Going across the first series of transition-metal carbides (groups IV to VIII) the C *K*-emission band changes from a single, nearly symmetrical peak, to a more complex structure similar to the C *K*-emission band of graphite. This effect is also observed for the second and third series of transition-metal carbides, shown in Figs. 13–15. The peaks of the carbide C *K*-emission bands correspond well with the humps labelled A, B, C, D, E, and F on the C *K*-emission bands of diamond, of graphite, and of the contamination carbon, and this indicates that these C *K* bands can be resolved into sub-bands.

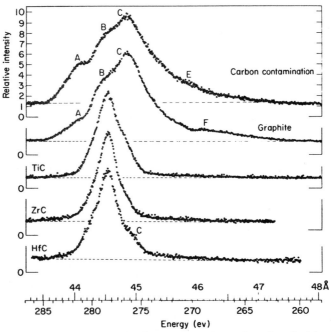

Fig. 13. Carbon *K*-emission bands: for contamination-carbon deposited by the electron beam; for graphite; and for the group-IV carbides, TiC, ZrC, and HfC. (The peaks are normalized to the peak C of the electron deposited carbon.)

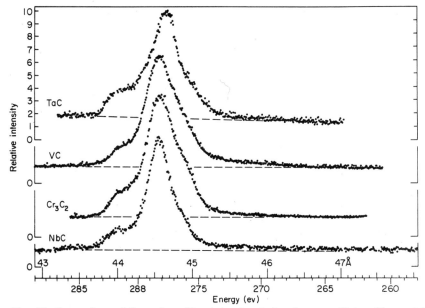

FIG. 14. Comparison of the carbon K-emission bands for the group-V transition-metal carbides TaC, VC, and NbC. Note the similarity in shape of the bands (the peaks have been normalized). The C K-emission band for Cr_3C_2 has been included with those for the group-V carbides because it has a similar shape. (Reproduced with permission from Holliday, 1967.)

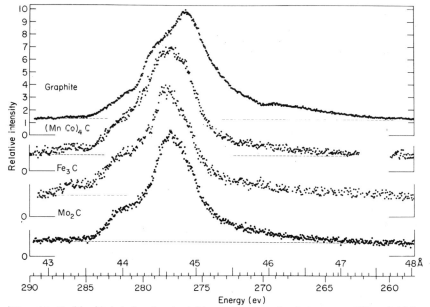

FIG. 15. Carbon K-emission bands (with peaks normalized) for group-VI and higher transition-metal carbides, compared to the C K-emission band for graphite. (Reproduced with permission from Holliday, 1967.)

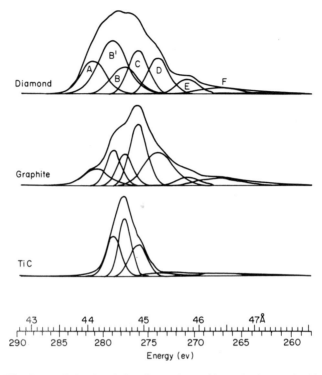

Fɪɢ. 16. The C K-emission bands for diamond, graphite and TiC, resolved into Gaussian curves A, B′, B, C, D, E, and F. These curves correspond to humps A, B′, B, C, D, E, and F on the respective emission bands (Figs. 12 and 18).

Figure 16 shows how the intensity distributions of the C K-emission bands from graphite and diamond have been resolved, using a *DuPont* curve-analyzer, by assigning the proper weighting to 7 Gaussian peaks that correspond in wavelength to humps A, B, C, D, E, and F, and to B′ (see Fig. 18). The intensity distribution of the carbide C K-emission from bands of TiC, VC and Fe_3C have been resolved by keeping the wavelength position of the sub-peaks A, B′, B, C, D, E, and F constant and varying the relative peak heights as shown in Fig. 16. Hereafter when the emission band is resolved into Gaussian peaks, which are indicated by the humps on the envelope of the band, they will be referred to as sub-peaks; this term is adopted because the peaks suggest sub-bands.

It was found necessary to have an additional sub-peak B′ between B and A, in order to account for all of the C K-emission band of Fig. 12. Confirmation for this sub-peak B′ was obtained when the C K-emission band from TiC was re-measured with the higher resolving-power spectrometer. A new peak is seen to the high-energy side of B between peaks A and B (Fig. 18). In addition, for the C K-emission band from TiC, the peak C is

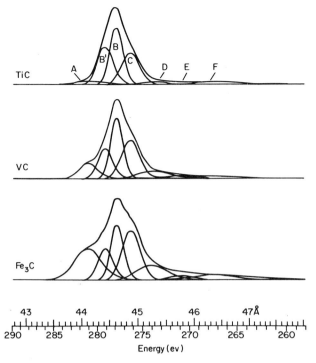

FIG. 17. The C K-emission bands for TiC, VC and Fe$_3$C resolved into Gaussian curves A, B', B, C, D, E and F.

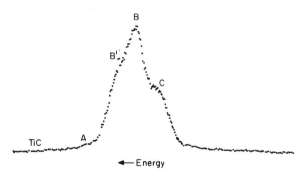

FIG. 18. The C K-emission band for TiC obtained with a higher resolving-power spectrometer.

much more prominent in Fig. 18 than in Fig. 12. Further evidence that these sub-peaks are real was provided by the fact that it was possible to reproduce the peak wavelengths of the C K-emission bands in the synthesized bands without changing the position of the sub-peaks.

For group-IV carbides (Figs. 13, 17) the Gaussian peak B is predominant, and less weighting is given to sub-peaks A, B', and C. The low-energy tail of the group-IV carbides is due to the peaks of very low intensity, D and E. For group-V carbides (Figs. 14, 17), more weight is given to peaks A and C, and for group-VI and higher (Figs. 15–17), the weighting is more equally divided among the various sub-peaks. Thus, the difference in peak wavelength shown for the C K-emission bands in Table II, and the changes in intensity distribution shown in Figs. 12–15 are due to changes in the weighting of the Gaussian sub-peaks.

TABLE II. Carbon K-emission for transition-metal carbides

Material	Peak-shift[†] Δ(ev)	$\Delta\lambda$(Å)	Peak wavelength	Halfwidth[‡] $W_{1/2}$ (ev)	Asymmetry[‡]	$\Delta H°_{298}$/C atom Enthalpy	Melting point[§] (°C)
Graphite	—	—	44·85 Å	6·0	0·83		
Diamond	+2·1	−0·33	44·52 Å	8·1	1·25		
			Group A				
TiC	+1·95	−0·31		3·0	1·1	−43	3200
VC	+1·8	−0·29		3·3	1·45	−28	2850
Cr_3C_2	+1·9	−0·30		3·3	1·6	−10·5	1870
ZrC	+2·05	−0·325		2·4	0·85	−44	3530
NbC	+1·9	−0·30		2·4	1·05	−33	3500
HfC	+2·0	−0·32		3·0	0·80		3800
TaC	+1·2	−0·19		3·0	0·80	−78	3880
TaC_5	+1·7	−0·27		2·7	1·7		3400
			Group B				
$(MnCo)_4C$	+1·2	−0·19		5·2	0·9		
Fe_3C¶ 80% ‖	+1·8	−0·29		4·4	1·25	+5·98	1650
Martensite 80% / Austenite 20%	+1·0	−0·105		5·0	0·7	+4·2	(1200)
Mo_2C	+1·2	−0·19		4·4	1·2	+4·2	2600
WC	+1·8	−0·29		6·7		+8·4	2850

† Peak shift relative to graphite.
‡ Not corrected for instrumental error.
‖ Fe–1·83 C alloy.

Mostly taken from Schaffer (1964) and from Darken and Gurry (1953).
¶ As second phase in Fe–1·83 C alloy.

The physical significance of these peaks is not entirely apparent. From a comparison of C K-emission band intensity distributions in Figs. 12–17, it can be seen that the greater the difference in electronegativity between the carbon and the metal the greater is the tendency for a single peak to predominate. An example of this is Y_2C, which has a greater electronegativity difference and produces a single-peak and narrower C K-emission band ($W_{1/2} = 2\cdot0$ ev) than the group-IV carbides (Table II).

Faessler (in studies such as in this Volume, p. 94) has found that a single resonant-peak predominates for the Si K-emission bands for a number of silicon compounds when the electronegativity difference is large. If electronegativity-difference is the controlling factor in the relative height of the sub-peaks then it would be expected that the C K-emission band for a given group would have the same shape. It will be seen from Figs. 13 and 14 that this is the case. The one exception to this rule is the C K-emission band of Cr_3C_2 where the intensity distribution matches that for the group-V carbides better than for the group-VI carbides.

It is of interest to note that the carbides for which a single resonant peak predominates in the C K-emission bands, have the highest binding energy as indicated by heat of formation and melting-point measurements. This is shown in Table II, where the carbides have been divided into two groups. Group A have high negative heats of formation and high melting points— much higher than for the respective metal. Group B have positive heats of formation with melting points which are approximately the same as for the metal. The C K-emission bands for group A have a predominant single resonant peak (peak B); while for group B, as for diamond, the intensities of the resonant peaks are more equal.

Further evidence that the intensity distribution in the second class of band

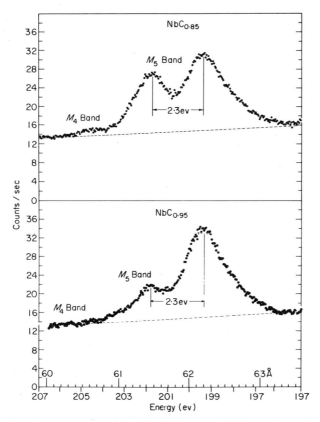

FIG. 19. The Nb $M_{4,5}$-emission bands for $NbC_{0.85}$ and $NbC_{0.95}$, obtained with a blazed platinum grating (2160 lines/mm).

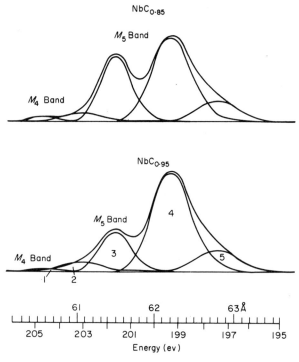

FIG. 20. Nb $M_{4,5}$-emission bands for $NbC_{0.85}$ and $NbC_{0.95}$ resolved into 5 Gaussian curves. The only difference in the Gaussian curves for the emission band for $NbC_{0.85}$, is that the height of peak 3 is increased relative to that for $NbC_{0.95}$.

changes is due to a change in weighting of the sub-peaks is shown for the metal emission bands of the carbides. Figure 19 shows the Nb $M_{4,5}$-emission bands (Holliday, 1966) from $Nb–C_{0.95}$ and $Nb–C_{0.85}$. The Nb M_5-emission band from $Nb–C_{0.95}$ has been resolved into two predominant peaks as shown in Fig. 20. The intensity distribution from $Nb–C_{0.85}$ was obtained simply by increasing the height of sub-peak 3, without changing the height-$W_{1/2}$ or the peak wavelength of any of the other peaks. The Mo M_5-emission band from Mo_2C is also composed of two predominant symmetrical peaks (Holliday, 1966).

The Ti $L_{2,3}$-emission bands from $Ti–C_{0.95}$ and from titanium metal are shown in Fig. 4. There is an increase in satellite intensity on the low-energy side of the band for $Ti–C_{0.95}$ at the same wavelength as in that for titanium metal. The wavelength limits of the L_2-emission and L_3-emission bands from titanium metal are approximately the same as those from $Ti–C_{0.95}$ but the peak for the L_3-emission from the carbide has shifted to lower energy relative to that for the L_3-emission from the pure metal. This shift for the Ti L_3-emission peak, is in a direction opposite to that found for titanium oxides, and has been discussed as evidence in favour of a "carbon donor" theory (Holliday, 1967c). However, since the shift does not involve the whole emission band which is the case for titanium oxides, the ionic character of the bond in TiC may not be the same as in titanium oxides.

On the high-energy side there is an increase in the width of the edge and the $W_{1/2}$ of the Ti L_3-emission band from TiC in relation to that from Ti metal. It was shown earlier that self-absorption does not affect the shape of the L_3-emission band. Thus, the change in intensity distribution of the L_3-emission band is due to chemical effects. However, as indicated earlier, there was a reduction in the L_2 to L_3 ratio due to self-absorption. Thus, the reduction in the L_2 to L_3 ratio for Ti–$C_{0.95}$ in Fig. 4 is probably due to self-absorption. However, other evidence indicates that self-absorption may not be the only effect causing the reduction in the L_2 to L_3 ratio for Ti–$C_{0.95}$. For the L_2-emission from Ti–$C_{0.95}$ in Fig. 4 there is a rise in intensity just before the L_2 band, beyond the peak, up to the L_3-emission edge. The effect of self-absorption, curve B in Fig. 3, is to eliminate this rise in intensity just before the Ti L_2 band and to cause a reduction in intensity beyond it.

It was shown earlier that the Fe $L_{2,3}$-emission band from pure iron is composed of two peaks. The Fe $L_{2,3}$-emission from 50%–50% Fe–Fe_3C in Fig. 5 shows that the low-energy peak has been depressed relative to the high-energy peak, but the wavelength limits of the L_2 and L_3 bands have not changed. In addition, there is no change in the L_2 to L_3 intensity-ratio. Conclusive evidence is not obtained from the Fe $L_{2,3}$-emission bands of Fig. 5 because the Fe_3C target contains 50% Fe. It is not possible to obtain Fe_3C in large quantities because of its instability if isolated from the Fe matrix.

From the study of the second class of emission-band changes it appears that the emission band is made up of a number of sub-bands with varying oscillator strengths, and in the case of the carbide C K-emission bands the oscillator strength appears to be related to the electro-negativity difference of the carbon and metal atoms. For the metal emission bands from the carbides, there appear to be two main peaks and the intensity distribution is determined by the relative weighting of these. An insufficient number of metal emission bands have been measured to determine whether a relation exists between the relative weighting of the sub-peaks and the electro-negativity difference.

A possible relation between the sub-peaks in the emission bands and band theory is to be found in the band calculations of Coulson and Taylor (1952), who divided the graphite K valence-band into one π-sub-band and three σ-sub-bands. In the synthesized graphite C K-emission band (Fig. 16) peak A corresponds in energy to the π-band, and peaks (B'B), C and D to the three σ-bands. The π-electrons are regarded as responsible for electrical conduction and the σ-electrons for bonding. The difficulty in relating the emission-band sub-peaks with the Coulson and Taylor valence sub-bands is that the least number of π-electrons and σ-electrons occur in those carbides with strongest bonding and greatest electrical conductivity. As stated earlier, there is some evidence that carbon is donating electrons to the metal d-band in the group-IV and group-V transition-metal carbides, and it could be argued that some π-electrons and σ-electrons are transferred to the metal atom in these carbides. If this is the explanation, then one has still to explain why some of the electrons are being transferred from the bottom of the valence band. The fact that there is a relation between the relative weighting of the C K-emission band sub-peaks and bonding in the carbides, indicates that it may be possible to relate these sub-peaks to the resonant-bond theories proposed by Pauling (1960).

(ii) Emission-band Changes of Class 3

The third of the classes, into which we have divided these changes in emission bands with chemical combination, is characterized by a shift of the entire band and by large changes in shape. Compounds with substantial ionic-bond character manifest these emission-band changes as illustrated in the spectra from Ti–$O_{1.02}$ and titanium metal (Fig. 21). The wavelength limits of the Ti L_2-emission and L_3-emission bands from TiO have shifted 1·1 ev relative to those for these emission bands from Ti metal. It will also be noted that there is a change in intensity distribution of the Ti L_3-emission

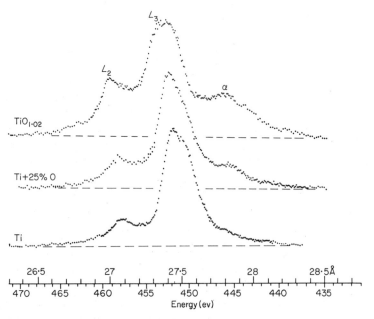

FIG. 21. The Ti $L_{2,3}$-emission bands (with peaks normalized) for $TiO_{1.02}$, Ti+25% O, and Ti. The peak of the Ti L_3 band for Ti+25% O has shifted 0·1 ev, and for TiO 1·1 ev, relative to the L_3 peak for Ti metal, towards higher energies.

band from TiO relative to that from Ti metal. The results in Fig. 3 show that these changes in relative intensity cannot be due to self-absorption. There is also a significant increase in satellite intensity to both the high-energy and low-energy side of the band. As already noted the satellite, labelled peak α, corresponds in wavelength to the satellite on the low-energy side of the Ti L_3-emission band of titanium and Ti–$C_{0.95}$ (Fig. 4).

In Fig. 22 it can be seen that the α to L_3 intensity-ratio increases with the degree of oxidation; this is also the case for the L_3-emission peak-shift which is plotted for TiO_x as a linear function of x in Fig. 23. It is generally accepted that an increase in the oxidation number x results in an increase in the number of electrons transferred and in the ionic character of the bond. Thus, the Ti L_3-emission band-shift is a direct function of the number of electrons transferred to the oxygen atom, or of the charge on the Ti atom. Since the α to L_3 intensity-ratio also increases with x, it appears that this ratio is also a function of the charge on the Ti atom.

The Ti $L_{2,3}$-emission bands from $SrTiO_3$ and $TiO_{1.97}$, in Fig. 24, provide further evidence that the Ti L_3-emission band-shift and the α to L_3 intensity-ratio are more directly functions of the charge distribution than of composition or structure. Although $SrTiO_3$ has a different atomic percentage of oxygen, contains a heavier element and has a different crystal structure ($SrTiO_3$ is cubic and TiO has a rutile structure), both have the same L_3-peak wavelength and α to L_3 intensity-ratio. The one factor common to both compounds is the titanium valence, approximately $+4$. This result shows also that neither wavelength-shift nor satellites are functions of structure; which is evidenced again in Fig. 23, where a linear relation is observed between peak-shift and oxygen composition (oxides with $x = 0.75$–0.17 have a NaCl structure, those with $x = 1.5$ are trigonal and those with $x = 1.97$ have a rutile structure).

Even though wavelength shift is not sensitive to structure, it is sensitive to chemical change. A break in the curve is seen when the limit of oxygen solubility is reached. Figure 21 shows the Ti $L_{2,3}$-emission band from Ti $+25\%$ O compared to Ti metal. We observe an increase in the L_2 to L_3 intensity-ratio, a slight shift in energy of the L_3 peak relative to the L_3 peak of Ti, and an increase of the satellite α. These changes are in the same direction as observed for TiO_x. However, there is no entire band-shift but the L_3-peak shifts 0.1 ev to higher energy. This is a similar type of shift observed for the Ti L_3-emission band from Ti–$C_{0.95}$. The changes in the Ti L_3-emission band for Ti $+25\%$ O indicate that these fit into class 2 better than in class 3. The break in the curve in Fig. 23, at the oxygen solubility limit, shows that there is a large change from the solution to the compound in the degree of ionic character of the Ti–O bond.

It is of interest to look at the oxygen K-emission bands from titanium oxides. The O K-emission bands from Ti $+25\%$ O and Ti–$O_{1.02}$ are shown in Fig. 25. It appears that the O K-emission bands are composed of three

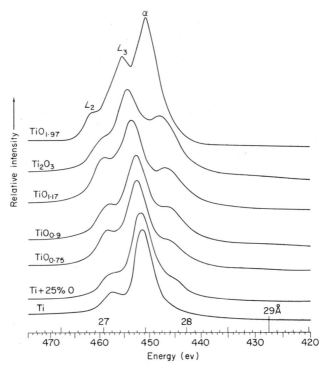

FIG. 22. Ti $L_{2,3}$-emission bands (with peaks normalized) for Ti, Ti + 25% O and TiO$_x$ oxides.

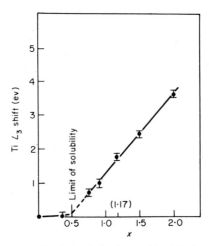

FIG. 23. The peak energy-shift of the Ti titanium oxides (TiO$_x$) and for Ti + 25% O, relative to the Ti band from titanium metal. Error bars indicate the experimental error.

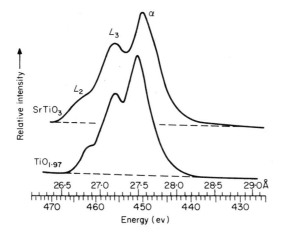

FIG. 24. The Ti $L_{2,3}$-emission bands for $SrTiO_3$ and for $TiO_{1.97}$; both L_3 peaks have the same wavelength (peaks normalized).

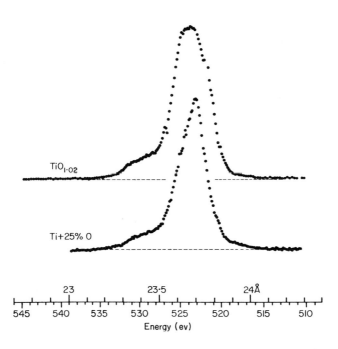

FIG. 25. Oxygen K-emission bands for $Ti + 25\%$ O, and for $TiO_{1.02}$. There is no significant shift in peak position (peaks normalized).

main peaks occurring at similar wavelengths. The high-intensity peak at 23·62 Å appears to be composed of two peaks and the broadening that occurs in this peak, going from the solution to the oxide is because the two peaks have more nearly the same intensity in the case of TiO. However, no significant shift in wavelength was observed. In addition, there was only slight shift in wavelength of the O K-emission band, going from $x = 0·75$ to $x = 1·97$. The obvious reason for the O K-emission band not changing in peak wavelength with x is that the $2p$-level in oxygen is rapidly filled, and the addition of oxygen to Ti has much more effect on the electronic structure of the Ti atom. One possible reason why there is little change in the O K-emission band wavelength, going from the solution to the compound, while there is a distinct change in the Ti L_3-emission band wavelength (Fig. 23), is that the oxygen atom in solution is obtaining "free" electrons from the metal, while in the compound it draws electrons which are partially bound to the Ti atom. This result shows that it is important to measure the emission bands from all the elements in a compound or alloy, before a complete picture of emission-band changes and bonding is obtained.

5. CONCLUSION AND SUMMARY

On the experimental side the above results show that much more care must be taken to separate surface effects from self-absorption effects when interpreting the changes in emission bands with excitation voltage or take-off angle. There appears to be nearly conclusive evidence that the majority of changes observed in the emission bands, with alloying or with chemical combination, are due to changes in the electronic structure of the solid and not due to self absorption.

From the above studies of a large number of emission bands of metals and compounds and the extensive study of alloys made by Curry, we have a clearer picture of the value of band theory to soft x-ray spectroscopy, and vice versa. We observe some correlations of pure-metal emission bands with valence-band theory and with specific-heat measurements. However, in the case of alloys and compounds band theory does not predict or explain the observed changes. Using the concepts of electro-negativity and bonding, it is possible to predict and qualitatively explain a number of the observed changes and there is sufficient agreement to encourage theoreticians to respond to the appeal by Ziman (1967) that they should study the changes in emission bands in terms of bonding. If this approach is taken, soft x-ray spectroscopy may be able to provide valuable information about such solid-state and metallurgical problems as binding energy, the stability of alloys and compounds, the rôle of resonance in bonding, and the understanding of electro-negativity.

Acknowledgement

The author wishes to thank Professor Faessler of the University of Munich for communicating his Si K-emission spectra and for his discussion of the relation between emission-band intensity distribution and the electro-negativity difference for silicon compounds prior to their publication. At this laboratory the author wishes to acknowledge the work of W. H. Hester and E. P. R. Reiche on the design improvements and alignment of the spectrometer.

References

Bearden, J. A. (1964). "X-ray Wavelength". U.S. Atomic Energy Commission Contract AT(30–1), p. 2543.

Bonnelle, C., Wuilleumier, F. and Senemaud, C. (1966). *In* "Röntgenspektren und Chemische Bindung". Physikalisch-Chemisches Institut der Karl-Mark-Universitat, Leipzig.

Cheng, C. H., Gupta, K. P., van Reuth, E. C. and Beck, P. A. (1962). *Phys. Rev.* **126**, 2030.

Chopra, D. and Liefeld, R. (1964). *Bull. Amer. Phys. Soc.* **9**, 404, (BG2–3).

Chopra, D. (1964). Ph.D. Thesis; New Mexico State University.

Clift, J., Curry, C. and Thompson, B. J. (1963). *Phil. Mag.* **8**, 593.

Coulson, C. A. and Taylor, R. (1952). *Proc. Phys. Soc.* **A65**, 815.

Darken, L. S. and Gurry, R. W. (1953). "Physical Chemistry of Metals", p. 364. McGraw-Hill, New York.

Faessler, A. (1962). *In* "Colloq. Spectr. Intl. 19th; Univ. Maryland", ed. by E. R. Lippincot and M. Margoshes, p. 307. Spartan Books, Washington, D.C.

Fischer, D. W. and Baun, W. L. (1967a). *J. Appl. Phys.* **38**, 229.

Fischer, D. W. and Baun, W. L. (1967b). *Norelco Reporter* **14**, 92.

Holliday, J. E. (1960). *Rev. Sci. Inst.* **31**, 891.

Holliday, J. E. (1962). *J. Appl. Phys.* **33**, 3259.

Holliday, J. E. (1966). *In* "The Electron Microprobe", ed. by T. D. McKinely, K. F. V. Heinrich and P. W. Wittry, p. 3. Wiley, New York.

Holliday, J. E. (1967a). *In* "Handbook of X-Rays", Chap. 38, ed. by E. F. Kaelbe. McGraw-Hill, New York.

Holliday, J. E. (1967b). *Norelco Reporter* **14**, 84.

Holliday, J. E. (1967c). *J. Appl. Phys.* **38**, 4720.

Lukirskii, A. P. and Brytov, I. A. (1964). *Akad. Sci. USSR Bull. Phys. Ser.* **28**, 749; *Izv. Akad. Nauk USSR* **28**, 841.

MacRae, A. U. (1964). *Surface Science* **1**, 319.

Mueller, W. M., Mallett, G. R. and Fay, M. J. (eds.) (1966). "Advances in X-ray Analysis", Vol. 9, pp. 329–397. Plenum Press, New York.

Parratt, L. G. (1959). *Rev. Modern Phys.* **31**, 616.

Pauling, L. (1960). "The Nature of the Chemical Bond". Cornell University Press, Ithaca.

Rooke, G. A. (1968). *J. Phys.* C **1**, 767.

Schaffer, T. B. (1964). "High Temperature Materials". Plenum Press, New York.

Shaw, C. H. and Jossem, E. L. (1959). U.S. Atomic Energy Commission Rep. No. AT(11–1)–191, p. 50.

Skinner, H. W. B. (1940). *Phil. Trans.* **A239**, 95.

Skinner, H. W. B., Bullen, T. G. and Johnston, J. E. (1954). *Phil. Mag.* **45**, 1070.

Tomboulian, D. H. (1957). *In* "Handbuch der Physik", ed. by S. Flugge, Vol. 30, p. 246. Springer-Verlag, Berlin.

Van Ostenburg, D. O., Lam, D. J., Shmizu, M. and Katsuki, A. (1963). *J. Phys. Soc. Japan* **18**, 1744.

Wiech, G. (1964). Dissertation zur Erlangung der Doktorwürde, der Ludwig-Maximilians Universität, München.

Ziman, J. M. (1967). *Int. Sci. and Tech.*, No. 67, p. 44.

Soft X-Ray Emission Spectra at Threshold Excitation

ROBERT J. LIEFELD

New Mexico State University, Las Cruces, New Mexico, U.S.A.

ABSTRACT

L-Series emission spectra, "self-absorption spectra", excitation curves and continuous spectra from nickel, copper, and zinc are presented, as recorded with a unique two-crystal vacuum spectrometer. A brief discussion is given of the several corrections to be made to observed valence-band emission spectra; and spectra recorded are used to show how the grossly distorting effects of anode self-absorption and satellite emissions may be eliminated by exciting the emission with electrons of near-line threshold energy. It is then shown that a background-free line contour can be obtained by recording the valence band line-plus-continuum with a near line threshold energy and subtracting from this the line-free continuum. The latter is recorded with a sub-line threshold energy and is shifted by the difference in the excitation energies.

1. INTRODUCTION

A. Corrections to Observed Emission Spectra

In his review article Parratt (1959) listed the corrections to be applied to observed x-ray valence-band emission spectra prior to attempts at detailed interpretation regarding solid-state energy-band structures. A slightly expanded version of that list includes corrections for (1) non-linear intensity response of the detector, (2) scattered background intensities, (3) variation of detector sensitivity with photon energy, (4) finite resolving power (spectral window), (5) self-absorption in the emitting source, (6) continuous spectrum background, (7) satellite emissions, (8) energy width and shape of the initial state, (9) excitation spectra, and (10) variations in transition probabilities. Application of all but numbers (9) and (10) of these corrections is the responsibility of the experimentalist. Due to difficulties associated with the acquiring of necessary information for making the corrections, most published spectra have been presented uncorrected or only partially corrected.

The non-linear-response correction (1) is frequently small for photon counting detector systems, and it is not difficult to make the measurements required for its assessment; but correction (2) for background is often quite significant, especially for one-crystal and grating instruments, and it is very difficult to make the necessary measurements for its estimation. For two-crystal instruments the scattered background is usually small but it contains an inherent variation with the spectral intensity incident at the second crystal.

A large number of careful auxiliary experiments are required to establish the detector-sensitivity and spectral-window corrections, (3) and (4) respectively, especially if the primary spectra cover a broad energy-range. Though frequently of major importance, correction (5) for self-absorption has been generally ignored or been regarded as insignificant for the need of accurate information about the distribution of true source within the emitting sample and about the pertinent absorption spectra.

Correction (6) for continuum background is always important; however, it is usual to resort to an arbitrary best-estimate determination of background because it is nearly impossible to establish how much of the background is continuum and how much is scattered intensity, and also the extent to which these are modified by crystal reflectivity, by window transmissions and the like. The correction (7), for satellites, is even more difficult to effect because often these emissions are sufficiently coincident with the parent line or band that they are not even recognized. Even when they are identified, their removal is highly problematical because their intensity distribution is still an unknown. Additional experiments are required to determine the inner-level width required for correction (8). Only corrections (9) and (10) are relatively simple for the experimentalist; because they should be left to the theoreticians.

B. Background to the Present Investigation

The work presented in this paper is part of a long-term effort to acquire accurate L-series valence-band emission spectra of zinc, copper, nickel, cobalt, and iron, and to make all of the necessary measurements pertinent to application of corrections (1) through (8). These spectra have all previously been recorded, and in some cases have been studied by several investigators.

In the published record to date there is little agreement among the results and some very striking disagreements. None of the major corrections had previously been attempted.

C. The Instrument

The spectrometer used for the L-spectra measurements was designed by J. A. Soules and R. J. Liefeld (to be published), as a versatile one-crystal or two-crystal instrument featuring kinematic (ball and plate) axes, and "drum and tape" drive-couplings. Figure 1 shows the basic arrangement. A vertical divergence-limiting collimator is directed towards the drum-mounted first crystal; the second crystal, on an identical drum is attached to a table which rotates about the first-crystal axis; and the detector is a gas-flow propor-

FIG. 1. A view showing basic arrangement of the two-crystal (KAP) spectrometer.

tional counter detector which rotates about the second-crystal axis. A second counter, mounted on the collimator, is sometimes used as a source-intensity normalizing detector. The first-crystal table is driven by a micrometer lead-screw tangent-arm arrangement and the platform carrying the second crystal and the detector is independently driven by a worm gear and worm drive. With these drive mechanisms, and crossed tape linkages, the elements of the spectrometer may be positioned for each data point so as to eliminate beam walking at the crystals.

The instrument has been equipped with potassium–acid–phthalate (KAP) crystals, which have a spacing $(2d)$ of 26·6 Å, high reflectivity, and adequate resolving power for recording the spectra in question. A reasonable measure of the energy resolution obtainable with these crystals is given by their $(1, -1)$-position rocking-curve widths at half maximum. Figure 2 shows typical results at several wavelengths in the range of interest. Of particular note is the fact that the $(1, -1)$ curve-widths are nearly constant at about 0·65 ev, from below 10 Å to above 20 Å. This is evidence that the central portion of the spectral window for the instrument has an essentially constant width over this region. The application of the correction for finite instrumental resolution is consequently much simpler. The KAP crystals also have a rather slowly varying reflectivity over the wavelength region of interest and this assists in making correction (4) tractable.

Considerable effort has been expended to insure the "integrity" of the materials studied. Polished bulk-metal samples have been used, at high temperatures when possible, in a working vacuum of about 10^{-9} torr provided by titanium-ion and sublimation pumping. The x-ray tube has several ribbon-filaments positioned in front of the anode so that the spectrometer

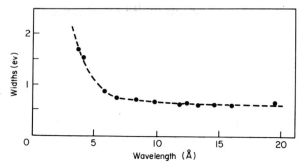

FIG. 2. The $(1, -1)$-position rocking-curve widths at half-maximum of potassium acid phthalate crystals.

views the latter through the spaces between the filaments. This geometry effects a nearly normal angle of incidence for the electron-beam, and a normal photon take-off angle, helping to minimize absorption in the source.

An example of the performance of the full-tracking two-crystal (KAP) spectrometer is presented in raw-data form in Fig. 3. This is a complete *L*-emission spectrum of nickel and was recorded with 3 kv and 15 mA target-input, in about 8 h running time. The spectrum extends from below 13 Å to above 17 Å, and covers a Bragg-angle range of about 8°.

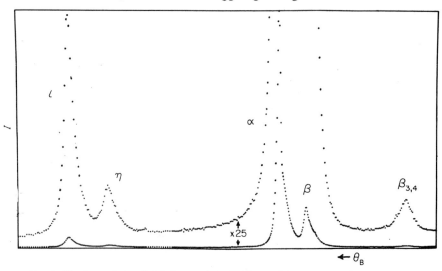

FIG. 3. The *L* spectrum of nickel, recorded with the tracking two-crystal spectrometer.

2. SELF-ABSORPTION

A. Line-shape Dependence on Electron Energy

Almost as soon as it began, the effort to record *L* spectra of the elements from zinc through iron was diverted to an attempt at checking the results of Van den Berg (1957), who showed that the Ni and Cu L_α-line (L_3-valence) shapes and positions are functions of the electron-energies used to excite the spectra. Although we have not obtained the same result as Van den Berg in all situations, and cannot agree with his interpretations, his principal result is certain; these lines do vary in shape and position with the energy of the exciting (or incident) electrons.

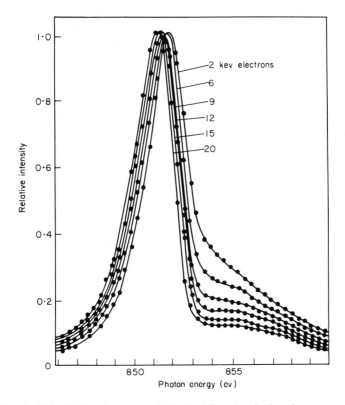

FIG. 4. Nickel L_α-line shapes recorded with different excitation electron-energies.

Figure 4 shows typical raw data for nickel. It can be seen that the line shift is considerable (about 1 ev over the range studied), and that the change of line-shape is also drastic. However, the changes are regular and suggest the possibility that self-absorption is responsible for the effect.

B. Self-absorption Spectra

If self-absorption is the explanation for the effects described above, then for two curves which have different amounts of self-absorption (this being their only difference), the point by point intensity-ratio of the two would be similar to the photon absorption spectrum of the anode material.

Such a comparison is presented in Fig. 5. In the upper part the Ni $L_\alpha L_\beta$-region emission spectra, for respectively 2 kv and 15 kv electron-bombard-ment, are reproduced from the data of Fig. 4 such that the line intensities match on the low-energy side of the L_α line. In the lower part of Fig. 5 the point by point ratio of the line intensities (for 2 kv and 15 kv excitation) is displayed above the photon absorption spectrum of nickel. The latter was recorded with the same instrument by the usual absorbing foil technique, using the 2 kv emission-spectrum from nickel as the incident continuum. The

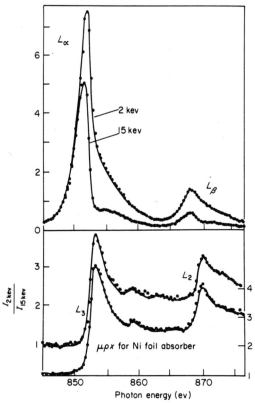

FIG. 5. *Above.* Nickel L_α-emission and L_β-emission spectra recorded for incident electrons with energies of 2 kev and 15 kev.

Below. The ratio of intensities for the 2-kev and 15-kev emission spectra (a "self-absorption" spectrum), compared with the photon-absorption spectrum of nickel.

agreement of the two curves in the lower half of Fig. 5 is very good and provides convincing evidence that the principal difference between the emission lines for 2 kev and 15 kev excitation is due to self-absorption in the emitting anode. We use the term "self-absorption spectra" for these point by point intensity-ratios of the emission lines, because for appropriate electron-energies the ratios are reasonable facsimiles of directly measured photon-absorption spectra. The technique is generally interesting because it permits observation of the absorption characteristics of bulk material samples.

The good agreement of the nickel self-absorption spectrum with the nickel photon-absorption spectrum is surprising. The familiar argument that the two absorbers might not be entirely similar could be advanced; in the present case the nickel anode was cleaned by heating it to red heat in the ultra-high vacuum of the x-ray tube, and the foil absorber was a self-supporting nickel foil, obtained from the Chromium Corporation of America, as electro-deposited nickel on copper which was chemically stripped from the nickel after mounting. No comparison of the composition or crystal structures of the samples has been attempted.

It must also be noted that in Fig. 5, the intensity ratio for the two lines has been compared with the foil absorption-coefficient, rather than the logarithm of the intensity ratio as might have been expected. This was because in this instance not taking the logarithm of the intensity ratio serves to account for the excess-thickness effect (Parratt et al., 1957) for the self-absorption spectrum in relation to the direct photon-absorption spectrum. In fact, both emission spectra should be subjected to corrections (1) through (4) prior to taking the ratio; and then the logarithm of the intensity ratio would be the appropriate form in which to display the result. However, our objective here is to demonstrate the magnitude of the distortion in the shape of the nickel valence-band line due to anode self-absorption, and for this objective the intensity-ratio of the observed lines suffices.

C. Avoidance of Anode Self-absorption

The introductory list (p. 133), of pertinent corrections to observed valence-band emission spectra, includes a correction for self-absorption in the source. Such a correction, if it is to be accurate, requires knowledge of detailed and corrected absorption spectra of the material of the source, and precise knowledge of the distribution in the source of the line-emitting excitations and of the continuum. These are generally so difficult to obtain that it is clearly safer to avoid significant anode self-absorption by using sufficiently low-energy electrons for excitation, and an absorption-minimizing x-ray tube geometry. With the present geometry, of normal-incidence electron-beam and normal take-off angle for photons, the Ni L_α-line shape is the same, within the uncertainties of measurement for electron energies of from 2 to 2·5 kev. A detectable change (1–2%), due to self-absorption, occurs for incident-electron energies of about 3 kev. Since most x-ray valence-band studies have generally been carried out with higher energy electrons and with much smaller photon take-off angles, it is appropriate to reiterate the warning made by Hanson and Herrera (1957) that major distortions due to anode self-absorption probably exist in most of the published curves, and to call for remeasurement of many of the spectra in question.

3. SATELLITE EMISSIONS

A. Identification of Satellite Emissions

The shape of the Ni L_α line, for excitation conditions and x-ray tube geometry which preclude detectable self-absorption, is exemplified by the curve in Fig. 5 obtained using excitation-energies of 2 kev. Considerable emission intensity exists at energies higher than that of the L_3-absorption edge. Most, if not all, of this intensity is due to a complex of satellite lines derived from multiple-vacancy states set up by radiationless (Auger) transitions from L_1 and L_2 initial states. The only way in which the importance of these satellites can be determined is to study the shape of the lines for successively lower excitation energies. We have found in the case of nickel that the "excess" intensity, on the high-energy side of the line, decreases regularly with decreasing excitation-energy and apparently disappears for incident-electron energies of less than that required to produce L_2 initial states.

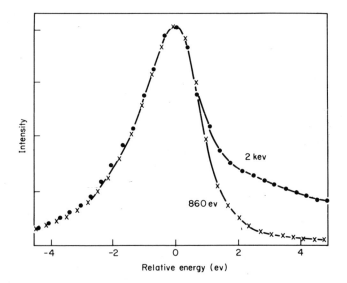

FIG. 6. A comparison of the nickel L_α-line shapes for incident electrons of 2 kev (with satellites) and 860 ev (without satellites).

Figure 6 shows a comparison of the shapes of the nickel L_α line for incident electrons with energies of 2 kev and 860 ev and shows that with 2 kev a large amount (25–30%) of the integrated line intensity is composed of the satellite emissions. In many previous investigations the satellites have been sufficiently suppressed by self-absorption that they have gone undetected, or have been regarded as small and insignificant.

B. Avoidance of Satellite Emissions

It is clear that the excitation of the copious L_α satellites must be avoided if one wishes to observe the intensity distribution of the simple L_3-state to valence-state transition. As already mentioned, the satellite intensity disappears when the L_1 and L_2 states are not produced in the emitting source[†]. In the case of the first transition-series and adjacent elements, incident-electron energies of just less than that required for L_2-excitation are considerably higher (16–17 ev for nickel) than is required for the production of L_3 states. One would expect this to be quite enough to produce multiple vacancies of the type L_3, $M_{4,5}$ by direct electron–atom interaction; but this is not a likely process. Richtmeyer (1937) explained the non-existence of the L_α satellites for elements of atomic number between 50 and 75, on the basis of Auger transitions being energetically impossible for this range of elements, and the evidently negligible production of L_3, $M_{4,5}$ states by the electron bombardment.

In order just to avoid the production of L_2 states in the source, it is necessary to use incident electrons of less than the L_2-excitation energy. This can be accomplished in either of two ways: by recording the L_β-line excitation curve and setting the anode potential at a value slightly less than the L_β-line excitation threshold, or by recording the L_α-line excitation curve and setting the anode potential at a value which is slightly less than the L_α-line excitation threshold plus the L_2–L_3 energy-difference. The second method is more reliable because the L_β line is less intense than the L_α line and yields a less precise determination of the threshold level.

C. Excitation Curves

To record the L_α-line excitation curve, the two-crystal monochromator is set at the position for the peak of the line and the intensity at that position is recorded as a function of the x-ray tube accelerating potential. Figure 7 shows a typical curve for nickel and indicates that the L_α line "appears" abruptly at a threshold voltage. This is presumably due to the large density of available states for electrons just above the Fermi-level in nickel, and a consequently large probability of excitation by incident electrons with energies just greater than threshold. The energy scale of Fig. 7 is the dial settings of the anode power-supply unit used for the experiment; corrections to account for the calibration of this power supply, and for the work-function of the cathode, have not been made.

It should also be pointed out that, because the energy of the nickel L_α line is very close to the L_3-state excitation energy, the high-energy end of the continuous spectrum moves into the spectral window of the instrument almost simultaneously with the "appearance" of the L_α line. The actual record is then a superposition of the L_α-line excitation curve and a continuum

---

[†] Some emission intensity still exists on the high-energy side of the L_α lines for sub-L_2-state excitation which can be ascribed to bound-electron excitation states. This is not measurable with our present sensitivity in the cases of Zn and Cu, and is barely detectable in the case of Ni; it is clearly observable for Co and Fe where it is sufficiently intense to prevent the observation of sharp emission edges.

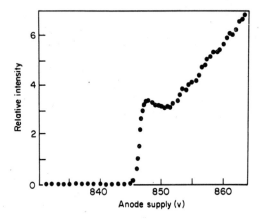

FIG. 7. The nickel L_α-line excitation curve recorded with nearly mono-energetic incident electrons ($\Delta E < 0\cdot2$ ev).

isochromat. Figure 8 shows a typical continuum isochromat for nickel, recorded with the monochromator set at a position about 20 ev less than the nickel L_3-state excitation energy. Such an isochromat should be subtracted from the curve to Fig. 7 to obtain the actual L_α-line excitation curve but, since the line to continuum intensity-ratio is about 20:1 at threshold, the distortion in Fig. 7 is not serious.

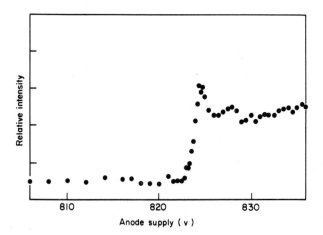

FIG. 8. A typical nickel-continuum isochromat.

The excitation-curve technique works well even in the case of a metal that has a density of available electron states just above the Fermi-level, which is much smaller than that for nickel. Figure 9 presents data for copper which show the copper L_α-line threshold occurring at an anode voltage of about 930·5, which is some 3 v above the start of the continuous-spectrum iso-

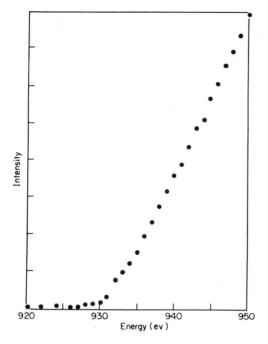

FIG. 9. The L_α-line excitation curve for copper.

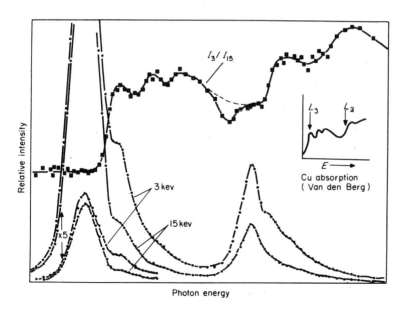

FIG. 10. The $L_{2,3}$ "self-absorption" spectrum of copper.

chromat. This observation indicates that the relative position of the lowest unfilled valence-levels is at an energy ~ 3 ev above that of the peak of the copper L_α line; information which is more conventionally obtained from the position of the pertinent absorption-edge relative to the emission line. Self-absorption spectra, such as that for copper shown in Fig. 10 are especially useful in this respect because, for a given material, they provide the emission-line and absorption-edge positions in one experiment and on a coherent energy scale.

4. THE BACKGROUND CONTINUUM

For the first transition-series and adjacent elements, excitation by electrons with insufficient energy to produce L_2 states gives a spectrum principally consisting of the L_α line, and a continuous spectrum which has its high-energy limit at an energy only slightly above the position of the L_α line. It is then possible to obtain a measure of the background continuum by recording the continuous spectrum with an electron-energy insufficient to excite the L_3 state. Figure 11 shows such spectra, as recorded from an

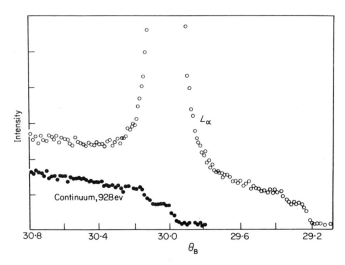

FIG. 11. The copper L_α line plus continuum for above-line threshold excitation, and the line-free continuum for below-line threshold excitation.

oxygen-free high-conductivity copper anode. The only difference in the experimental conditions for the two spectra was the anode potential. In Fig. 12 the continuum of Fig. 11 has been shifted to higher energies by the amount of the incident electron energy-difference to provide some estimate of the background continuum present in the recorded line-plus-continuum.

The structure at the high-energy end of the line-plus-continuum spectrum is clearly part of the continuum and not of the line spectrum. Another observation made from Fig. 12 is that the tailing of the line at high and low energies is apparently different. This may or may not be "real", and care is necessary because the data in Fig. 12 are raw data. Both spectra presented here should be subjected to the previously listed corrections, (1) through (5), before detailed comparison. The crystal reflectivity and the transmissions of detector and x-ray tube windows which are part of the spectral window of the instrument, all decrease with decreasing photon-energy; this is partly responsible for the observed intensity difference between the line-plus-continuum and the shifted continuum on the low-energy side of the line. Also, in the case of the recorded line-plus-continuum, those electrons in the incident beam which produce L_3-shell excitations do not produce continuum photons near the high-energy limit. For this reason, the continuous spectrum may be relatively weaker in the region between the line excitation energy and the high-energy limit observed when the line is excited. This effect would cause less apparent line-tailing to appear at the high-energy side of the line than really exists.

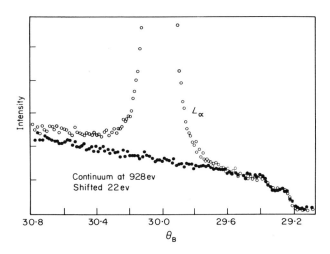

FIG. 12. The copper L_α line plus continuum, and the shifted line-free continuum.

However, this tailing discrepancy is insignificant with regard to the main line shape, because at the line peak the line to continuum intensity-ratio is more than 30:1. Since the line-plus-continuum and the continuum alone were recorded with the same instrument from the same sample, using the same x-ray tube current, a fairly good estimate of the L_α-line shape can be obtained by taking the difference, point by point, of the curves shown in Fig. 12. This method is probably superior to the arbitrary background estimation commonly used.

5. *L*-SERIES VALENCE-BAND SPECTRA FREE FROM SELF-ABSORPTION AND SATELLITE EMISSION AND CORRECTED FOR BACKGROUND CONTINUUM

The data for copper with the background subtracted are presented in Fig. 13. Again it must be argued that since some of the pertinent corrections (2 to 5, and 9) have not yet been applied to these data, interpretive speculation is hazardous. However, it is possible that this curve indicates some structure in the intensity distribution for the L_3–valence transition; particularly the termination of the filled valence levels at an energy some 2·5–3 ev above that

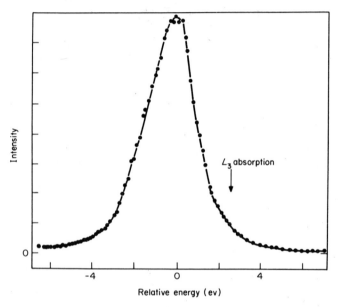

FIG. 13. Shape of the L_α line for copper, free from self-absorption and satellites, and corrected for the background continuum.

of the line peak. If this is real then it contradicts some earlier predictions based on hydrogenic-wave function transition probabilities which suggested that, in such a case, the *d–p* transition probability is so much greater than the *s–p* transition probability that the x-ray line will only contain information about the filled valence levels of *d*-type wavefunction.

In Fig. 14 the self-absorption-free, satellite-free, and continuum-free data for the line shapes for copper and for nickel are compared. The greater width of the copper line is readily observed, as well as a probably significant difference in the line shapes.

Further evidence that x-ray valence-band emission spectra obtained with these techniques may provide realistic information about the valence levels in metals is obtained from the *L* spectra of zinc. Figure 15 shows the zinc

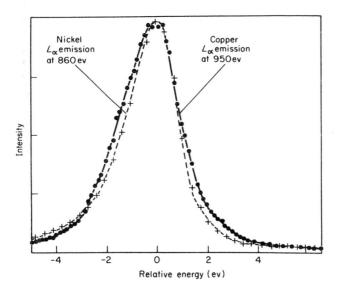

FIG. 14. $L\alpha$ lines, for copper and nickel, free from self-absorption and satellites, and corrected for background continuum.

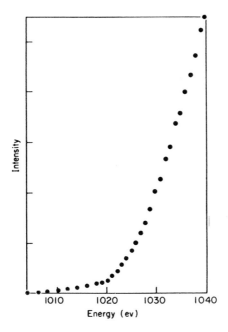

FIG. 15. The zinc L_α-line excitation curve.

L_α-line excitation curve, and indicates that the line-excitation threshold occurs at an anode voltage of 1020 v, which is some 10 v after the onset of the continuum isochromat. This indicates that the relative energy of the lowest unfilled valence-levels (i.e. available to electrons ejected from the L_3 shell) is about 10 ev above that of the zinc L_α line. Figure 16 presents the zinc self-absorption spectrum, which also locates the relative energy of the lowest unfilled valence-levels at about 10 ev above that of the L_α line. Figure 17 shows the self-absorption-free, satellite-free, and continuum-free data for the zinc L_α line. This spectrum is evidently composed of a broad low structure, $\sim$ 15 ev wide, and a tall narrow structure with a base-width of $\sim$ 4 ev; the results suggest that these structures may be associated with the 4s and 3d valence-bands of zinc.

6. CONCLUSION

The analysis of the spectra shown here is incomplete in that all of the pertinent corrections have not been made, and detailed interpretations cannot be advanced until they have. However, it is clear that elimination of the grossly distorting effects of anode self-absorption and of satellite emissions,

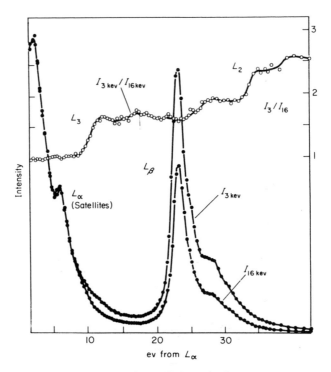

FIG. 16. A zinc "self-absorption" spectrum.

along with the nearly correct subtraction of the background continuum, produces spectral line shapes which are much more amenable to interpretation than those previously available.

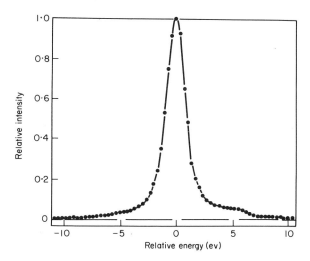

FIG. 17. The zinc L_α line, free from self-absorption and satellites and corrected for the background continuum.

Acknowledgement

This investigation was supported by the Physics Department of the New Mexico State University, and by the National Aeronautics and Space Administration.

References

Hanson, H. P. and Herrera, J. (1957). *Phys. Rev.* **105**, 1483.
Parratt, L. G. (1959). *Rev. Mod. Phys.* **31**, 616.
Parratt, L. G., Hempstead, C. F. and Jossem, E. L. (1957). *Phys. Rev.* **105**, 1228.
Richtmeyer, F. K. (1937). *Rev. Mod. Phys.* **9**, 391.
Van den Berg, C. B. (1957). Dissertation, University of Gröningen.

Soft X-Ray Spectra for Nickel and Nickel Alloys and Comparison with the Theoretical Densities of States

J. R. CUTHILL, A. J. MCALISTER, M. L. WILLIAMS and R. C. DOBBYN

Alloy Physics Section, Institute for Materials Research, National Bureau of Standards, Washington, U.S.A.

ABSTRACT

This paper reports emission spectra measured with a 2-metre grazing-incidence spectrometer, using photoelectric detection and digital recording. The $M_{2,3}$-emission spectrum of nickel was studied with the temperature held at $\sim 960°C$ to ensure an oxide-free surface, and a number of spectra were summed to resolve detail not previously observed. Calculated transition-probabilities appear to account for the observed L spectrum being always narrower than the M spectrum. A comparison of the aluminium $L_{2,3}$ and nickel $M_{2,3}$ overlapping spectra from NiAl$_3$, NiAl and Al–Ni solid solutions, with these spectra from the respective pure metals, reveals certain major features that are unrelated to crystal structure. The spectra suggest that the electronic structure of the aluminium undergoes a progressive change with increasing nickel content, while the high-energy emission edge of the L spectrum remains unshifted with nickel content up to at least 50 at. %. With Ti–Ni alloys, there is no overlap of the titanium $M_{2,3}$ spectrum with the nickel $M_{2,3}$. The nickel $M_{2,3}$-emission from the intermetallic compound TiNi exhibits a peak at the same energy as the peak in the pure Ni spectrum, but also exhibits a much larger peak at ~ 3 ev lower energy.

1. INTRODUCTION

Our long-range objective is to obtain information about the changes in the density-of-states on alloying, and on the formation of the intermetallic compounds; this might lead to a better understanding of the principles of alloying behaviour. We have done some preliminary work on alloys, which we discuss later, but at an early stage of the work it became evident that more detailed data on the pure metal components are needed, and to date most of our effort has been in this direction (Cuthill *et al.*, 1967).

2. THE SPECTROMETER

Figure 1 is a photograph of our 2-meter grazing-incidence spectrometer, and gives an indication of its size. Electronic recording is used and the data are obtained in digital form, both on punched paper-tape and on hard copy.

151

The accumulated count from the photo-electron detector is recorded at preselected time intervals, during a continuous wavelength-scan.

We record at 4-sec intervals which gives, at the drive-speed used, print-outs at about every 0·05 ev. A particular feature of this recording system is that virtually no counts are lost while the previously accumulated count is printed out. This accumulated count is transferred in 40 nsec to a storage

FIG. 1. Photograph of the soft x-ray spectrometer showing the vacuum chambers and magnetic drive.

register, and counting for the next interval starts immediately. Digital recording on punched paper-tape facilitates the summing of a large number of scans by computer. Summing in this manner increases signal-to-noise ratio and, coupled with the superior linearity of electronic recording, has enabled us to resolve new fine-structure in the Ni $M_{2,3}$ spectrum. A ratemeter-driven chart recorder is used for monitoring the spectrum during a scan.

Fig. 2. View of the grating in its mounting; with lead-screw and Rowland circle track. The lead-screw is $1\frac{1}{2}$ inch diameter.

The spectrometer is equipped with a Seigbahn grating, 30000 lines/in, ruled on glass. This is shown in its mounting in Fig. 2. The detector is a commercial magnetic electron-multiplier with the pre-amplifier mounted along with it in the vacuum chamber. A lead-screw, seen below the Rowland circle track in Fig. 2, drives the detector around this track and keeps it pointing at the centre of the grating. The lead-screw is driven by differential gearing (Fig. 3), situated below the track and on the extension of the vertical axis of the grating, so that the lead-screw axis can swing through a horizontal angle as it drives the photo-multiplier carriage around the Rowland circle.

FIG. 3. Differential gearing which drives the lead-screw, and is mounted on vertical axis of grating.

The lead-screw shaft (the upper in Fig. 3) is driven by a drive-shaft whose orientation remains fixed and is rotated through the end of the main chamber by a magnetic drive showing prominently on the end of the chamber in Fig. 1.

3. RESULTS AND DISCUSSION

A. $M_{2,3}$-Emission Spectra for Nickel

Our results for the nickel $M_{2,3}$ spectrum are shown in Fig. 4. The upper curve is the raw data and is the sum of 44 scans. The specimen temperature was maintained at $960 \pm 7°C$ by the heating produced by the electron beam, with a vacuum of 5×10^{-8} torr. The temperature was monitored with a platinum platinum–rhodium thermocouple, spot welded to the specimen in the uniformly heated zone, but just to one side of the area from which the characteristic x-radiation is sampled. The temperature was maintained at this high-value so as to be above the oxide decomposition-temperature on any crystal plane of nickel (Germer et al., 1963), thus ensuring an oxide-free surface. In fact, sufficient data were obtained at 832°C to show that there is a significant change in structure due to oxide formation even at this only slightly lower temperature. This is important because probably most of the nickel spectra recorded in the literature have been obtained at lower temperatures, and with poorer vacuum, giving results seriously affected by oxide contamination. Further, by holding the specimen at a known temperature, meaningful corrections could be made for temperature-broadening.

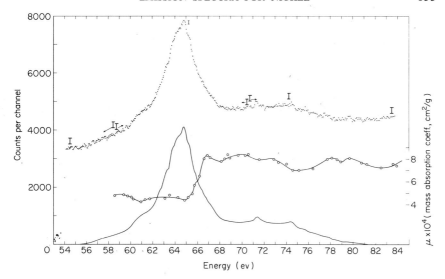

FIG. 4. Ni $M_{2,3}$-emission spectrum. *Above.* Sum of 44 scans, without background subtracted. *Below.* Same curve with background subtracted, and with absorption spectrum for Ni (Tomboulian *et al.*, 1957) superimposed. (Reproduced with permission from Cuthill *et al.*, 1967.)

The widths of both source and detector slits were 0·1 mm. The excitation voltage used, for each of the 44 scans summed, was 2500 v. Similarity with some spectra recorded at 1500 v suggests that self-absorption distortion is small. We attribute this to the use of a very shallow angle of incidence of the electron-beam (to reduce penetration depth), and an x-ray take-off direction normal to the specimen surface.

The lower curve in Fig. 4 is the same $M_{2,3}$ spectrum with the bremsstrahlung background subtracted-out (according to assumptions outlined by Cuthill *et al.*, 1967), and with the Ni $M_{2,3}$-absorption spectrum (Tomboulian *et al.*, 1957) superimposed.

The Ni M_3-emission was resolved from the observed $M_{2,3}$ spectrum by assuming that the M_2 and satellite components are identical to the M_3 except for scale and position. Two adjustable parameters are thus available: the ratio of peak heights, and the energy-shift of the peak with respect to the Ni M_3 spectrum. The values of these parameters for the M_2-emission and for the two high-energy satellites, were determined from a best-fit procedure by computer. This method of separation is described in detail in the paper already cited (Cuthill *et al.*, 1967).

The Ni M_3-emission spectrum is shown in Fig. 5, superposed upon Ni L_3 spectrum recorded at threshold excitation by Chopra and Liefield (1964). Such spectra purport to represent to a first approximation the density of states of the metal, but distorted by differing transition probabilities. It is a well-known observation for the transition metals that the L-emission spectrum always shows a narrower bandwidth than the corresponding *M* spectrum. Some estimate of transition-probability differences for the transition metals can be obtained from the augmented-plane-wave (APW) calculation made by Wood (1960) for iron. The mathematical expression required

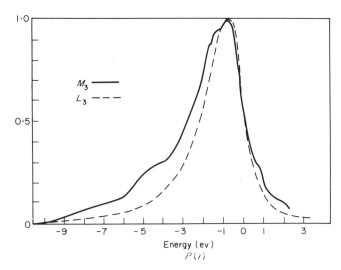

FIG. 5. Nickel M_3-emission spectrum, with the nickel L_3-emission spectrum superimposed. (Reproduced with permission from Cuthill *et al.*, 1967.)

involves the radial integral, which enters the dipole matrix-element and gives for the transition probability Γ, the relation

$$\Gamma = \left| \int P_a(r) \, r \, P_{3d}(r) \, dr \right|^2$$

where $P_a(r) = r R_a(r)$. The suffix a represents a 2p-orbital for L spectra and a 3p-orbital for M spectra; R is the radial part of the electron wavefunction.

For this calculation, the 2p and 3p atomic-wavefunctions (from Herman and Skillman, 1963) were used for the final states in respectively the L and M spectra. These two wavefunctions are shown in Fig. 6. In lieu of any available nickel 3d valence-band wavefunctions, we used those calculated by Wood (1960) for iron; these wavefunctions for electrons at the top and at the bottom of the 3d-band are those reproduced in Fig. 6. Note that the d-part of the wavefunction at the top of the band is atomic-like, while that at the bottom of the band is of bonding-type. Wood suggested that this same d-band character should prevail throughout most of the first transition-series, including copper. We integrated to a distance midway between the radius of the Wigner-Seitz cell and half the interatomic spacing. In this evaluation of Γ, a value is obtained which is only 10% larger at the top of the M valence band with respect to the bottom of the band. For the L spectra there is a 70% increase in the transition probability Γ at the top of

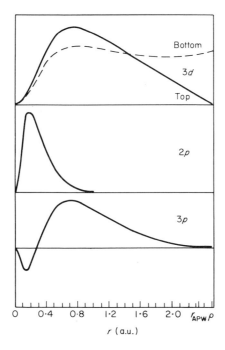

FIG. 6. Iron $3d$ wave-functions at the top and at the bottom of the valence band (Wood, 1960), and nickel $2p$ and $3p$ atomic-wave-functions (Herman and Skillman, 1963). (Reproduced with permission from Cuthill *et al.*, 1967.)

the valence band with respect to the bottom of the band. Thus the transition-probability distortion effect is more severe for L spectra, and accounts well for L spectra having narrower bandwidths than the corresponding M spectra.

A comparison of our observed M_3-emission spectrum with the theoretical density of states is shown in Fig. 7. Curve A is the theoretical density of states for paramagnetic nickel, recently calculated by E. C. Snow (unpublished) using a method that forms an extension to the calculations described by Snow *et al.* (1966). In order to take account of the effects of temperature (960°C) and of lifetime-broadening which will be present in the experimental spectrum, we have applied to the density-of-states curve a Fermi-distribution function and an energy-dependent Lorentzian smearing function. The details of the correction for the effect of lifetime-broadening are described elsewhere (Cuthill *et al.*, 1967). Curve B is the theoretical density of states modified in this manner. Curve C is our experimentally determined Ni M_3 spectrum.

We should like at this point to urge theoreticians who calculate energy bands to make available wavefunctions as well as density-of-states histograms; preferably indicating in the latter the percentage of s, p, and d electron-character.

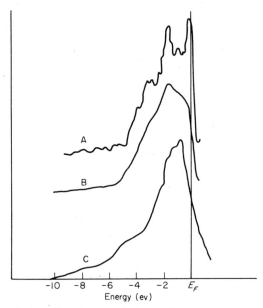

FIG. 7. A, Theoretical density of states obtained using an augmented-plane-wave (APW) calculation (E. C. Snow, unpublished). B, Theoretical density of states (Snow *et al.*, 1966) with Fermi-Dirac temperature correction (to 960°C) and with smearing-function applied to take account of the lifetime broadening present in the experimental M_3 spectrum. C, Nickel M_3-emission spectrum, experimentally determined at 960°C. (Reproduced with permission from ·Cuthill *et al.*, 1968b.)

B. Plasmaron Peak in the Al $L_{2,3}$-Emission Plasmon Satellite

In view of the present interest in electron interaction effects, we briefly describe a search for plasmaron-edge structure in the extreme low-energy tail of the Al $L_{2,3}$-emission band (Cuthill *et al.*, 1968a). The predicted edge (Hedin, 1967) is expected to occur at about 34 ev below the L-emission edge, and to be quite weak, no more than about 1% of the peak intensity in the band, and less if there is significant wave-number dependence in the matrix elements. (Hedin, this Volume, p. 337.) The investigation is further complicated by order overlap. Only if the observed spectrum can be corrected for this overlap can a meaningful inspection for the predicted edge be made. Figure 8 shows the low-energy spectrum (curve B), the corresponding first-order spectrum (curve A) and the estimated first-order background (curve C). The second-order grating efficiency was taken to be the ratio of the measured second-order and first-order edge discontinuities, and was assumed constant over this narrow spectral range. The difference between the measured first-order spectrum and background was multiplied by this ratio, and the product subtracted from curve B to yield the corrected curve D, which is shifted down in Fig. 8 for clarity. Error bars represent 70% statistical confidence level. Curve D shows a very weak edge, just within the random

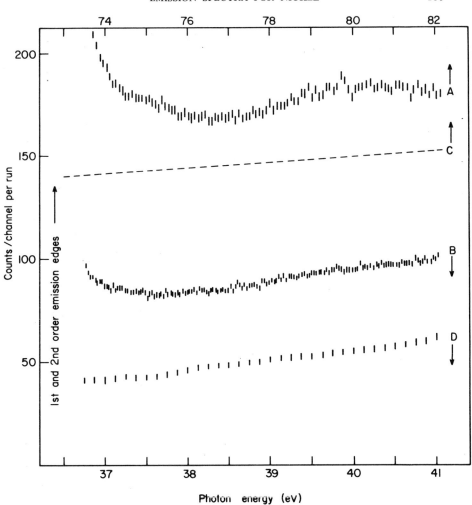

FIG. 8. Average counts per channel, $L_{2,3}$-emission spectrum of Al. A, Region above the first-order emission edge, plotted versus upper energy scale; B, region of the predicted edge; C, background above the first-order edge; D, corrected spectrum in the region of the predicted edge (shifted down arbitrarily for clarity). The vertical bars represent 70% confidence level. (Reproduced with permission from Cuthill *et al.*, 1968a.)

noise and quite near the expected location of 38·6 ev. The intensity change across this edge is approximately 0·3% of the peak parent-band intensity.

C. Emission Spectra from Al–Ni and TiNi

As already mentioned, our long-range interest is in alloys, and work on the Ni–Al system and TiNi is discussed in detail elsewhere (Cuthill *et al.*, 1968b).

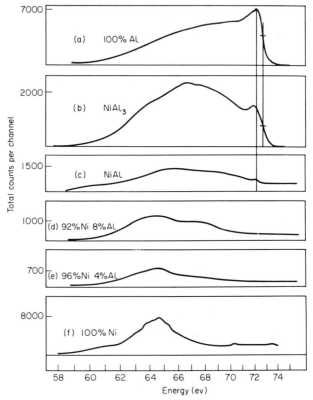

FIG. 9. Al $L_{2,3}$-emission and Ni $M_{2,3}$-emission spectra for Al–Ni alloys. (a) Al $L_{2,3}$-emission from pure aluminium; (b–e) component spectra (overlapping emission bands), respectively for Ni Al$_3$, Ni Al, Al(4%) in Ni, and Al(8%) in Ni; (f) Ni $M_{2,3}$-emission from pure nickel. (Reproduced with permission from Cuthill et al., 1968b.)

The first alloys we studied were from the aluminium–nickel system. In Fig. 9 we show the pure-aluminium L spectrum and pure-nickel M spectrum, together with the corresponding spectra from NiAl$_3$, NiAl and from the two nickel terminal-solid-solution alloys, of respectively 4% and 8% aluminium. The aluminium L and nickel M component-spectra overlap extensively, making the spectra of such a system difficult to analyse. However, some general observations may be offered. The sharp emission edge for L pure aluminium remains evident in the spectra from the alloys, up to at least 5% Ni composition, and its energy remains unshifted. The centre of the electron concentration moves to lower energy with increasing content of nickel. Assuming that alloying has no intrinsic significant effect on the relative transition-probabilities, virtually the entire change observed can be attributed to changes in the electronic structure of the aluminium; at the 2500 v accelerating potential used, the fluorescent-yield from pure aluminium is observed to be six times that from pure nickel, in the spectral range considered. This yield-ratio results in only 17% of the observed NiAl spectrum being attributable to the nickel, and only 3% of the observed NiAl$_3$ spectrum. Also, it can readily be shown that these alloy component-spectra are not simply a linear combination of the pure aluminium $L_{2,3}$ and pure nickel $M_{2,3}$ spectra, as though the alloys were simply a mechanical mixture. Further, we observe a rather systematic change in the spectra, in going from pure aluminium to pure nickel—even though there are many abrupt changes in crystal structure across the system.

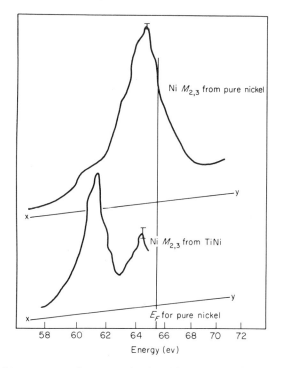

FIG. 10. Nickel $M_{2,3}$ spectrum from pure nickel, and from the TiNi intermetallic compound. (Reproduced with permission from Cuthill et al., 1968b.)

In contrast with the Al–Ni alloy system, we find that the nickel $M_{2,3}$-emission from TiNi can be compared with that from pure nickel without overlap problems. This is shown in Fig. 10. The spectrum splits into two peaks upon the formation of the intermetallic compound. The weaker of the two is not shifted in energy from that of the maximum in pure nickel; the stronger occurs at 3 ev lower than the maximum in pure nickel. It is interesting to compare this double-peak spectrum with the double-peak curve of electronic specific heat versus electron-to-atom ratio for combinations of the first transition series of bcc metals (Cheng et al., 1960). On the basis of regarding the area under the pure-nickel spectrum, up to the Fermi-energy, as corresponding to an e/a ratio of 10, the areas up to respectively the low-energy peak, the valley between the peaks and the high-energy peak, correspond to e/a values of 3·9, 6·0 and 7·2; which compare with the respective e/a values of 4·5, 6·0 and 6·3 at the corresponding points on the Beck plot.

Acknowledgement

We should like to express our thanks to R. E. Watson for advice and instruction in the theoretical aspects of this work. We are grateful also to E. C. Snow for calculating the percentage of s, p and d character in the APW density-of-states curve for nickel.

References

Cheng, C. H., Wei, C. T. and Beck, P. A. (1960). *Phys. Rev.* **120**, 426.

Chopra, D. and Liefeld, R. (1964). *Bull. Amer. Phys. Soc.* **BG3**, 404.

Cuthill, J. R., McAlister, A. J., Williams, M. L. and Watson, R. E. (1967). *Phys. Rev.* **164**, 1006.

Cuthill, J. R., Dobby, R. C., McAllister, A. J. and Williams, M. L. (1968a). *Phys. Rev.*, in press.

Cuthill, J. R., McAlister, A. J. and Williams, M. L. (1968b). *J. Appl. Phys.* **39**, 2204.

Germer, L. H., Hartman, C. D., MacRae, A. U. and Schreibner, E. J. (1963). Bell Telephone Syst. Tech. Publ. Monograph No. 4364.

Hedin, L. (1967). *Solid State Commun.* **5**, 451.

Herman, F. and Skillman, S. (1963). "Atomic Structure Calculations". Prentice-Hall, New Jersey.

Snow, E. C., Waber, T. J. and Switendick, A. C. (1966). *J. Appl. Phys.* **37**, 1342.

Tomboulian, D. H., Bedo, D. E. and Neupert, W. M. (1957). *J. Phys. Chem. Solids* **3**, 282.

Wood, J. H. (1960). *Phys. Rev.* **117**, 714.

Distributions d et f de quelques Métaux et Composes obtenues par Spectroscopie X

C. Bonnelle

Laboratoire de Chimie Physique de la Faculté des Sciences de Paris, France

ABSTRACT

Experimental results for $L_{2,3}$-emission and absorption spectra of copper and various transition elements, for the metal and for the oxide, are discussed, as well as the $M_{4,5}$ spectra of elements that have an incomplete f-shell. Most of the results chosen for discussion can be compared to the theoretical calculations of density of states. To enable this comparison, the theoretical $N(E)$ curve is convoluted with a Lorentz distribution with a width at half-height equal to that of the appropriate internal level. The width of the d-band and the position of its maximum with respect to the Fermi-level are given for Cu, Ni, Co, Fe, Cr and Pd. The distribution of f-states for gadolinium, and the structure of its emission spectrum are discussed.

1. INTRODUCTION

Nous présenterons les résultats obtenus dans notre laboratoire pour les spectres d'émission et d'absorption L_3 du cuivre, de quelques métaux de transition de la première série et du palladium, ainsi que pour les spectres $M_{4,5}$ du gadolinium.

Les transitions mettant en jeu des niveaux internes à caractère p permettent d'atteindre principalement la distribution des états d. Pour les éléments étudiés, et contrairement à ce qui a lieu pour les spectres M_2 et M_3, les spectres L_2 et L_3 sont suffisamment séparés en énergie pour être parfaitement résolus. Leur observation est donc particulièrement souhaitable pour l'étude des distributions d.

Nos conditions d'observation ont été décrites (Bonnelle, 1966). Nous nous sommes attachées à définir l'état physicochimique de l'échantillon à l'étude; les spectres des oxydes ont été étudiés dans chaque cas comparativement à ceux du métal. L'influence de la réabsorption dans la cible émissive a été mise en évidence et discutée. La tension d'excitation a été variée dans de larges limites; son rôle sur l'intensité de certaines structures secondaires a permis, dans quelques cas, de différencier les structures propres à la bande x principale de celles attribuables à des émissions satellites. Nos courbes représentent la variation de l'intensité émise I et du coefficient d'absorption

μ en fonction de la fréquence v. La résolution varie de 0·1 à 0·3 ev suivant le domaine spectral.

2. DISCUSSION DES RESULTATS

A. Generalités

Les courbes $I(v)$ et $\mu(v)$ résultent, sous certaines réserves, de la composition de la courbe de densité d'états $N(E)$ avec la distribution énergétique du niveau profond, ici L_3, et les probabilités de transition $p(v)$ depuis ou vers ce niveau.

Les probabilités de transition dépendent de la fréquence, dont nous pouvons négliger la variation dans le petit domaine spectral couvert par nos courbes expérimentales, et d'un élément de matrice que l'on peut de même supposer constant, en première approximation. Il est nécessaire, par contre, de tenir compte de la composition de $N(E)$ avec la distribution du niveau profond pour l'interprétation de certains de nos résultats. Cette distribution peut être assimilée à une courbe de Lorentz dont la largeur à mi-hauteur varie de 0·9 à 1·45 ev pour nos différents cas d'expérience (cf. Tableau I, et Bonnelle, 1966). Nous avons, en particulier, montré que, pour les métaux, la composition avec une telle courbe déforme d'autant plus la distribution $N(E)$ que celle-ci est plus abrupte et donc que la densité d'états au niveau de Fermi est plus grande. La précision sur la position du niveau de Fermi s'en trouve limitée dans certains cas.

TABLEAU I †

	$\mathscr{L}_{L_3}$	$3d-E_F$	$\mathscr{L}_{\text{à mi-hauteur}}$	$\mathscr{L}_{\text{totale}}$
Cu	0·9	2·6	3·2	6·0
Ni	0·9$_7$	0·8	2·6	5·0
Co	1·0	1·1$_5$	3·6	6·5
Fe	1·5	2·5	6·0	7·0
Cr	0·8$_5$	2·8	4·5	5·7
Pd	1·3	1·4	4·4	6·4

† Valeurs en ev.

Nous avons choisi de rapporter ici préférentiellement les résultats pour lesquels une comparaison avec des calculs théoriques de densité d'états est actuellement possible.

Dans la Fig. 1 on présente comparativement les émissions L_α observées pour Cu, Ni, Co et Fe à l'état métallique, rapportées au niveau de Fermi correspondant et à une même échelle d'énergie. La Fig. (2) représente la variation du coefficient d'absorption au voisinage de la discontinuité L_3 observée pour Ni, Fe et Cr, rapportées au maximum d'intensité principal de l'émission $L\alpha$. Elles mettent en évidence les changements de la distribution

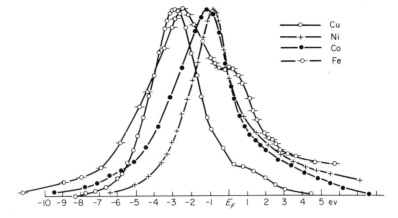

FIG. 1. Emissions L_α de Cu, Ni, Co et Fe à l'état métallique.

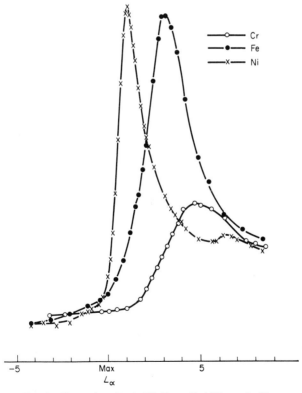

FIG. 2. Absorptions L_3 de Ni, Fe, et Cr à l'état métallique.

des niveaux d, occupés et inoccupés, en fonction de leur remplissage. Les largeurs à mi-hauteur et les largeurs totales des émissions L_α sont portées Tableau I, ainsi que la largeur de la distribution du niveau L_3 et la distance de L_α au niveau de Fermi, notée ici E_F.

La comparaison entre nos résultats expérimentaux et les courbes de densité d'états théoriques a été faite en tenant compte de la composition avec le niveau L_3. En effet, comme le montrent les données du Tableau I, la largeur de L_3 ne peut être négligée devant la largeur des bandes d, en particulier pour le cuivre et le nickel.

Cette composition a pour effet d'étouffer certaines des structures présentes dans les courbes de densité d'états. De plus, pour discuter de la forme des bandes, il faudrait pouvoir s'affranchir complètement des émissions satellites présentes vers les grandes énergies des émissions L_α dans tous les cas étudiés. Nous ne donnerons ici pour les métaux que des courbes expérimentales obtenues à des énergies d'excitation suffisamment basses pour minimiser le plus possible l'intensité relative de ces émissions satellites par rapport à la bande principale. Mais celles-ci, bien que d'intensité faible, restent généralement observables.

Deux données physiques, importantes à connaitre, peuvent être déduites à partir de nos spectres : ce sont la largeur de la bande d et la position de son maximum par rapport au niveau de Fermi. Nous en discuterons plus spécialement ici.

B. Les Metaux

Pour le cuivre, on dispose d'un grand nombre de calculs théoriques effectués à l'aide de diverses approximations. Le Tableau II résume les différentes valeurs obtenues, comparativement aux valeurs expérimentales.

L'accord le meilleur est obtenu avec les calculs effectués par des méthodes "self-consistantes," que ce soit la méthode des fonctions de Green (Wakoh, 1965) ou l'approximation des ondes planes progressives (APW) (Snow et Waber, 1967). Avec cette dernière, la position par rapport au niveau de Fermi de la bande d, calculée en utilisant comme terme d'échange 2/3 du terme de Slater, est en excellent accord avec notre mesure expérimentale.

TABLEAU II

	Largeur bande d (ev)	$3d$–E_F (ev)
Segall (1962) (fonctions Green)	4·1	4·0
Burdick (1963) (APW)	3·4	3·4
Wakoh (1965) (fonctions Green H.F.S. self-consistant)	3·3	2·7
Snow et Waber (1967) (APW self-consistant)		
avec terme échange de Slater	2·6	4·0
avec 2/3 terme échange de Slater	3·5	2·7
Emission L_3 (largeur à mi-hauteur non corrigée)	3·2	2·6

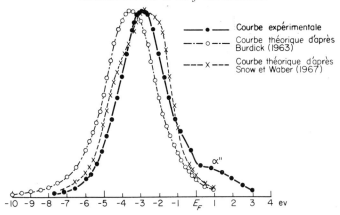

FIG. 3. Emission L_α du cuivre et courbes déduites de calculs théoriques de densité d'états.

Les courbes de densité d'états théoriques d'après Burdick (1963) et Snow et Waber (1967) composées avec une distribution lorentzienne de largeur à mi-hauteur de 0·9 ev (cf. Tableau I) sont portées Fig. 3 comparativement à notre courbe expérimentale. La forme de la courbe déduite d'après Burdick (1963) est tout à fait analogue à celle de notre courbe expérimentale, mais comme on pouvait s'y attendre étant donné l'approximation utilisée, sa position par rapport au niveau de Fermi est peu précise.

La Fig. 4 représente de même pour le nickel la comparaison entre l'émission L_3 et les courbes théoriques obtenues en effectuant la composition de différentes courbes de densité d'états avec une courbe de Lorentz représentant la distribution du niveau L_3; nous y avons porté la courbe obtenue par G. Allan (communication privée) à l'aide d'une méthode de liaisons fortes qui introduit le couplage spin-orbite et celles que nous avons déduites pour le métal ferromagnétique à partir des courbes de densité d'états donnés, pour les deux directions de spin, à l'aide de calculs "self-consistant"

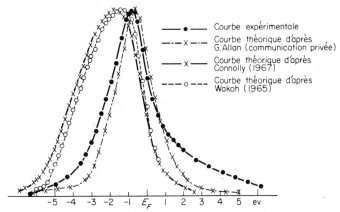

FIG. 4. Emission L_α du nickel et courbes déduites de calculs théoriques de densité d'états.

par Wakoh (1965) (approximation des fonctions de Green) et par Connolly (1967) (approximation APW avec encore un terme d'échange égal à 2/3 du terme de Slater). L'accord le meilleur avec l'expérience est obtenu pour la courbe de Allan dont la largeur à mi-hauteur est de 2·5 ev.

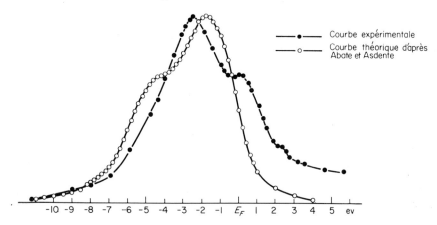

FIG. 5. Emission L_α du fer et courbe déduite de calculs théoriques de densité d'états.

La Fig. 5 représente la comparaison entre notre courbe expérimentale pour le fer et la courbe théorique déduite, pour le métal ferromagnétique, des calculs de Abate et Asdente (1965) effectués à l'aide d'une méthode de liaisons fortes. L'accord entre théorie et expérience reste ici très qualitatif, mais les largeurs totales des bandes expérimentales et calculées sont du même ordre de grandeur. L'une et l'autre courbes présentent des structures marquées, décalées les unes par rapport aux autres dans l'échelle des énergies. D'après différents contrôles expérimentaux, la structure observée au voisinage du

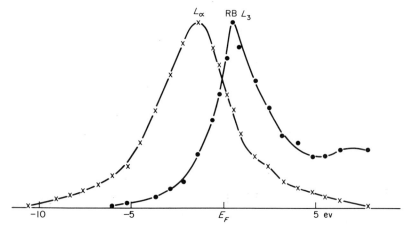

FIG. 6. Emission et absorption L_3 du palladium métallique.

niveau de Fermi semble appartenir à la bande principale du métal, mais les dépouillements de nos spectres sont difficiles dans ce cas par suite de la présence éventuelle de fer dans les différents cristaux analyseurs.

Les courbes d'émission et d'absorption L_3 du palladium sont portées Fig. 6. Leur comparaison avec les résultats obtenus pour le nickel met en évidence le rôle de l'extension des sous-couches d sur leur forme et leur distribution. Bien que la résolution soit inférieure, dans ce domaine spectral, à celle atteinte pour les spectres du nickel, on peut conclure que la distribution d se trouve nettement élargie dans son ensemble, ce qui se traduit donc aussi par un déplacement du niveau de Fermi vers les grandes énergies par rapport au maximum de la bande d.

C. Les Oxydes

Les spectres d'émission et d'absorption de différents oxydes, Cu_2O, CuO, NiO, Co_3O_4 et Fe_3O_4 ont été observés. En les comparant au spectre du métal correspondant, nous avons mis en évidence le rôle de la liaison sur certains caractères spectraux, tels que position en énergie, largeur, structures

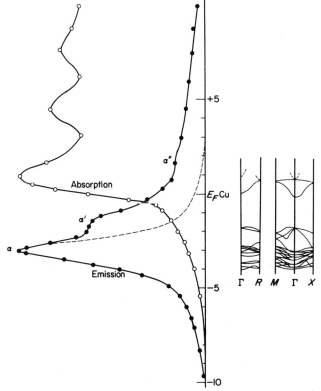

FIG. 7. Emission et absorption L_3 du cuivre dans Cu_2O et structure de bandes d'après Dahl et Switendick (1966).

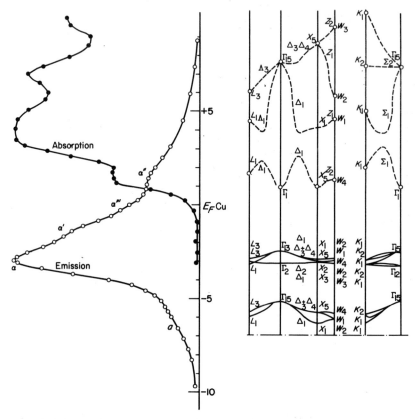

FIG. 8. Emission et absorption L_3 du cuivre dans CuCl et structure de bandes d'après Song (1967).

secondaires, etc. Nous ne rapporterons pas ici l'ensemble de ces résultats qui ont déjà été discutés par ailleurs (cf. principalement, Bonnelle, 1966, 1968). Nous présenterons seulement Figs. 7 et 8 nos courbes expérimentales pour Cu_2O et CuCl comparativement aux calculs de bandes effectués respectivement par Dahl et Switendick (1966) (approximation APW) et par Song (1967) (approximation des liaisons fortes pour les états occupés et approximation des ondes planes orthogonalisées pour les états inoccupés). Une discussion de ces résultats a été donnée (Bonnelle, 1967). Rappelons qu'il est possible d'après nos mesures expérimentales d'obtenir, pour ces deux composés, la largeur de la bande interdite et de situer les unes par rapport aux autres les différentes orbitales occupées et inoccupées. En particulier, l'observation de "transition croisée" dans le spectre d'émission de CuCl nous a permis de situer la bande $3p$ du chlore à environ 3.5 ev en dessous de la bande $3d$ du cuivre.

Il faudrait là encore pouvoir s'affranchir complément des émissions satellites présentes vers les grandes énergies de la bande principale α pour discuter en toute rigueur des formes et des largeurs des différentes distributions d'états électroniques. Soulignons que nos spectres permettent cependant, en faisant certaines hypothèses pour décomposer l'émission globale observée, d'obtenir avec une approximation acceptable les largeurs de ces distributions.

Une étude des spectres d'émission et d'absorption du gadolinium dans le métal, l'oxyde et différents hydrures est en cours dans notre laboratoire. L'émission et l'absorption M_5 du métal sont données Fig. 9. L'émission présente différentes structures dont les intensités relatives varient très fortement avec les conditions d'excitation. Les trois plus intenses ont été notées 1, 2 et 3. L'émission 1 doit correspondre à la bande proprement dite, c'est à dire ici principalement à la distribution des états f. Sa forme et sa position varient lorsque l'on passe du spectre du métal à celui de l'oxyde.

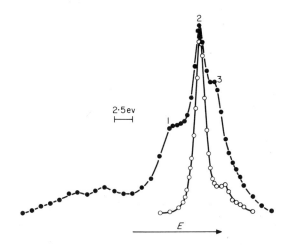

FIG. 9. Emission et absorption M_5 du gadolinium.

L'émission 2 coincide à la précision expérimentale avec la raie d'absorption. En accord avec l'interprétation proposée par Stewardson et Wilson (1956) pour les émissions M_4 et M_5 de l'europium, nous interprétons cette émission par des transitions d'électrons excités, présents dans la bande f normalement inoccupée, vers le niveau interne M_5. L'émission 3 est proche de la structure prisente vers les grandes énergies de la raie d'absorption et pourrait lui correspondre, en première approximation. Pour l'émission M_4, on retrouve de même une composante analogue à 1 et des composantes analogues à 2 et 3, mais l'émission 1 est alors d'intensité très supérieure à celles des émissions 2 et 3. Elle est donc beaucoup mieux résolue. Sa forme et sa position varie très nettement avec la liaison. L'exposé détaillé des résultats sera donné dans une prochaine publication (C. Bonnelle et R. C. Karnatak, à paraître).

Signalons pour conclure que nous avons observé dans les spectres M_4 et M_5 des éléments $5f$, des émissions qui peuvent être de même interprétées

comme des transitions "en résonance" avec les raies d'absorption correspondantes (C. Bonnelle et G. Lachere, à paraître).

Acknowledgement

J'adresse mes vifs remerciements àu Dr. G. Allan et au Dr. G. Leman qui m'ont obligeamment communiqué les résultats de leurs calculs avant publication.

Bibliographie

Abate, E. et Asdente, M. (1965). *Phys. Rev.* **140**, A1303.
Bonnelle, C. (1966). *Ann. Physique* **1**, 439.
Bonnelle, C. (1967). *Journ. de Phys.* **28**, C3, 65.
Bonnelle, C. (1968). "Physical Methods of Inorganic Chemistry", Ch. 2. Wiley, New York.
Burdick, G. A. (1963). *Phys. Rev.* **129**, 138.
Connolly, J. W. D. (1967). *Phys. Rev.* **159**, 415.
Dahl, J. P. et Switendick, A. C. (1966). *J. Phys. Chem. Solids* **27**, 931.
Segall, B. (1962). *Phys. Rev.* **125**, 109.
Snow, E. C. et Waber, T. J. (1967). *Phys. Rev.* **157**, 570.
Song, K. S. (1967). *Journ. de Phys.* **28**, C3–43.
Stewardson, E. A. et Wilson, J. W. (1956). *Proc. Phys. Soc.* **A69**, 93.
Wakoh, S. (1965). *J. Phys. Soc. Japan* **20**, 1894.

Soft X-Ray Emission Spectra of Alloys and Problems in their Interpretation

C. CURRY

Physics Department, University of Leeds, England

ABSTRACT

A general view is taken of the state of progress in the field of soft x-ray emission spectroscopy of binary alloys. Separate emission bands correspond to the filling of core vacancies in each type of constituent atom. Even when these vacancies are in similar shells, say L_3, the two L_3-emission bands tend to be very different in shape and width. The changes of shape are sometimes considerable and take place in a progressive manner with concentration. Some problems of experimental work on alloys are reviewed, and new results for some binary alloys, Cu–Si, Al–Cu, Al–Mn and Mg–Si are presented.

1. INTRODUCTION

The aim of this paper is to present a general view of experimental findings with alloys as distinct from pure metals. The total amount of experimental information is not extensive, and clearly care is needed in generalizing from limited information. Nevertheless, it seems desirable to take a broad view— so far as this is possible—quite as much as to consider the details of any particular spectrum.

While attempting to take this general view of experimental facts, the substance of the paper centres mostly around work carried out at Leeds in recent years. The latter part reports recent results which help to fill out the picture to some extent. Most other results, however, fit into the pattern which will be outlined, and could also be used to illustrate the general statements made. Some of this material is referred to as the paper proceeds.

The discussion is limited to emission spectra. Absorption results are scarce in the 100 Å region, which is where many of the experiments are done.

2. SIMPLE EXPECTATIONS

In a pure metal a typical emission transition begins with an empty core state, and after the transition an empty valence-electron state is formed. The initial state is often well-defined in energy so that the spectra indicate the range of energies of filled valence-electron states, from which electrons may be supplied for transitions to the core vacancies.

Passing now to the binary alloy, the existence of two types of constituent atom means that two separate emission bands are to be expected for each kind of vacancy (K, L_2, L_3, etc.). These correspond to the filling of core vacancies of the appropriate kind in each type of atom and are, of course, observed; the interest centres on the shape of these band spectra in relation to the environment of the atoms in which the emission transitions take place.

Attention is drawn, in the first place, to very broad features of results of experiments on these emission bands in alloys. These are as follows. (1) Distinctions between pure-metal band spectra tend to be preserved on alloying (though, generally, not perfectly—and sometimes in fact they are accentuated). In some instances no perceptible changes occur; but more often substantial changes do occur. (2) Spectra-shape changes can be considerable and may be progressive with concentration. A pattern of change, adhered to by many results, will be indicated later; at this stage it may be pointed out that the emphasis is on *shape* changes. Width and position changes appear to be relatively minor in character. (3) There is some evidence that structure of an alloy is secondary to concentration in respect of the effect on band spectra.

Of the many possible examples of these statements two are now chosen which show these effects boldly.

Firstly the $L_{2,3}$-emission spectra for aluminium–magnesium ordered alloys, and for the pure metals, are shown in Fig. 1. These particularly illustrate the spectra-shape changes which occur on alloying.

It will be noted that, starting from either of the pure-metal spectra, progressive shape changes occur on dilution with the other metal. In this

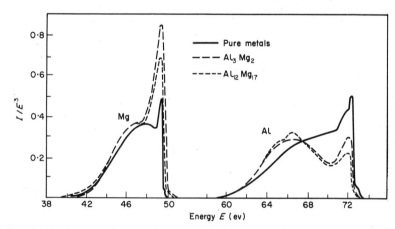

FIG. 1. Aluminium and magnesium $L_{2,3}$-emission spectra from the pure metals and the alloys Al_3Mg_2 and $Al_{12}Mg_{17}$. (Reproduced with permission from Appleton and Curry, 1965.)

example the differences between the pure-metal band spectra are accentuated by the alloying process. The emission bandwidths remain constant (to a few tenths of an electron volt), and the high-energy edges do shift and broaden, but only by quite small amounts. The smallness of these changes seems remarkable, in view of the substantial structure changes involved.

Secondly, the Zn $M_{2,3}$-emissions from pure zinc and from α-brass (containing 30 atomic % Zn) are shown in Fig. 2. No significant difference between the two band-spectra is observed. In this example the pure-metal emission bands are closely preserved on alloying.

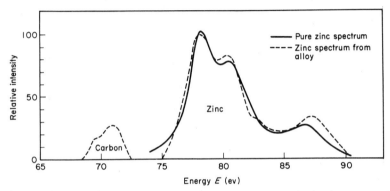

FIG. 2. Comparison of the pure-zinc $M_{2,3}$-emission spectrum with the zinc $M_{2,3}$-emission from an alloy containing 30 atomic % Zn in Cu. (Reproduced with permission from Clift *et al.*, 1963b.)

A similar result was found throughout the whole range of copper-nickel alloys (Clift *et al.*, 1963a), but the result with α-brass is more interesting still, since structure changes are involved. The environment of the emitting zinc atoms is, in one case, the hexagonal-close-packed lattice of neighbouring zinc atoms, and in the other, a face-centred-cubic structure with copper atoms predominant.

One or two experiments have shown no measureable emission-band change with order-disorder transformation. Catterall and Trotter (1962) showed this with Cu M-emission from Cu_3Au; Appleton and Curry (1967) observed it with Al L-emission from $AlFe_3$. The general fact will emerge that alloy concentration affects the emission bands in a way which has some system about it, even though substantial structure change may also be involved. This also indicates that in respect of its effect on the spectra, structure is not so substantial a factor as might have been anticipated.

3. LINES OF EXPLANATION OF OBSERVATIONS

Changes in the pure metal wavefunctions, on alloying, must underlie the observed effects. Two aspects of these changes will be considered in turn.

(1) The wavefunction for a particular electron energy-state may increase near one type of atom and decrease near the other; i.e. a restricted sharing of the electron between the atoms takes place. This will depend, in degree, on the kinds of atoms involved (especially their valences), on arrangement of the atoms, and also on the energy of the state. (2) The symmetries of the wavefunctions, as seen by the two kinds of atom, may differ, and may change on alloying. This will affect the emission transition probabilities and their energy dependence.

Both of these aspects have significance for the emission-band forms, and changes in the band spectra will arise on account of changes of both kinds. It seems clear that effects from wave-symmetry alone would have to be spectacular to account entirely for what is observed (i.e. clear differences of bandwidth, and near constant bandwidths for separate constituents). The pattern of change with alloy composition, yet to emerge, appears to agree in trend with that anticipated from changes in wavefunction of type (1) not entirely masked by changes of type (2). The results, as a whole, indicate enhanced low-energy emission in the band-spectrum due to the higher-valence constituent, and *vice versa*. This corresponds to a pile-up of the wavefunction around excess positive nuclei, and a thinning of the wave-function elsewhere.

4. GENERAL COMMENTS ON EXPERIMENTS ON ALLOYS

For the purpose of discussing experimental work on soft x-ray emission from alloys, it is convenient to make the distinction between dilute alloys and more concentrated ones. There are points in favour of pursuing work with both.

With dilute alloys, only minor changes take place in the solvent band-spectra. More substantial changes are apparent in the solute spectra; but since solute spectra are comparatively weak they are generally difficult to measure. Some solute emission spectra are available but these, though interesting, are rather ill-determined. With solvent spectra it is possible to look for small changes in such key features as zone-boundary discontinuities, high-energy peaks and edge-widths.

Since detailed changes are under investigation it is necessary, in the experiments, to have good reproducibility and stability, etc., during scanning. In pure-metal spectra, zone-boundary discontinuities are quite minor features; and detection of changes in such features on alloying, presents experimental problems of a further order of difficulty.

With more concentrated alloys, it is possible to cover a whole range of environments for the emitting atom; starting from solid solutions at one end of an alloy range, and passing through substantial structure-changes as one proceeds across the concentration range. The effects of order–disorder transformation and other changes of structure may be sought. The changes in emission spectra seen with concentrated alloys are generally much more substantial, and are much easier to observe with confidence.

Against these experimental advantages, of the use of concentrated alloys, must be set the greater theoretical problems arising with the non-dilute alloy—especially if this is also disordered. The solvent spectra of dilute alloys are probably the most amenable to theoretical consideration; there are more problems of interpretation for the solute spectra, and also for the solvent spectra from less-dilute alloys where the approximation of non-interacting solute atoms ceases to be applicable.

The results to be described have been obtained with concentrated alloys. Results of other experimentalists, some working on the same and others on different alloy systems (including both concentrated and dilute alloys), show emission-band changes which follow the same trends as observed in our laboratories.

5. RESULTS OF EXPERIMENTS ON ALLOYS

Results of experiments on the following systems are now described: Al–Mg alloys (Appleton and Curry, 1965): Ni–Zn, Al–Fe and Mg–Ni alloys (Appleton and Curry, 1967): Al–Mn, Al–Cu, Cu–Si and Mg–Si alloys (C. Curry and R. Harrison, unpublished). For the latter series, the results are more tentative than the rest, and more experiments have to be performed to reduce the uncertainties in the spectra to a minimum. However, the general features seem sufficiently clear to allow these results to be reported at the present stage. It may be said again that the *shapes* of these emission spectra are comparatively easy to determine; the changes on alloying, in the form of any one spectrum, are much greater than the uncertainties in the individual spectra. This is true in all cases, apart from Ni, Cu and Zn emission spectra, where no perceptible changes of form are observed.

The order of presentation departs from the chronological order of the experiments, so that the features of the results as a whole may be more apparent.

A. $M_{2,3}$-Emission from Nickel–Zinc Alloys

$M_{2,3}$-Emission spectra for α-phase (36 atomic % Zn), and β-phase (48 atomic % Zn) Ni–Zn alloys are shown in Fig. 3. This is envisaged as an extension of the work on Ni–Cu and Cu–Zn alloys (Clift et al., 1963a,b), and the results are similar to those found earlier. The spectra are not different from those of the pure metals. The differences between the curves are not significant, being just about equal to the mean deviations for each averaged curve.

The last result is not too surprising when it is recalled that transitions of $3d$-electrons to inner vacancies are being observed, and that these electrons

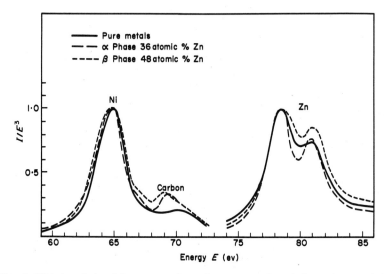

FIG. 3. Nickel and zinc $M_{2,3}$ spectra from the pure metals and the α-phase and β-phase alloys. (Reproduced with permission from Appleton and Curry, 1967.)

are fairly tightly bound. Their wavefunctions may not be greatly affected by the substitution of a foreign atom as neighbour, especially if that atom is not very different in atomic number.

The $M_{2,3}$ spectra of these pure metals are themselves not easy to interpret. The main peak arises from transitions of 3d-electrons to the M_3 shell ($3P^4_{3/2}$). In the isolated atom the 3p-wavefunction is approximately at the same mean distance from the nucleus as the 3d-wavefunction. Thus, although an energy difference of several tens of volts exists between them, the broadening of the atomic levels may not be very different in the solid. The slight broadening of the level of lower energy could therefore mask any abrupt high-energy limit in the spectrum. Any small difference which might otherwise be seen on alloying could also be blurred for this reason.

B. Aluminium $L_{2,3}$-Emission from Aluminium Alloys

The forms of the Al L-emission bands from the alloys Al–Mg, Al–Mn, Al–Cu and Al–Fe are respectively presented in Figs. 1, 4, 5 and 6.

(i) *Al–Mg*. Several features of Fig. 1 have already been pointed out. Note also, that in the Al L-emission band the ratio of the intensity in the low-energy part of the emission band to that at higher energies (near the Fermi-edge) increases progressively with dilution of the Al in Mg, for these ordered alloys.

(ii) *Al₃Mn*. Features observed in the Al L-spectrum from Al₃Mn (Fig. 4), are as follows. (a) The sharp peak at the Fermi-edge, characteristic of the pure Al emission-band, is no longer evident. (b) The maximum of the

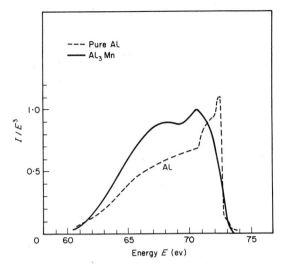

FIG. 4. Aluminium $L_{2,3}$-emission from Al_3Mn and from pure Al.

emission-band intensity is at lower energy relative to the Fermi-edge. (c) The bandwidth is rather less (0·4 ev) than that from the pure metal. (d) The edge position is unchanged within the accuracy of measurement. (e) The low-energy emission is higher in intensity, relative to that at higher energies, than for the pure aluminium spectrum.

(iii) Al_2Cu. The alloy Al_2Cu gives an Al L-spectrum (Fig. 5) with some striking features, and attention is drawn to the following points. (a) The

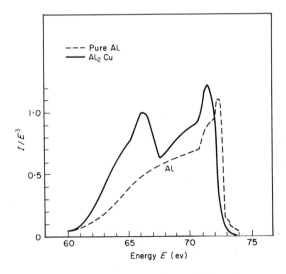

FIG. 5. Aluminium $L_{2,3}$-emission from Al_2Cu and from pure Al.

high-energy peak comes at rather lower energy than in the pure-aluminium emission band, and a second prominent peak appears near 66 ev. Spurious experimental causes for this sharp peak have been eliminated. (b) The whole of the Al $L_{2,3}$-emission band is shifted towards lower energies by rather less than 0·5 ev. The bandwidth remains virtually unchanged. (c) The high-energy edge is broadened a little (to about 0·4 ev). (d) The low-energy emission is again increased in intensity relative to that at higher energies, as compared with the pure-aluminium spectrum.

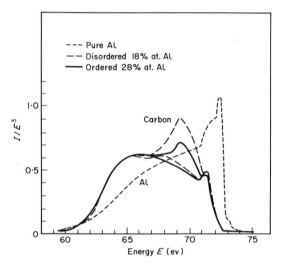

FIG. 6. Aluminium $L_{2,3}$-emission spectra from ordered and disordered α-phase Al–Fe alloys, and from pure Al. (Reproduced with permission from Appleton and Curry, 1967.)

(iv) *Al–Fe*. Results of experiments on ordered and disordered alloys, respectively containing 28 and 18 at % Al, appear in Fig. 6. Elimination of carbon *K*-emission was difficult with these alloys; the feature in the region 67–71 ev is due to carbon, and can vary considerably with the amount of carbon contamination. A carbon *K*-emission curve may be subtracted in the right proportion to remove the prominent peak at 69 ev. Such a procedure leaves essentially the same Al *L*-spectrum from different measurements for more than one sample of the same alloy. The features of the spectra are as follows. (a) Within experimental uncertainties the same spectral shape is obtained from both alloys. (b) The edge-width is about 0·4 ev, rather larger than for pure aluminium. (c) The edge shift is about 1 ev towards lower energy. (d) The bottom of the emission band shifts about 0·5 ev, also to lower energy, so that the bandwidth is less than for pure aluminium by about 0·5 ev. (e) The small hump just below the Fermi-edge appears to be real. It is sufficiently removed from the carbon *K*-emission band for this conclusion to be formed. (f) The low-energy emission intensity relative to that at higher energies is again increased for the alloys, compared with pure aluminium. The behaviour here is similar to the behaviour of the Mg $L_{2,3}$ spectrum in Mg–Li alloys observed by Catterall and Trotter (1959). There, the Mg high-energy peak was considerably reduced in the alloys, and the lower-energy part rather more dominant; the general change in the band being very similar to the present observations.

C. Magnesium $L_{2,3}$-Emission from Magnesium Alloys

(i) *Mg–Al*. Returning to Fig. 1 and noting the changes in the magnesium L-emission spectrum, the following points are observed. (a) The changes in the emission band are of the opposite kind to those noted with aluminium. The intensity in the high-energy part of the emission band, relative to that in the low-energy part, increases on dilution of Mg with the higher valency Al. (b) The edge-shift is positive and progressive, though small; the larger of the displacements is less than 0·2 ev. (c) The edge-widths also increase by a small amount, though the bandwidths are not measurably changed.

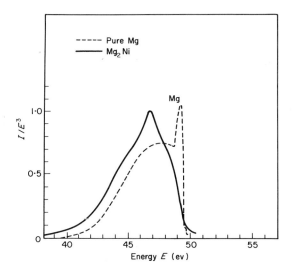

FIG. 7. Magnesium $L_{2,3}$-emission from Mg$_2$Ni and from pure Mg. (Reproduced with permission from Appleton and Curry, 1967.)

(ii) *Mg$_2$Ni*. Figure 7 shows the Mg $L_{2,3}$-emission band from this alloy. The following may be noted. (a) The peak at the high-energy limit of the pure Mg spectrum has completely disappeared in the alloy spectrum. (b) The bottom of the band has moved to lower energies (by about 1 ev). (c) The high-energy cut-off takes place in two stages; a small step in the emission, at 49·5 ev, which is the position of the edge in the pure-metal spectrum, and a further edge at 48·9 ev, where most of the intensity fall-off occurs. (d) The peak of the emission is fairly prominent and comes at 46·9 ev. This is almost certainly real; other possible causes having been eliminated. (e) The behaviour of the spectrum follows the same trend of dependence on concentration as observed for the aluminium L-emission band (and for the Mg L-emission band from Mg–Li alloys); though in a rather special way. In the alloy the low-energy part of the emission band is intensified relative to the high-energy part.

The peak in the emission may be connected in some manner with the high density of d-states contributed by the nickel. The empty $3d$-states and some of the $4s$-states may be filled in such a way that the top of the d-band comes at the lower of the two edges, i.e. about 0·6 ev below the Fermi-energy. The peak in the Al L-emission from Al_2Cu, at about 66 ev, could be associated with the d-electrons contributed by Cu in a similar manner.

D. $M_{2,3}$-Emission Bands from Al–Fe and Mg–Ni Alloys

The Fe and Ni $M_{2,3}$-emissions, from the alloys Al–Fe and Mg–Ni, have also been investigated. In keeping with the results for Cu–Ni, Cu–Zn and Ni–Zn alloys, these emission bands do not change significantly from the corresponding pure-metal spectra.

E. Silicon $L_{2,3}$-Emission

Figures 8 and 9 show Si L-emission spectra from Cu_3Si and Mg_2Si, with the pure-silicon L spectrum for comparison.

Both of these emission bands show a tendency to peak sharply in the low-energy region, and to exhibit fall-off in intensity (compared with the

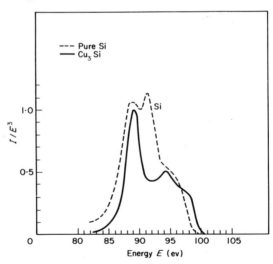

FIG. 8. Silicon $L_{2,3}$-emission from Cu_3Si and from pure Si.

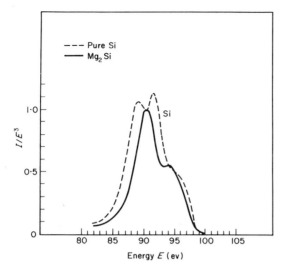

FIG. 9.   Silicon $L_{2,3}$-emission from Mg_2Si and from pure Si.

spectrum for pure silicon) in the upper part. The same trend, as noted for an element of higher valency when diluted with one of lower valency, is apparent. The low-energy peaks are at about 89·5 ev in Cu_3Si and 90·5 ev in Mg_2Si. In Cu_3Si the emission band is of much the same width as that for pure silicon, but has shifted to higher energy by about 1 ev. For Mg_2Si the bandwidth decreases by 1 ev or more; this arises from a shift of the lower energy end of the emission band.

6. SUMMARY OF TRENDS IN EMISSION-BAND FORMS ON ALLOYING

Although it is not possible to predict in detail how the spectrum of a metal will change on alloying, the following rules (Appleton and Curry, 1967) seem to apply. (a) Emission spectra of transition metals and of copper and zinc (M spectra) do not change perceptibly on alloying. (b) The spectra of other metals change in a manner such that the intensity at the high-energy end of the emission band increases relative to that at the low-energy end when the metal is alloyed with one of greater valency, and vice versa. (c) High-energy edge-shifts are small and negative when the intensity of the high-energy emission is reduced, and positive when it is increased. (d) Bandwidths rarely change by more than 10% on alloying. There is little tendency towards equality in width and shape, as expected from a simple common-valence-band model.

For rule (b) to be consistent, low valences must be attributed to transition metals, in the sense that they can contribute only a fraction of a 4s-electron to the band, and contain vacancies in the $3d$-band which can be considered to accept electrons. In addition to references already quoted, the following papers also supply evidence for this rule: Crisp and Williams (1960), Das Gupta and Wood (1955), Gale and Trotter (1956), Skinner and Johnston (1938).

7. CONCLUSIONS

The results available seem generally to indicate a strong tendency for preferential grouping of valence-electrons around the two species of atom in an alloy. The changes in the spectra are broadly consistent with what might be expected in the light of the kinds of atom concerned. No evidence is found that can certainly be attributed to bound states; but in some cases prominent, rather narrow peaks appear towards the low-energy part of the emission bands.

Two features, regarded as rather surprising in nature, are shown fairly generally by the results: the nearly fixed bandwidths and the fact that these widths seem to be related to the structure and concentration of the alloys to a smaller extent than might be anticipated.

References

Appleton, A. and Curry, C. (1965). *Phil. Mag.* **12**, 245.
Appleton, A. and Curry, C. (1967). *Phil. Mag.* **16**, 1031.
Catterall, J. A. and Trotter, J. (1959). *Phil. Mag.* **4**, 1164.
Catterall, J. A. and Trotter, J. (1962). *Proc. Phys. Soc.* **79**, 691.
Clift, J., Curry, C. and Thompson, B. J. (1963a). *Phil. Mag.* **8**, 593.
Clift, J., Curry, C. and Thompson, B. J. (1963b). *Phil. Mag.* **8**, 639.
Crisp, R. S. and Williams, S. E. (1960). *Phil. Mag.* **5**, 1205.
Das Gupta, K. and Wood, E. (1955). *Phil. Mag.* **46**, 77.
Gale, B. and Trotter, J. (1956). *Phil. Mag.* **1**, 759.
Skinner, H. W. B. and Johnston, J. E. (1938). *Proc. Camb. Phil. Soc.* **34**, 109.

Comment

On the Interpretation of **X**-Ray Band Spectra from Alloys

G. A. ROOKE†

1. INTRODUCTION

In the foregoing paper Curry has reviewed the spectra recorded from alloys and in Part 3 some ideas for interpreting alloy spectra are presented by Stott and by Harrison; the following comments are therefore brief. In view of the general and consistent nature of the results from alloys and of the apparent success that can now be achieved in the interpretation of pure-metal spectra, I feel that renewed efforts should be made to interpret alloy spectra. This may mean that some existing ideas on alloys have to be greatly modified and new models and concepts formed.

2. DISCUSSION

I shall comment on the alloy spectra under three classes: (a) alloys which contain metals having d-electrons in the valence band and which produce spectra almost identical with the pure-metal spectra; (b) alloys which contain light metals and produce spectra with bandwidths similar to the pure-metal spectra; (c) alloys which contain light metals and produce spectra with bandwidths different from the pure-metal spectra.

Alloys of class (a). These alloys are formed from metals, containing d-electrons in the valence bands, which in general produce spectra that show only contributions from the d-band. It is expected that the d-bands will not be greatly changed on alloying (see Harrison in Part 3 of this Volume). In ordered structures, the d-bands may be narrower than the pure-metal d-bands, due to the reduced overlap of the electrons of like atoms, but they will not be appreciably shifted in energy (for example see Arlinghaus, 1965). This narrowing of the d-bands has not been observed in x-ray band spectra but this may be due to the disordering effects of the electron bombardment. Disordered structures will show d-bands which are even more like the pure-metal bands due to the greater overlap. It is therefore not surprising that the spectra are all rather similar to the respective pure-metal spectra.

Alloys of class (b). This class contains most of the other alloys whose spectra have been recorded. The distinguishing feature of these alloys is that

† Present address: Metallurgy Department, University of Strathclyde, Glasgow, U.K.

the component bandwidths are similar to the pure-metal bandwidths and not to each other.

It is possible that these results may be due to clustering; however, they have been obtained consistently from several different alloy systems and at several different concentrations and they have also been observed for alloys produced both by evaporation and in the solid form. There has not been sufficient clustering to show up either in powder patterns or as an effect on the plasmon frequency in cases where these have been used to check the order of the system. The spectra have often shown shapes which are markedly and consistently different from the corresponding pure-metal spectra; this may be due to the absorption of clusters. All this evidence suggests that, if they exist, clusters lie in a very restricted range of sizes. It would be a major coincidence if this has occurred for several different alloys and therefore it seems likely that these are genuine effects of alloying.

If it is assumed that alloying and not clustering causes the above spectral characteristics, then our ideas on alloys must be considerably changed. For example, consider Al–Mg alloys, where a rigid-band model would predict a bandwidth lying between those for the two pure metals: the spectra suggest that the bandwidth must be almost as wide as that of the pure aluminium band, and they also indicate that the states with lowest energy have wave-functions of almost zero magnitude in the region of the magnesium atoms.

We might draw a model density of states as shown in Fig. 1. This can be approximately separated into two partial densities of states, namely

$$N_x(E) = \int_{E \,=\, \text{const}} \frac{\psi^*\psi\,\mathrm{d}\tau}{\mathbf{V}_k(E)}\,\mathrm{d}^2k$$

where the integral over x implies an integration over only those cells containing atoms of type x.

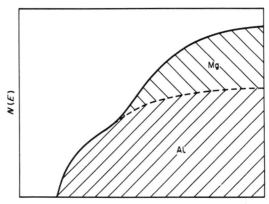

FIG. 1. Density of states for proposed model of an alloy containing aluminium and magnesium. The shaded portions show respectively the separate partial densities-of-states for aluminium and magnesium.

It is not suggested that the effects are due to transition probabilities, but that they are features existing in the unperturbed valence-band and affect such properties as the total electronic energy and heat of formation.

We can offer the following simple arguments that suggest this as a possible model. The pseudo-atom model (Ziman, 1967) requires that local electron densities remain similar to the pure-metal electron densities provided that most of the electronic density in the pseudo-atom lies close to the ionic centre. Band-structure calculations for β-brass (Arlinghaus, 1965) show that in this alloy the s,p-electrons retain the pure-metal electronic densities which the metallurgist uses to represent a completely disordered alloy. The metallurgist uses the concept of an ideal solution which has no heat of formation, and therefore either has no change in the electronic structure around each atom; or has changes that cause any energy differences to cancel fortuitously, and this seems unlikely. A Wigner–Seitz calculation, which assumes that the electronic density in each cell is pure-metal like, must predict that the energy of the lowest state in each cell is similar to that of the pure metal.

Hence the spectra suggest an alloy model that is considerably different from those previously postulated. There is very little experimental evidence for the nearly-rigid-band models, and theoretically these all depend on perturbation methods. If a "two band" model is plausible it would indicate that the difference in potential from cell to cell in an alloy is large, and that perturbation techniques are no longer applicable. It may also be possible to extend such a model to give theoretical justification to the "two band" model successfully used by Varley (1954) to calculate heats of formation of alloys.

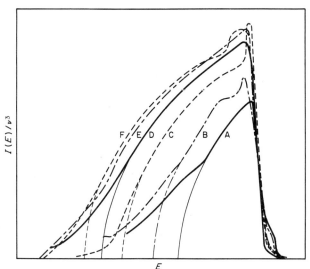

FIG. 2. A series of sodium $L_{2,3}$-emission spectra taken from sodium–indium alloys, of compositions assumed to progressively increase in sodium from A to F.

Alloys of class (c). As an example of this class, some results from a series of sodium–indium alloys are presented here. The sodium $L_{2,3}$ spectra are shown in Fig. 2. Samples were produced by placing separate pieces of the two metals together in a furnace and evaporating a thin film onto a copper target; a spectrum was recorded and then the target scraped clean. This process was repeated until all the metal had been evaporated. It was assumed that the alloy composition changed with each evaporation and that the change was always such that the percentage of one metal increased at the expense of the other.

A second series of spectra was recorded, but in this case all were of nearly the same shape as that for pure sodium which approximately corresponds to the widest spectrum (F) in Fig. 2. There was no apparent reason why alloying had not occurred in this second series and therefore the results should be treated with caution until they can be more carefully repeated.

In general, alloys with the B32 crystal structure are formed from a monovalent metal and a trivalent metal, with ~ 50 atomic % composition. This structure forms, it is assumed, when the monovalent atom gives its valence electron to the trivalent metal which can then form a diamond-type lattice. The monovalent ions are then held in the interstaces by ionic attraction and form an interlocking diamond-type lattice.

The present spectra for sodium–indium alloys have been normalized so that the bottom of each band has the same intensity, and in Fig. 3 they have been shifted until the estimated energy of the bottom of each band is

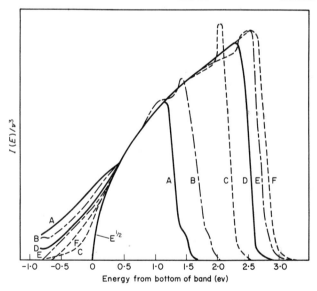

FIG. 3. Sodium $L_{2,3}$-emission spectra (recorded from sodium–indium alloys) adjusted on the energy-scale so that the bottoms of each band (found by the parabolic extrapolation) coincide; i.e. each is at zero energy.

zero and will coincide. The spectra can then be interpreted by assuming that the sodium atoms have contributed the largest percentage of their electrons to the indium for alloy A, and that for successive spectra the sodium atoms in the alloy have contributed less, presumably because the sodium concentration was increasing with each evaporation.

If this is right, and if the theory for alloys with the B32 crystal structure is correct, then these results provide further evidence that the density of states when measured locally, may be different for each type of atom in the system.

There are indications of a tail on the low-energy side of these spectra which increases as the band becomes narrower. This may be due either to local clusters with higher sodium concentration or to some transitions from the electron distribution on the neighbouring indium atoms being involved in transitions to the sodium core-states.

3. CONCLUSION

Spectra from alloys in class (a) are not anomalous provided it is assumed that those from the pure metals that have d-bands partly or completely filled show contributions almost entirely from the d-electrons. Some spectra from alloys in classes (b) and (c) are entirely inconsistent with the normal rigid-band theories. The results in all cases may be due to clustering, but consistency in the changes of shape of the spectra and considerations of the possibility of clustering both suggest that the effects may be real. Then a two-band theory might be a reasonable approximation, and this possibility should be considered.

Acknowledgement

The author is indebted to the University of Western Australia for a research studentship during the tenure of which the experimental results were recorded, to the U.K.A.E.A. for a Fellowship during which many of the foregoing ideas were formulated, and to the Science Research Council for a Fellowship during which this discussion was written. He also wishes to thank D. J. Fabian for help in preparing the manuscript.

References

Arlinghaus, S. (1965). Quart. Progr. Rep. Solid State and Molecular Theory Group of M.I.T., No. 56, 4.
Varley, J. H. O. (1954). *Phil. Mag.* **45**, 887.
Ziman, J. M. (1967). *Proc. Phys. Soc.* **91**, 701.

Review of Optical Properties of Metals and Alloys due to Interband Transitions

F. Abelès

Institut d'Optique† et Faculté des Sciences, Paris, France

ABSTRACT

It is usually possible to distinguish absorption due to interband transitions from that which is due to free carriers (intraband absorption). This review is limited to the low-frequency interband structure associated with energy levels not very far (a few electron volts) below the Fermi-surface. The following classes of metals are considered: (1) alkalis for which there are still unresolved experimental and theoretical problems; (2) noble metals for which the theory is still at a difficult stage, but there are no experimental difficulties; (3) normal polyvalent metals; (4) ferromagnetic metals taking nickel as an example; (5) noble-metal with normal-metal alloys, and noble-metal with transition-metal alloys, in order to show a clear distinction between the optical properties of these two types of alloy.

1. INTRODUCTION

Our objective, in this paper, is to review that part of the optical properties of metals and alloys which is related to interband transitions. The effect of intraband transitions is deliberately omitted from the discussion. In order fully to examine interband absorption, it is necessary to separate it from intraband absorption, but this is a problem which can be solved if the measurements have been conducted far enough into the infrared; that is for low enough energies.

The optical properties of metals and alloys have been extensively studied during the last few years and the present review is by no means intended to be exhaustive. In fact, it will partly serve to indicate the difficulties and unsolved problems. It will be limited to photon energies of a few electron volts, that is to the not very far ultraviolet. For a more complete review of the status of our knowledge up to 1965 see Abelès (1966).

After a brief survey of the methods for computing the complex dielectric constant of metals, the optical properties of the following classes of materials will be discussed: alkalis, noble metals, polyvalent metals, ferromagnetic metals and alloys.

† Laboratoire associé au Centre National de la Recherche Scientifique.

2. CALCULATION OF THE DIELECTRIC CONSTANT

Let $\varepsilon = \varepsilon_1 + i\varepsilon_2$ be that part of the dielectric constant which is due to interband transitions. It is very easy to show that ε_1 and ε_2 verify a Kramers–Kronig relation. Indeed, the real and imaginary parts of the complex dielectric constant and that part of it which is due to intraband transitions satisfy such a relation, so ε_1 and ε_2 must satisfy it also. For example

$$\varepsilon_1(\omega) = \frac{1}{\pi\omega} \int_0^\infty \frac{d}{d\omega'}(\omega'\varepsilon_2(\omega')) \ln \left| \frac{\omega+\omega'}{\omega-\omega'} \right| d\omega'. \tag{1}$$

Thus it is sufficient to compute one only of these quantities (either ε_1 or ε_2) from first principles; the other is then obtained by a straightforward mathematical calculation. It is definitely simpler to compute ε_2, so that in what follows we limit our discussion to the variation of ε_2 with energy $\hbar\omega$. The study of the variation of ε_1 with energy would not result in any new information.

In the one-electron approximation, the contribution of transitions between two bands in a single region of the Brillouin zone is given by (see Ehrenreich and Cohen, 1959):

$$\varepsilon_2 = \frac{4\pi^2 e^2 h^2}{3m^2\omega^2} \sum_{m,n} \int \frac{2}{(2\pi)^3} dk \, |M_{mn}|^2 \, \delta(E_n - E_m - \hbar\omega) \tag{2}$$

where m and n respectively label occupied and unoccupied states, and where

$$M_{mn} = \frac{\hbar}{i} \int \psi_{km}^* \nabla \psi_{kn} \, d\tau \tag{3}$$

is the interband oscillator strength, the integration being performed over the volume of the unit cell. Temperature and damping effects have been neglected. If $|M_{mn}|^2$ may be considered constant, then ε_2 is readily seen to be inversely proportional to $(\hbar\omega)^2$ and proportional to the joint density of states

$$\frac{2}{(2\pi)^3} \int dk \, \delta(E_n - E_m - \hbar\omega)$$

for vertical transitions between bands m and n. If E_F is the Fermi-energy, $E_m < E_F$, and $E_n > E_F$, these restrictions being necessary to insure that the transitions occur only when the lower state is occupied and the upper state is empty.

The calculation of ε_2 according to Eq. (2) is a very difficult task and it has been successfully done so far only for the two semiconductors germanium and silicon (Brust, 1964; Dresselhaus and Dresselhaus, 1967). Figures 1 and

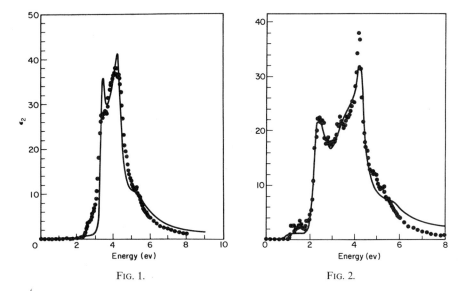

FIG. 1.   FIG. 2.

FIGS. 1 and 2. Frequency dependence of the imaginary part ε_2 of the dielectric constant for silicon (Fig. 1) and germanium (Fig. 2). The solid curve represents the experimental results of Philipp and Ehrenreich (1963), while the dotted curve shows the results calculated by Dresselhaus and Dresselhaus (1967) with a Monte-Carlo procedure using 10 000 points.

2 show the measured (solid lines) and calculated values (dotted lines) of ε_2 for silicon and germanium, after Dresselhaus and Dresselhaus (1967). The computation uses a Fourier expansion for the electronic energy bands and takes into account damping effects by introducing a relaxation time $\tau = 2 \times 10^{-14}$ sec, which is independent of the band indices m and n and of the photon energy $\hbar\omega$, and is the same for both materials. The agreement between the calculated and experimental curves is very good indeed. Unfortunately, this method has not yet been applied to metals, for which different approximations have been used; all these approximations have proved to be more or less unsatisfactory and the discrepancies between computed and experimental curves may be rather large. We shall discuss these approximations when examining the different types of metals.

It is important to note that Eq. (2) has been derived using a one-electron approximation. The neglect of electron interaction is certainly one of the shortcomings of the theory, but it is difficult at present even to estimate its importance. The calculations of $\varepsilon_2(\omega)$ for silicon and germanium, already mentioned (Dresselhaus and Dresselhaus, 1967), take account of certain many-body effects. These workers contend that they have retained the first term in the expansion of the Landau theory of the Fermi liquid, and this is probably one of the reasons for the greater success encountered in the computation of the dielectric constant of silicon and germanium.

3. EXPERIMENTAL RESULTS AND DISCUSSION FOR VARIOUS CLASSES OF METALS

A. The Alkali Metals

An experimentalist trying to measure ε for the alkali metals is faced with a very difficult task. The results which we shall discuss were obtained in H. Mayer's laboratory (Clausthal) and by J. N. Hodgson. Because of the reactivity of these metals to air, all the measurements must either be done in vacuum or on well protected surfaces.

The studies by Mayer and his co-workers were performed on free surfaces prepared in the ultra-high vacuum in which the measurements were made. The alkali metals used were carefully prepared by reduction of very pure chlorides in vacuum and then purified and degassed by multiple fractional-distillation in ultra-high vacuum. Finally, the mirror-like surfaces were prepared by distillation of the prepared metals into a cup-container of glass within an ultra-high-vacuum chamber provided with suitable windows for the entrance and exit of light (Mayer and Hietel, 1966). The measurements were of the polarimetric type (oblique incidence).

Hodgson (1966), in his investigations, used thick films deposited by vacuum evaporation on to one face of a silica-glass prism, with pressures during the deposition of $\sim 10^{-7}$ torr. Immediately after deposition, the exposed surface of the film was covered with a layer of vacuum grease and the optical constants were measured in air by internal reflection from the glass–metal surface, using a polarimetric technique.

The results obtained by Mayer et al. were surprising and are still puzzling to the theoreticians. In principle the alkalis are simple metals and their optical properties should be easily understood in the framework of pseudo-potential theory. The first calculations by Butcher (1951), before the results obtained by Mayer et al. were available, were made in order to interpret earlier results of Ives and Briggs (1936 and 1937) in the near ultraviolet and visible region of the spectrum. This calculation makes use of an adjustable parameter, the Fourier coefficient V_G of the pseudo-potential; G represents one of the twelve vectors associated with the [110] system for the alkalis. Returning to Eqs. (2) and (3), it can be shown that M_{mn} is proportional to V_G, so that ε_2 or σ is proportional to the square of this quantity. In his calculations, Butcher uses V_G as an adjustable parameter and he was able to account conveniently for the results obtained by Ives and Briggs; but nowadays, V_G is known from de Haas-van Alphen measurements and from theoretical calculations (Ham, 1962; Ashcroft, 1965; Lee, 1966) and this leads to a large discrepancy between measured and computed values of ε_2 or $\sigma = \varepsilon_2 \omega / 4\pi$.

Figure 3 shows the results obtained by Mayer and co-workers for Na, K, Rb and Cs at 20°C. The results for Rb were tentative only; the recent results obtained by Aufschnaiter (1966) are shown in Fig. 4. In Fig. 3, the peaks corresponding to the lower energies, respectively 0·825 ev for K, 1·2 ev for Cs and 1·8 ev for Na, are unexpected features. They are extra peaks, unpredicted by theory and not attributable to direct interband transitions. Figures 5 and 6 show the temperature variations of σ for Cs and Na (Mayer and Hietel, 1966). The increase of σ when the temperature is lowered, in the vicinity of 1·2 ev for Cs, is also difficult to explain.

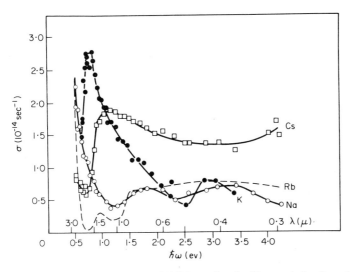

FIG. 3. $\sigma(\omega)$ for Na, K, Rb and Cs at 20°C (reproduced with permission from Mayer and Hietel, 1966). The results for Rb were tentative only; more recent results for this metal are shown in Fig. 4.

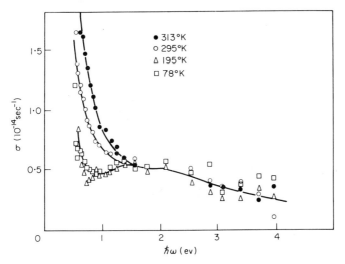

FIG. 4. $\sigma(\omega)$ for Rb. (Reproduced with permission from Aufschnaiter, 1966.)

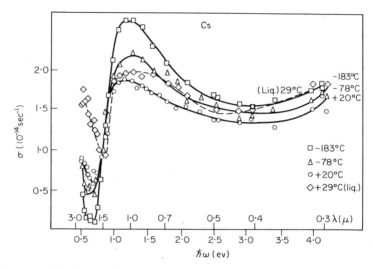

FIG. 5. $\sigma(\omega)$ for Cs, for different temperatures. Note that this metal is in the liquid state at +29°C. (Reproduced with permission from Mayer and Hietel, 1966.)

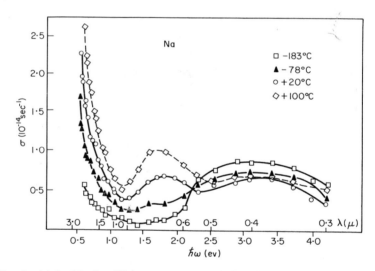

FIG. 6. $\sigma(\omega)$ for Na, for different temperatures. Sodium is in the liquid state at +100°C. (Reproduced with permission from Mayer and Hietel, 1966.)

A few suggestions have been made, to explain the unexpected absorption bands below the interband edge, but none of them are really satisfactory (Cohen and Phillips, 1964; Overhauser, 1964; Cohen, 1966; Ferrell, 1966). However, Professor Mayer has informed us that the measurements are being carefully repeated and it is advisable to leave discussion of these extra absorption bands until after the completion of this new series of measurements. We shall therefore discuss here only the normal interband-absorption.

Let us first examine the case of sodium. To fit the theoretical curve to the experimental data of Mayer and Hietel (1966), we have to choose unrealistically large values of $V_G = V_{110} = 0.475$ ev. The value of V_{110} deduced from experiments is of the order of 0.2 ev. The calculations by Butcher were improved by Miskovsky and Cutler (1967) who, instead of using a single OPW for the conduction and first excited bands, used a linear combination of seven orthogonalized-plane-waves. The work of Miskovsky and Cutler showed that an analogous calculation performed by Appelbaum (1966) was in error because the wavefunctions used by this author were not properly orthogonalized. Figure 7 shows that the use of a single OPW leads to an absorption, σ, approximately three orders of magnitude too small, the maximum for σ is definitely outside the visible region. Using the seven-OPW calculation, we essentially recover the result first reached by Butcher, i.e. a value is obtained for σ which is approximately one order of magnitude too small.

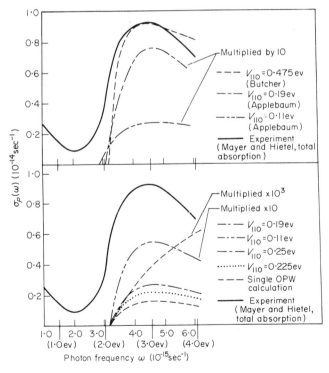

FIG. 7. $\sigma(\omega)$ for Na. The upper curves show the experimental results, together with the theoretical results of Butcher (1951) and Appelbaum (1966). The lower curves show the importance of using a linear combination of seven OPW's in place of a single OPW. (Reproduced with permission from Miskovsky and Cutler, 1967.)

Another way of overcoming the difficulty has been suggested by Animalu (1967); he showed that V_G must be replaced by an "optical" matrix element $\tilde{V}_G$. If $\sigma_0(\omega)$ denotes the Butcher result, which is proportional to $|V_G|^2$, the true $\sigma(\omega)$ is given by

$$\sigma(\omega) \simeq \sigma_0(\omega) \frac{|\tilde{V}_G|^2}{|V_G|^2} = \sigma_0(\omega)\, F(\omega) \tag{4}$$

$\tilde{V}_G$ differs from V_G by the extra non-negative term $\hbar\omega P_G$, P being the projection operator of pseudopotential theory and is always positive. We thus have two distinct cases:

$$V_G > 0 \text{ (Li and Na)}, \; \tilde{V}_G > V_G \quad F(\omega) > 1$$

$$V_G < 0 \text{ (K, Rb and Cs)}, |\tilde{V}_G| < |V_G| \quad F(\omega) < 1.$$

Figures 8, 9 and 10 show $\sigma(\omega)$ for respectively sodium, potassium and lithium. The calculations by Animalu gave results which are about 50% of the experimental values. The curve for sodium marked Overhauser (Fig. 8),

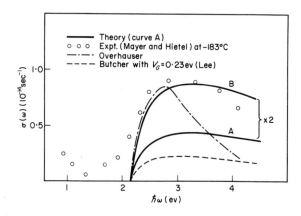

FIG. 8. $\sigma(\omega)$ for Na, showing a comparison of the experimental results with those of three different theoretical treatments: Butcher (1951), Animalu (1967) and Overhauser (1967). Note the theoretical curve A has been multiplied by a factor of 2 to obtain curve B. (Reproduced with permission from Animalu, 1967.)

shows the results he obtained by incorporating exchange into the framework of a self-consistent, time-dependent Hartree–Fock theory, deliberately omitting the screening of the exchange interaction (Overhauser, 1967). It should be noted that a comparison between theory and experiment is particularly difficult for potassium, where the unexplained peak at the lower frequencies probably has a strong influence on the interband absorption. For lithium, the agreement between theory and experiment is probably much better. It should be remembered that the results (Hodgson, 1966), are

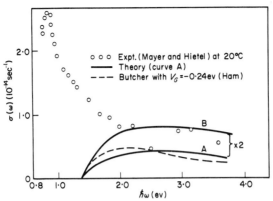

FIG. 9. $\sigma(\omega)$ for K: comparison of the experimental results with theory (Butcher, 1951; Animalu, 1967). Note the theoretical curve A has been multiplied by a factor of 2 in order to obtain curve B. (Animalu, 1967.)

obtained with protected surfaces. Unfortunately, they do not go far enough into the ultraviolet to allow a detailed comparison with theory.

Mahan (1967a) has shown that many-body processes involving plasmons enhance the theoretical value for σ by an additional 50%, and therefore the theoretical values of σ are now about $\frac{3}{4}$ of the experimental values. In effect one has the Butcher–Animalu result, to which is added a final-state interaction caused by the dynamically screened Coulomb-interaction between the electron and the hole, $4\pi e^2/q^2\varepsilon(q, \omega)$. This contains two contributions: one term corresponding to the normal Coulomb-interaction which is screened by the low-frequency dielectric function, $\varepsilon(q, \omega \simeq 0)$ [the intensity at the threshold, ω_t, is $0.05\,\sigma_0(\omega_t)$ for Na]; and a second term, arising from the poles, of $1/\varepsilon(q, \omega)$, which corresponds to an interband transition where the electron and hole scatter by the virtual exchange of a plasmon [intensity, $0.40\,\sigma_0(\omega_t)$ for Na].

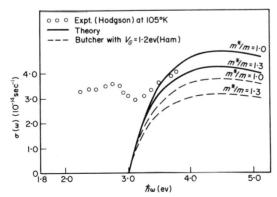

FIG. 10. $\sigma(\omega)$ for Li (Animalu, 1967). Note the actual theoretical curve corresponds to $m^*/m = 1.0$.

B. Noble Metals

As we have seen, the position regarding alkali metals is rather disappointing. Thus, for these simplest of metals we have a very uncertain knowledge of the dielectric constant as a function of frequency up to 5 ev. On the theoretical side, many problems have been raised but no clear answer given. Compared with the alkalis, the noble metals have a more complicated electronic structure due to the mixing of the valence d-bands with those of the s,p-type electrons. There is in fact a hybridization of the valence s,p-bands with the d-bands. The band structure for Cu is shown in Fig. 11, from the calculations by Segall (1962). Results very similar to those of Segall were obtained by Burdick (1963), and there is little doubt concerning the overall validity of Fig. 11.

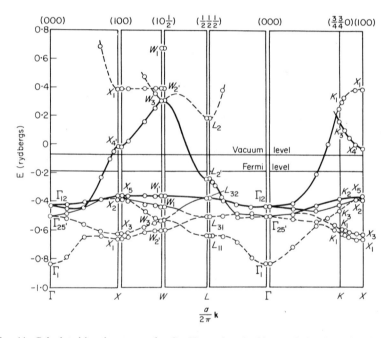

FIG. 11. Calculated band structure for Cu. (Reproduced with permission from Segall, 1962.)

The band shapes for Ag are qualitatively similar (Segall, 1961), but from the optical measurements it can be concluded that the d-bands lie about 2 ev lower (relative to the Fermi-energy) than those in Cu. For metallic gold band energies have been calculated for only a few symmetry points (Segall, 1961), but one expects the band structure to be quite similar to that of Cu. In fact, from various experimental measurements, it can be concluded that the band structure for Au is more like that for Cu than for Ag (Cooper *et al.*, 1965).

From the experimental viewpoint, the situation is much more satisfactory for noble metals than for the alkalis. Measurements in various laboratories have given similar results, and there is also very little difference between the values of the dielectric constant obtained from measurements on thin films and those obtained from measurements on the bulk metal. Moreover, the optical properties of noble metals are known from the infrared to the far vacuum ultraviolet. The frequency variations of ε for Cu, Ag and Au have been measured (Ehrenreich and Philipp, 1962; Beaglehole, 1965; Cooper et al., 1965).

The theoretical position is by far less satisfactory. Let us discuss the first absorption peaks only: their assignment to a definite transition is still a matter of controversy. A careful calculation by Cooper et al. (1965) shows that the observed structure associated with the low-energy peak can be conveniently represented in the form shown in Figs. 12–14. The peak is attributed to transitions $L_{32} \to L_{2'}$ (d-states to Fermi-level); but there is a discrepancy of a factor of the order of 3 for the three metals. Unfortunately, this is not the only possible assignment for the first interband absorption peak in the three metals. Mueller and Phillips (1967), using energy levels and model wavefunctions from a combined interpolation scheme devised by Mueller (1967), have calculated $\varepsilon(\omega)$ for Cu in the random-phase approximation. Their result is shown in Fig. 15 together with the experimental points of Ehrenreich and Philipp (1962) and of Beaglehole (1965). Here, the first

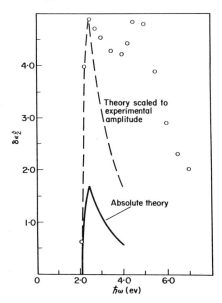

FIG. 12. $\delta\varepsilon_2^b$ versus ω for Cu; $\delta\varepsilon_2^b$ is that part of ε_2 which is due to interband transitions only, i.e. after subtraction of the contribution of the intraband transitions. (Reproduced with permission from Cooper et al., 1965.)

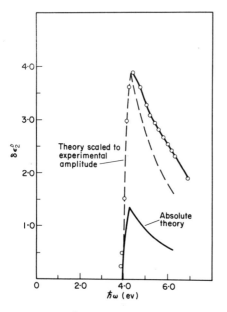

FIG. 13. $\delta\varepsilon_2^b$ versus ω for Ag (Cooper *et al.*, 1965).

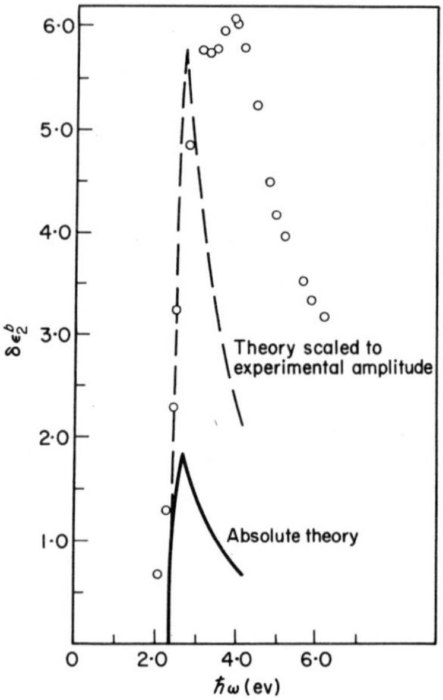

FIG. 14. $\delta\varepsilon_2^b$ versus ω for Au (Cooper *et al.*, 1965).

interband absorption peak, near 2 ev, is very poorly rendered by the calculation. The authors suggest that it should be identified as a virtual exciton-resonance.

This brings us to the exciton effects in metals; a problem that has been discussed by Mahan (1967b) but is still rather controversial. Recently piezo-optical measurements by Gerhardt *et al.* (1967) have been used to identify electronic transitions in copper, and the interpretation of the 2 ev edge in ε_2 for copper is the same as that given by Cooper *et al.* (1965); i.e. transitions from the top of the d-bands to the Fermi-level. Gerhardt *et al.* (1967) find a previously unresolved edge at 3·9 ev, which they assign to $X_5-X_{4'}$ transitions; it might be expected that a very careful measurement of ε would reveal it, but the region near 4 ev is experimentally difficult, because the copper oxides complicate the interpretation of the measurements.

The third absorption-edge in copper is 4·3 ev. Here again we have two different interpretations. Cooper *et al.* (1965) assign this absorption-edge to a $d-p$ transition ($X_5-X_{4'}$), whereas Mueller and Phillips (1967) and Gerhardt *et al.* (1967) consider it to be due to a conduction-band conduction-band transition ($L_{2'}-L_1$, or $p-s$). It is probable that the last interpretation is the correct one especially in view of the quantitative analysis for the piezo-optical constants (Gerhardt *et al.*, 1967). The same conclusion can be drawn from the study of the effect of hydrostatic pressure on optical properties of Cu (Zallen, 1966).

This brief discussion shows how difficult is the interpretation of the optical absorption at relatively low energies (a few electron volts only) for the noble metals.

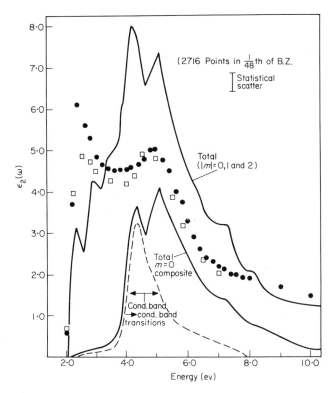

FIG. 15. $\varepsilon_2(\omega)$, due to interband transitions only, for Cu. The boxes show the experimental data of Ehrenreich and Philipp (1962) and the dots those of Beaglehole (1965). (Reproduced with permission from Mueller and Phillips, 1967.)

C. Polyvalent Metals

These are relatively simple metals and their properties are well understood by using the pseudopotential concepts introduced by Harrison (1959, 1966a). The first thorough study of the optical properties of a polyvalent metal, aluminium, was made by Ehrenreich et al. (1963). These workers noted that absorption peaks are likely to occur not only at symmetry points but also whenever two energy bands are running nearly parallel, and they showed that this indeed occurred for aluminium. Such parallel bands may be found for most polyvalent metals, and the correlation has been examined by Harrison (1966b), who gives a table showing for Be, Mg, Ca, Ba, Zn, Cd, Al, In, Tl and Pb the positions of the predicted absorption-edges together with the strengths of absorption. As expected, the agreement was good for aluminium, where a peak is predicted at 1·59 ev and is observed at 1·55 ev.

The same problem has been discussed in more detail by Golovashkin *et al.* (1967a). They find that for a family (G) of Bragg-reflection planes (Brillouinzone planes), the conductivity $\sigma_{(G)}$ is given by

$$\sigma_{(G)} = \frac{e^2}{12\pi\hbar} n_G k_G \frac{\omega_G^2}{\omega(\omega^2 - \omega_G^2)^{1/2}} \tag{5}$$

where n_G is the number of physically equivalent G-planes, k_G is their reciprocal-lattice vector and $\hbar\omega_G = 2|V_G|$. The total optical conductivity is obtained by summation over the various families (G).

Equation (5) shows that σ becomes infinite for $\omega = \omega_G$. This is due to the neglect of relaxation processes, and to the use of a linear combination of plane waves to describe the electron wavefunctions. A crude approximation, in which the δ-functions in Eq. (2) for ε_2 are replaced by Lorentz functions of widths γ, leads to a value of σ which is finite for all frequencies.

As already mentioned, the best understood polyvalent metal is aluminium. Its energy bands are represented in Fig. 16, which is taken from Ehrenreich *et al.* (1963). These authors have made a careful investigation of the absorption peak at 1·5 ev and have shown that it can be attributed to transitions from the regions around W and Σ. The Σ_3 and Σ_1 bands are quite parallel and

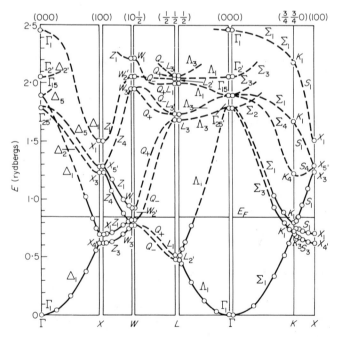

FIG. 16. Calculated energy bands of Al along its crystal symmetry axes. (Reproduced with permission from Ehrenreich *et al.*, 1963.)

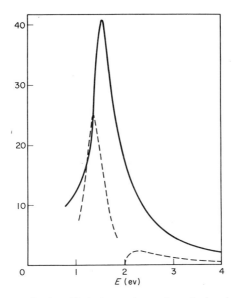

FIG. 17. Calculated contributions (dashed curves) to ε_2 from the interband transitions around W and Σ. The first peak is the sum of the contributions from the W_2-W_1 and $\Sigma_3-\Sigma_1$ transitions; the second is due to the W_3-W_1 transition. The full curve shows the experimental values (Ehrenreich et al., 1963).

Ehrenreich and co-workers have found that their contribution to ε_2 is about 50% larger than that from the W region. The result for $\varepsilon_2(\omega)$ is shown in Fig. 17 as the dashed peak between 1 and 2 ev. The second dashed curve, from 2 ev to 4 ev, is an estimate of the possible contribution from W_3-W_1 band-transitions to ε_2.

Golovashkin et al. (1967a) have analysed the first two absorption-bands of aluminium using their theoretical expression for σ (Eq. 5) and the experimental results of Shklyarevskii and Yarovaya (1964). They find $V_{200} = 0.75$ ev compared with the value 0.76 ev deduced from de Haas-van Alphen measurements. Furthermore, they find another maximum in $\sigma(\omega)$ at 0.44 ev, which they attribute to a $2V_{111} = 0.44$ ev energy-gap comparing favourably with the value 0.48 ev deduced from de Haas-van Alphen measurements.

The same analysis can be applied for indium. Figure 18 shows the results obtained by Golovashkin et al. (1967b); the values of σ, due to interband-transitions only, are plotted for 295°K and 4·2°K, in order to show the importance of low-temperature measurements. The peak at 0.8μ (1·5 ev) is attributed to transitions near the (111) planes and the second one at 2.1μ (0·6 ev) to transitions near (200) planes. Thus, by making small corrections, Golovashkin et al. (1967a) find $|V_{111}| = 0.70$ ev and $|V_{200}| = 0.28$ ev, compared with the values $|V_{111}| = 0.365$ ev and $|V_{200}| = 0.46$ ev indicated by Harrison (1966b). There is now a serious discrepancy between theory and experiment; although the heights of the two peaks are in the ratio 39:30, or 1·3:1, which compares not too badly with the 1·7:1 predicted by Harrison.

Another point to be noted is the displacement of the absorption peaks with temperature; in fact it is difficult to draw any definite conclusion from Fig. 18, because there is an overlap of the peaks which is particularly important at room temperature. A first approach would indicate opposite displacements for the two peaks in indium.

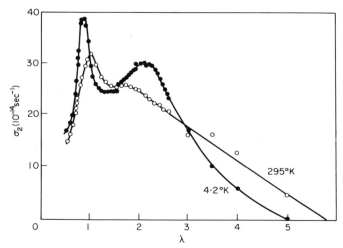

FIG. 18. $\sigma(\lambda)$ in indium, at room and liquid-helium temperatures. Wavelengths are in microns (Golovashkin *et al.*, 1967).

Thus, for aluminium at least, the simple parallel-band-effect theory gives accurate values of V_G, which are in very good agreement with values of the same quantity deduced from other experiments. But the magnitude of the computed peak in ε_2 is smaller than the experimental by a factor of about two (see Fig. 17). This may be due to inadequacies in the calculation method, but from the discussion of Ehrenreich *et al.* (1963), it is probable that an important cause is the neglect of many-electron effects.

It would be interesting to have experimental measurements for other simple polyvalent metals in order to be able to test the theoretical results. In indium too, Golovashkin *et al.* (1967b) find that the experimental values of σ are twice as large as those computed. A careful examination of optical interband absorption in these metals might facilitate our understanding of many-body effects in metals.

D. A Ferromagnetic Metal: Nickel

Nickel may be considered as an example both of transition and of ferromagnetic metals. The characteristic here is that the Fermi-level intersects the d-bands, and this has important implications for the physical (electrical, magnetic and thermal) properties. The optical properties of noble metals are also influenced by d-band excitations, and this leads us to expect that these and the transition metals could be treated in the same way. In ferromagnetic metals, we must also take into account the splitting of the energy bands corresponding to majority and minority spins. There is no well established energy-band structure for nickel to-date; Hodges *et al.* (1965) have used a pseudo-potential method for the calculation of the energy bands of nickel and they estimated a ferromagnetic splitting of 0·35 ev.

It is very difficult to separate inter-band and intra-band contributions to the dielectric constant for nickel. Indeed the interband-character of the fine-structure in measured σ, in the infrared region, remained obscure until the results of ferromagnetic-Kerr-effect measurements, due to Krinchik and his collaborators, were shown to provide such interpretation (see Krinchik, 1966). Magneto-optical effects in ferromagnetic materials result from the spin–orbit interaction, and from the fact that the electronic structure and distribution for majority and minority spins is different in such materials. The intraband transitions are thus definitely weaker than the interband transitions.

It has been recently shown by Hanus *et al.* (1967) that low-energy inter-band transitions can be conveniently detected by thermally modulated reflectivity. Figure 19 shows the frequency variation of the reflectivity R, the pulsed-current modulated reflectivity $\Delta R/R$, from 0·1 to 10 ev. In the low-energy region, from 0 to 2 ev, we can see two peaks at 0·25 and 0·4 ev and a

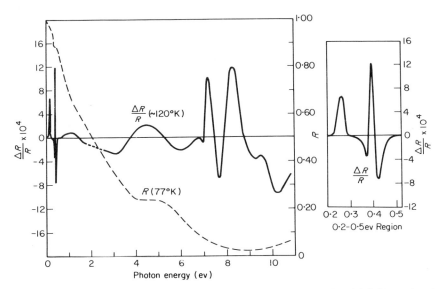

FIG. 19. Reflectivity R, and temperature-modulated reflectivity $\Delta R/R$ for nickel. (Reproduced with permission from Hanus *et al.*, 1967.)

shoulder at 1·3 ev. The measurements of the ferromagnetic Kerr-effect by Krinchik and Nurmukhamedov (1965) and by Martin *et al.* (1964) gave qualitatively different results: 0·2 to 1·5 ev, as seen from Fig. 20(a). The

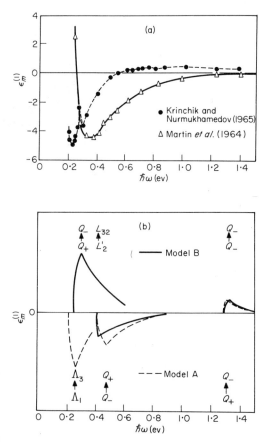

FIG. 20. Spectral dependence of the absorptive part of the off-diagonal dielectric constant $\varepsilon_m^{(1)}$ of nickel: (a) Experimental results; (b) Spectral dependence deduced from models A and B. The intensity of each of the three contributions reflects only the $1/\omega^2$ dependence, but the matrix elements have not been calculated (Hanus *et al.*, 1967).

observed structure is clearly due to transitions near the point L in the Brillouin zone, but there has been much controversy concerning the structure of the *s,p*-like valence-band at $L_{2'}$ and the *d*-band at L_{32}, with respect to each other and to the Fermi-level. The early ordering model (A) and newly suggested ordering scheme (B) are shown in Fig. 21. They are quite distinct, and even the ordering of the paramagnetic bands $E(L_{2'})–E(L_{32})$ is different. The spectral dependences of the absorptive part of the off-diagonal dielectric constant, respectively deduced from the two models are shown in Fig. 20(b).

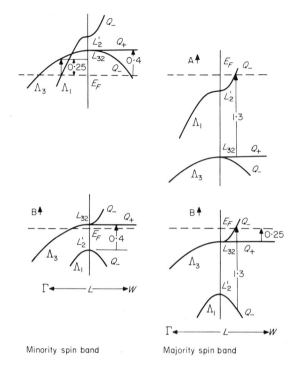

FIG. 21. Band structure of nickel near L, for models A and B, with energies in electron volts (Hanus *et al.*, 1967).

Further study will be necessary in order to have a definite idea of the energy-band structure in nickel together with a confident assignment of the elementary excitations to interband transitions.

E. Alloys

The theory of the physical properties of alloys has reached a less advanced stage than that pertaining to the properties of metals. We have seen that it is not easy to interpret the optical properties of the latter, and we may expect even more difficulties when trying this interpretation for alloys, this indeed proves to be the case. Moreover, there is a finite number of metals, whereas metal alloys form an unlimited group, for which there are but very few thorough and useful experimental studies of optical properties.

We shall examine here two types of noble metal alloy: those with a normal polyvalent metal and those with a transition metal. An example of the first type is provided by the α-brasses. They have been studied at liquid-helium temperatures using a calorimetric technique, by Biondi and Rayne (1959). They obtained the absorptivity as a function of frequency, but not the dielectric constant. Nevertheless, the experimental results have been submitted to a theoretical analysis by Lettington (1965).

These results are summarized in Fig. 22. The line A illustrates the displacement of the absorption edge of copper on alloying. If it is taken to represent the energy difference between the top of the d-band and the Fermi surface near L, as discussed above, then there must be a slight increase of this difference with increasing zinc concentration. The second maximum in the absorption (line B) moves in the opposite direction. If we accept the interpretation of Cooper *et al.* (1965), then this peak is due to a d–p transition, from the top of the d-band at X_5 to the valence band at $X_{4'}$. The points X_5 and L_{32}, which correspond to the top of the d-band, have almost identical

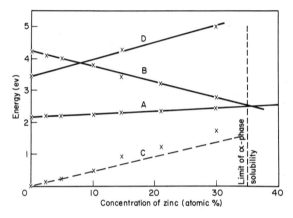

FIG. 22. Spectral changes with composition for copper–zinc alloys: A absorption-edge; B absorption peak; C movement of the Fermi-level relative to the bottom of the valence band; D a second absorption peak. (Reproduced with permission from Lettington, 1965.)

energy. Hence it is easy to bring about the shift of the Fermi-level, with respect to the bottom of the valence band, shown by the crosses in Fig. 22, C.

The dotted line denoted by C in Fig. 22 is the movement predicted by the rigid-band model, assuming a parabolic band with a unit effective mass. The line D shows the displacement of another weak absorption maximum, which might be the one seen at 3·9 ev by Gerhardt *et al.* (1967). These authors assigned it to a X_5–$X_{4'}$ transition, whereas Lettington thinks that it might give the energy difference between the states $L_{2'}$ and L_1. In their assignments for the two peaks at the higher energies, the respective interpretations of Lettington (1965) and of Gerhardt *et al.* (1967) do not agree.

The main point in studying the optical properties of these alloys is to discern what conclusions can be drawn concerning the theory of alloying. Lettington finds that the description of alloy behaviour given by Friedel (1954), which takes into account screening effects, is in general agreement with the experimental results of Biondi and Rayne (1959). Figure 23 shows the density of states in pure copper (full line) and in the alloy containing

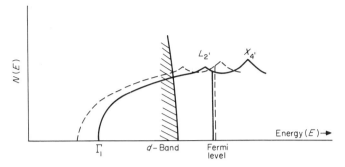

FIG. 23. Density of states for pure copper (full line), and for an alloy containing 30% zinc (dashed line). (Reproduced with permission from Lettington, 1965.)

30% zinc (dotted line). The results already indicated are now more obvious: the top of the d-band is undisplaced, the Fermi level is displaced only slightly and the bottom of the valence band at Γ is more seriously displaced.

Copper–germanium alloys containing up to 7% germanium have been measured by Rayne, but the results obtained in the ultraviolet, that is in the region of interband absorption, are difficult to analyse, mainly because the experimental curves are too entangled (see Fig. 1 of Rayne, 1961). It would be interesting to have more results for this type of alloy.

Let us examine now the situation where the normal polyvalent metal is replaced by a transition metal, that is the noble-metal transition-metal alloys. The electrical, thermal and magnetic properties of transition impurities dissolved in a noble metal can be explained by using the concept of virtual bound states, introduced by Friedel (1956). Broadly, this is a resonance phenomenon analogous to those which arise in classical mechanisms when systems with similar characteristic frequencies interact. The essential feature is the appearance, at an energy-value E_0, of a narrow band

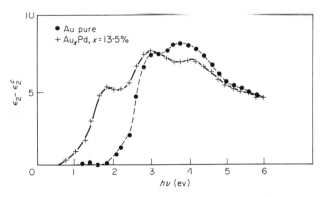

FIG. 24. Absorption, due to interband transitions only, in pure gold and in a 13·5% Au–Pd alloy (Abelès, 1966).

of levels of width 2Δ. It is then conceivable that a new absorption band might arise when the energy of the incident photon is $\hbar\omega \simeq E_F - E_0$; E_F is the Fermi-energy. This has been verified for Au–Ni and Au–Pd alloys. The latter series is the easiest to interpret. Figure 24 (Abelès, 1966) shows the interband absorption $\varepsilon_2 - \varepsilon_2^{(c)}$, in pure gold and in an alloy containing 13·5% Pd ($\varepsilon_2^{(c)}$ is the intraband absorption). In addition to a difference in the steepness of the absorption edge, a new absorption band is clearly seen near 1·7 ev.

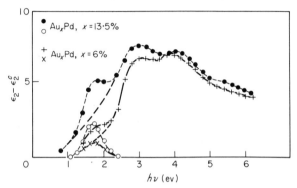

FIG. 25. The extra absorption bands due to virtual bound states for two Au–Pd alloys (Abelès, 1966).

Figure 25 summarizes the results corresponding to two Au–Pd alloys, containing respectively 6 and 13·5%. The new absorption bands have been resolved from the absorption-edge. It is clearly seen that both are centred at the same energy. Moreover, their integrated strength is proportional to the palladium concentration. This definitely shows the connection between these extra bands and the palladium impurities introduced in gold.

An analysis of the infrared measurements performed on the same alloys gives the variation of the effective mass and of the resistivity as a function of Pd concentration. From this, using a very crude model (Caroli, 1963), it is easy to extract the values of $E_F - E_0 = 1\cdot65$ ev and $\Delta = 0\cdot18$ ev. The extra bands in Fig. 25 are centred at 1·7 ev, which is in good agreement with $E_F - E_0$, and their width corresponds to a value of Δ of about 0·25 to 0·30 ev. Thus, there is good agreement between the values of $E_F - E_0$ and Δ, resulting from an analysis of intraband transitions (infrared) and those corresponding to the extra absorption band. This is a justification of the use of the concept of virtual bound states for the interpretation of the optical spectra of noble-metal transition-metal alloys.

3. CONCLUDING REMARKS

The above analysis of the interband absorption in metals and alloys is certainly far from a full description of all the available results. A selection has been made, in order to discuss the more salient features. In some instance various interpretations are offered for given experimental results. These are cases where the writer prefers to leave the reader to confront the difficulties. In fact, the optical properties of metals are far from being completely understood. We have attempted to indicate where the complete solution might come from a thorough study of electron–electron interactions. Whereas for x-ray emission this problem has been fully investigated, as one may see from the results presented elsewhere in this Volume, the situation is much more confused for the optical region.

We have discussed in greater detail the alkali metals, because they are the simplest but, paradoxically, we are faced there with the greatest difficulties. Hopefully, these are of experimental origin only, and therefore we are waiting with interest for the results of Mayer's research group. The noble metals are experimentally well studied but the analysis of their properties is still controversial. We hope soon to have good results for more polyvalent metals, where the use of pseudopotential calculations gives a good first approximation. The ferromagnetics are still in need of a thorough study. We have not mentioned the rare-earth metals, because their experimental investigation is to-date only very sketchy. A recent account of the present knowledge of the optical properties of rare-earth metals can be found in a useful article by Schüler (1966). The alloys could be given a longer survey, but it is felt that the experimental results so far available are still too meagre.

Acknowledgement

The author is indebted to Miss M. L. Thèye for her help with the written version of this review.

References

Abelès, F. (1966). *In* "Optical Properties and Electronic Structure of Metals and Alloys", ed. by F. Abelès, p. 553. North-Holland Publishing Company, Amsterdam.
Animalu, A. O. E. (1967). *Phys. Rev.* **163**, 557.
Appelbaum, J. A. (1966). *Phys. Rev.* **144**, 435.
Ashcroft, N. W. (1965). *Phys. Rev.* **140**, A935.
Aufschnaiter, S. (1966). Dissertation, Clausthal.
Beaglehole, D. (1965). *Proc. Phys. Soc.* **85**, 1007.
Biondi, M. A. and Rayne, J. A. (1959). *Phys. Rev.* **115**, 1522.
Brust, D. (1964). *Phys. Rev.* **134**, A1337.
Burdick, G. A. (1963). *Phys. Rev.* **129**, 138.
Butcher, P. N. (1951). *Proc. Phys. Soc.* **A64**, 765.
Caroli, B. (1963). *Phys. Kondens. Materie* **1**, 346.
Cohen, M. H. (1966). *In* "Optical Properties and Electronic Structure of Metals and Alloys", ed. by F. Abelès, p. 66. North-Holland Publishing Company, Amsterdam.
Cohen, M. H. and Phillips, J. C. (1964). *Phys. Rev. Letters* **12**, 662.
Cooper, B. R., Ehrenreich, H. and Philipp, H. R. (1965). *Phys. Rev.* **138**, A494.
Dresselhaus, G. and Dresselhaus, M. S. (1967). *Phys. Rev.* **160**, 649.
Ehrenreich, H. and Cohen, M. H. (1959). *Phys. Rev.* **115**, 786.

Ehrenreich, H. and Philipp, H. R. (1962). *Phys. Rev.* **128**, 1622.

Ehrenreich, H., Philipp, H. R. and Segall, B. (1963). *Phys. Rev.* **132**, 1918.

Ferrell, R. A. (1966). *In* "Optical Properties and Electronic Structure of Metals and Alloys", ed. by F. Abelès, p. 78. North-Holland Publishing Company, Amsterdam.

Friedel, J. (1954). *Adv. Phys.* **3**, 446.

Friedel, J. (1956). *Can. J. Phys.* **34**, 1190.

Gerhardt, U., Beaglehole, D. and Sandrock, R. (1967). *Phys. Rev. Letters* **19**, 309.

Golovashkin, A. I., Kopeliovich, A. I. and Motulevich, G. P. (1967a). *Soviet Phys. JETP* **24**, 91.

Golovashkin, A. I., Levchenko, I. S., Motulevich, G. P. and Shubin, A. A. (1967b). *Soviet Phys. JETP* **24**, 1093.

Ham, F. S. (1962). *Phys. Rev.* **128**, 82.

Hanus, J., Feinleib, J. and Scouler, W. J. (1967). *Phys. Rev. Letters* **19**, 16.

Harrison, W. A. (1959). *Phys. Rev.* **116**, 555.

Harrison, W. A. (1966a). *In* "Pseudopotentials in the Theory of Metals". Benjamin, New York.

Harrison, W. A. (1966b). *Phys. Rev.* **147**, 467.

Hodges, L., Ehrenreich, H. and Lang, N. D. (1966). *Phys. Rev.* **152**, 505.

Hodgson, J. N. (1966). *In* "Optical Properties and Electronic Structure of Metals and Alloys", ed. by F. Abelès, p. 484. North-Holland Publishing Company, Amsterdam.

Ives, H. E. and Briggs, H. B. (1936). *J. Opt. Soc. Amer.* **26**, 238.

Ives, H. E. and Briggs, H. B. (1937). *J. Opt. Soc. Amer.* **27**, 181, 395.

Krinchik, G. S. (1966). *In* "Optical Properties and Electronic Structure of Metals and Alloys", ed. by F. Abelès, p. 484. North-Holland Publishing Company, Amsterdam.

Krinchick, G. S. and Nurmukhamedov, G. M. (1965). *Soviet Phys. JETP* **21**, 22.

Lee, M. J. G. (1966). *Proc. Roy. Soc.* **295A**, 440.

Lettington, A. H. (1965). *Phil. Mag.* **11**, 863.

Mahan, G. D. (1967a). *Phys. Letters* **24A**, 708.

Mahan, G. D. (1967b). *Phys. Rev. Letters* **18**, 448.

Martin, D. H., Doniach, S. and Neal, K. J. (1964). *Phys. Letters* **9**, 224.

Mayer, H. and Hietel, B. (1966). *In* "Optical Properties and Electronic Structure of Metals and Alloys", ed. by F. Abelès, p. 47. North-Holland Publishing Company, Amsterdam.

Miskovsky, N. M. and Cutler, P. H. (1967). *Phys. Letters* **24A**, 611.

Mueller, F. M. (1967). *Phys. Rev.* **153**, 659.

Mueller, F. M. and Phillips, J. C. (1967). *Phys. Rev.* **157**, 600.

Overhauser, A. W. (1964). *Phys. Rev. Letters* **13**, 190.

Overhauser, A. W. (1967). *Phys. Rev.* **156**, 844.

Philipp, H. R. and Ehrenreich, H. (1963). *Phys. Rev.* **129**, 1550.

Rayne, J. A. (1961). *Phys. Rev.* **121**, 456.

Schüler, C. C. (1966). *In* "Optical Properties and Electronic Structure of Metals and Alloys", ed. by F. Abelès, p. 221. North-Holland Publishing Company, Amsterdam.

Segall, B. (1961). General Electric Research Lab., Report No. 61.RL.2785G (unpublished).

Segall, B. (1962). *Phys. Rev.* **125**, 109.

Shklyarevskii, I. N. and Yarovaya, R. G. (1964). *Opt. Spectrosc.* **16**, 45.

Zallen, R. (1966). *In* "Optical Properties and Electronic Structure of Metals and Alloys", ed. by F. Abelès, p. 164. North-Holland Publishing Company, Amsterdam.

Comment

The Role of Electron-emission Spectroscopy

Derek J. Fabian

1. INTRODUCTION

In concluding the experimental sections to this Volume it is clearly of value to compare the information made available by soft x-ray spectroscopy with that obtained from optical and electron-emission studies. In the foregoing paper the optical properties of metals and alloys has been adequately reviewed by Abelès; the present discussion covers photo-emission, and the electron emission that accompanies ion-neutralization by solids.

Both of these are experimental techniques which show considerable promise in the search for accurate band-structure information and density-of-states measurements. We therefore outline here the methods employed, some of the results obtained and the correlation of these results with soft x-ray band spectra.

2. PHOTO-ELECTRON SPECTROSCOPY

Photo-emission is a two-stage process, consisting of firstly an electron transition from the valence band to an unoccupied state above the Fermi level, and secondly the escape of the electron from the metal. The electron excitation is caused by photons striking the solid emitter.

The occurrence of nondirect transitions (which involve a change in k) as well as direct transitions has been discussed by Berglund (1966). By varying the photon energy in photo-emission from copper and silver, Berglund and Spicer (1964) showed that it is possible to differentiate between these two kinds of transitions and concluded that, below the Fermi-level, nondirect transitions predominate.

Nondirect transitions imply that conservation of the wave vector k is not an important selection rule; also the electron wavefunctions are not well represented by Bloch functions. For a given photon energy $\hbar\omega$, the optical or relative density-of-states typified by nondirect transitions is proportional to $N_c(E)N_v(E-\hbar\omega)$, where $N_c(E)$ is the conduction-band density of states and $N_v(E-\hbar\omega)$ is the electron density of the valence band at $E-\hbar\omega$. In addition, the inelastic scattering is assumed to be such that the squared matrix elements $|M(\omega)|^2$ are constant throughout the range of the experiment.

Electrons escape from the metal only when the photon energy exceeds the vacuum level, which is defined as the sum of the Fermi energy and the work function. In order to lower the work function, the experiments can be performed with a monolayer of caesium on the emitter (Spicer, 1966b). Basically the experiment measures the kinetic-energy distribution of electrons emitted from the material. Analysis of the results is complicated by electron–electron and electron–phonon scattering. These effects tend to broaden any peaks that are present, and therefore are analogous to Auger-broadening effects in soft x-ray spectra. The scattering effects may be accounted for by the factor $L(E)$ which is the scattering length; $L(E)$ can be determined fairly precisely from experiment.

For details of the experimental technique see Spicer and Berglund (1964) and Krolikowski (1967), who also present detailed analyses of the interpretation of the experimental data and show different ways of extracting the optical density of states from photo-emission data. From these data, which are usually in the form of a photo-electron energy-spectrum, it is necessary first to remove $L(E)$ and also the escape probability (a function that can usually be taken as constant when E, the energy to which the electrons are excited, is more than 1 or 2 ev above the vacuum level), and then to unfold the densities of initial and final states—respectively, $N_v(E-\hbar\omega)$ and $N_c(E)$. The unfolding is straightforward when there is pronounced structure in the valence band (below the Fermi-level) and an approximately constant density of final-states (in the conduction band, above the vacuum level). When this approximation cannot be made, the densities of initial and final states are generally unfolded from energy distributions obtained with different photon-energies.

From the studies of a variety of metals it appears that the peaks below the

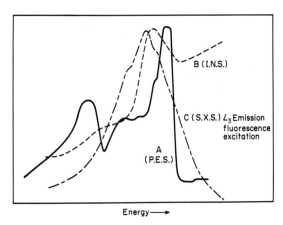

FIG. 1. Density of states for copper: A, obtained from photo-emission measurements (Krolikowski, 1967); B, unfolded from ion-neutralization data (Hagstrum and Becker, 1966); C, measured by the soft x-ray emission spectrum (Nemnonov et al., 1966).

Fermi-level in the optical density of states can generally be associated with nondirect transitions. The optical density of states for copper, which has been unfolded from photo-emission data by Krolikowski (1967), is shown in Fig. 1 (see also Spicer, 1966a). For comparison, the x-ray fluorescence L_3-emission spectrum measured by Nemnonov et al. (1966) and also the ion-neutralization spectrum measured by Hagstrum and Becker (1966) have been superimposed. The optical density of states for aluminium has not been derived, but the energy-distribution curves have been measured by Wooten et al. (1966); a few of these are shown in Fig. 2 superimposed with the soft x-ray $L_{2,3}$-emission spectrum for aluminium obtained by Rooke (1968).

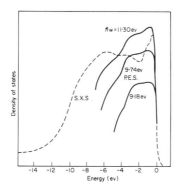

FIG. 2. Energy-distribution curves for photo-electron emission from aluminium (Wooten et al., 1966) compared with the $L_{2,3}$-emission spectrum (Rooke, 1968).

3. ION-NEUTRALIZATION SPECTROSCOPY

A. Principle of the Method

Ion-neutralization by solids was introduced as a new spectroscopy for studying densities of states in metals by Hagstrum and Becker (1966).

A specific crystallographic face of a target metal or material is bombarded with low-energy ions (4 to 10 ev), formed into a focused mono-energetic beam. The ions used are commonly the inert-gas ions He^+, Ne^+ or Ar^+. Density-of-states distributions for the solid can be extracted from the measured energy-spectrum of the electrons ejected in the process of ion-neutralization at the solid surface.

The electrons emitted are detected by a spherical collector which surrounds the target. A negative retarding potential is applied to this collector, and the kinetic-energy distribution for the ejected electrons can then be obtained by differentiation of the collector current. With the equipment used by Hagstrum and Becker, the surface of the material studied can be made atomically clean by ion-sputtering and annealing in vacuum, $\sim 10^{-10}$ torr, and may be examined by low-energy electron diffraction in the same apparatus.

B. Unfolding the Density of States

The theory of the ion-neutralization process, and of the method used for extracting density-of-states information from the energy-spectrum of the ejected electrons, is described by Hagstrum (1966). The neutralization of an ion at or very near to a solid surface will cause the emission of an electron from the surface when the process is accompanied by an Auger transition. The process is then radiationless and involves two electrons in the valence band of the solid: one which tunnels through the potential barrier at the surface and drops into the vacant atomic level of the ion to be neutralized; and a second which takes up the energy released in the transition made by the first, is thereby excited to an energy above the valence band and if suitably directed may escape from the solid.

The process involved is depicted in Fig. 3, where the two valence-band electrons involved, 1 and 2, are respectively termed (by Hagstrum) the *down* and *up* electrons. The atomic ground state is at an energy $E_i(s)$ below the zero or vacuum level; $E_i(s)$ will be equal to the free-space ionization energy of the atom less the interaction energy of the ion with the solid, and therefore varies with the ion–solid separation s.

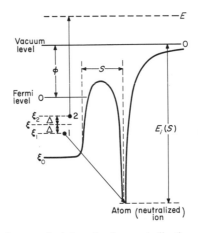

Fig. 3. Energy-level diagram depicting the ion-neutralization process; the electron-pair transition of electrons 1 and 2 respectively neutralize the ion and produce an excited Auger-electron above the valence band.

The final kinetic-energy E of the *up* electron, measured from the vacuum level, is clearly dependent on the initial energy-states of the two electrons in the band. For the transition depicted in Fig. 3, the *down* electron gives up an amount of energy $E_i(s) - \phi - \xi_1$, where $E_i(s)$ is evaluated for the ion–solid separation at which the neutralization occurs, and the Fermi-level for the solid is at an energy ϕ below the zero or vacuum level. ξ_1 is the initial-state energy for electron 1, measuring positively downwards from the Fermi-level.

The *up* electron takes up an amount of energy $E+\phi+\xi_2$, which must equal that released by the *down* electron. Hence

$$E = E_i(s) - 2(\xi + \phi) \tag{1}$$

where the initial-state energies for electrons 1 and 2, respectively ξ_1 and ξ_2, are symmetrically disposed with respect to ξ. A band of excited electrons is formed, and for those that surmount the surface potential barrier and escape from the solid the kinetic-energy distribution is determined by, among other factors, the valence-band energy distribution. The measured energy distribution for the external electrons (those that have escaped) is $X(E)$, and broadening effects—chiefly initial-state lifetime broadening and non-adiabatic broadening which is due to the motion of the ions and is always appreciable due to the ion-beam focusing fields—must first be removed to give $X_0(E)$.

The detailed procedure for unfolding the density of valence-band electron states $N(\xi)$ from $X_0(E)$ is given by Hagstrum (1966). In outline, the internal energy-distribution of *up* electrons excited out of the valence band, $F(E)$ is first obtained from $X_0(E)$ using

$$X_0(E) = F(E)P(E) \tag{2}$$

where $P(E)$ is the probability function for an excited electron escaping through the surface potential barrier. An example of the function $P(E)$, for He^+ ions at the (110) face of copper, is shown in Fig. 4. The function has an approximately flat region for $E > 4$ ev, and the form of $P(E)$ can be taken as essentially the same for different ions except for a normalizing factor.

$F(E)$, the energy distribution of excited *up* electrons, must depend on $N(E)$ the density of final-states at E, and on $F'(\xi)$ which is a function giving the total probability of an ion-neutralization Auger transition (or of all electron-pair transitions involving two electrons with energies symmetrically disposed about ξ):

$$F(E) = N(E) F'(\xi) \tag{3}$$

The density of final-states is approximately independent of E and can be taken as constant, hence $F'(\xi)$ can be written $F(\xi)$; with ξ related to E by Eq. (1). $F(\xi)$ is then the probability per unit energy of an electron-pair transition occurring to produce an excited electron at E, and must involve both initial-state energies ξ_1 and ξ_2. It can be expressed as

$$F(\xi) = \int_0^\xi U(\xi + \Delta) \, U(\xi - \Delta) d\Delta \tag{4}$$

Within the integral the product of the probabilities of transitions from the two initial-state energies gives the probability of the "pair" transition; these individual probability functions, $U(\xi)$, are taken as the same for the two initial states. This approximation is probably valid due to indistinguishability of the electrons in the initial states. The product is integrated with respect to

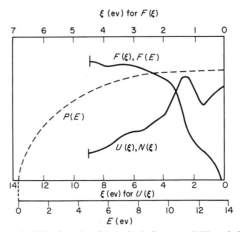

FIG. 4. The escape-probability function for excited electrons $P(E)$ and the ion-neutralization functions $F(\xi)$ and $U(\xi)$ (see text) for the [110] face of copper bombarded with low-energy He^+ ions. (Reproduced with permission from Hagstrum, 1966.)

$d\Delta$ (where $\Delta = \xi_2 - \xi = \xi - \xi_1$) because the two initial-state energies must be symmetrically disposed about ξ.

A further gross assumption made is that the matrix elements (or transition-probabilities) are independent of ξ and that therefore $U(\xi)$ depends solely on the density of states in the valence band, $N(\xi)$. Equation (4) then becomes

$$F(\xi) = \int_0^{\xi} N(\xi + \Delta)\, N(\xi - \Delta) d\Delta \qquad (5)$$

Hence $N(\xi)$ can be unfolded from $F(\xi)$, and therefore from the experimentally measured $X(E)$. The functions $F(\xi)$ and $U(\xi)$, for the electron emission from copper when bombarded by He^+ ions, are shown also in Fig. 4.

4. COMPARISON OF PHOTO-ELECTRON AND ION-NEUTRALIZA-TION SPECTROSCOPIES WITH SOFT X-RAY SPECTROSCOPY

From the comparisons made in Figs. 1 and 2 it is clear that photo-emission and ion-neutralization studies overlap usefully with soft x-ray emission measurements. From resolving-power considerations, more detailed information is to be expected from photo-electron spectra than from ion-neutralization spectra (for which the maximum resolution is a few tenths of an ev). For light metals, soft x-ray spectra are probably superior in resolution to both; although photo-electron spectroscopy is likely to give more detail, except where electron-scattering effects are important. Lifetime broadening may be expected to increase in the order soft x-ray emission, photo-emission, ion-neutralization electron emission (i.e. with the number of initial and final band-states involved, as distinct from "core" states).

With the heavier metals, and for example transition metals, comparatively more information can be expected from photo-emission and ion-neutralization studies (in that order) than from soft x-ray measurements; although difficulties still exist in the interpretation of photo-emission data (Phillips, 1965). Some of these points are illustrated by the results for nickel, cited by Cuthill *et al.* (1967) and shown in Fig. 5.[†]

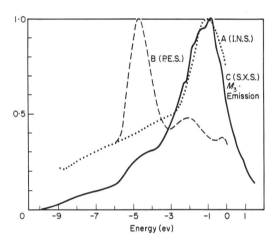

Fig. 5. Density of states for nickel: comparison of the data obtained from A, ion-neutralization, B, photo-electron emission, and C, the soft x-ray M_3-emission. (Reproduced with permission from Cuthill *et al.*, 1967.)

The $M_{2,3}$-emission spectrum and the ion-neutralization data (Hagstrum and Becker, 1966) closely agree, but a very large low-energy peak is observed in the density of states extracted from photo-electron spectra (Blodgett and Spicer, 1965) explanations suggested for this peak include many-body resonances and multiple d-band ionizations, but its interpretation is not clearly understood. Hagstrum (1966) shows that when the crystal surface is atomically clean the ion-neutralization function $U(\xi)$ will reflect the bulk density of states, and observes that the states in the s,p band are favoured by transition-probabilities for the tunnelling process.

The density of final-states, which in principle can be directly obtained from soft x-ray absorption spectra, and less directly from photo-emission measurements with varying photon-energies, does not readily unfold from ion-neutralization data. This is probably due to the limited number of ion-neutralization energies available. The neutralization energy, E_i–ϕ, is equivalent to the photon energy of photo-electron experiments, and because of the possibility of adsorption of the parent gas at the target surface, ion-neutralization studies have been limited to the noble-gas ions.

A disadvantage of both electron-emission techniques is that the data have to be separated from the escape-probability function $P(E)$. This can be a more serious problem with photo-electron emission when windows (e.g. lithium fluoride) are employed, because the measurements are then limited to low-energy photons. We find an additional disadvantage of ion-neutralization measurements where the unfolding of $U(\xi)$, for the individual *up* and *down* processes, must depend upon the validity of the assumptions made regarding the respective matrix or transition-probability factors, and will limit the effective energy-resolution; photo-electron spectroscopy is less limited by the assumptions used to unfold the densities of initial and final states.

An important practical feature of electron-emission experiments is their convenient application to non-metals, including semiconductors and insulators, materials which can only be readily studied by soft x-ray emission if fluorescence excitation is used. The problems of interpretation of electron-emission data are being steadily overcome with the help of accurate band-structure calculations and other experimental measurements, and the three types of spectroscopy are very likely to prove mutually helpful as techniques improve.

Acknowledgement

The author wishes to thank C. A. W. Marshall and G. A. Rooke for their invaluable help in the preparation of this discussion.

References

Berglund, C. N. (1966). *In* "Optical Properties and the Electronic Structure of Metals and Alloys", ed. by F. Abelès, p. 285. North-Holland Publishing Company, Amsterdam.

Berglund, C. N. and Spicer, W. E. (1964). *Phys. Rev.* **136**, A1030, A1044.

Blodgett, A. J. and Spicer, W. E. (1965). *Phys. Rev. Letters* **15**, 29.

Cuthill, J. R., McAlister, A. J., Williams, M. L. and Watson, R. E. (1967). *Phys. Rev.* **164**, 1006.

Hagstrum, H. D. (1966). *Phys. Rev.* **150**, 495.

Hagstrum, H. D. and Becker, G. E. (1966). *Phys. Rev. Letters* **16**, 230.

Krolikowski, W. F. (1967). Photoemission studies of the noble metals, cuprous halides and alkali halides. T. Rept. 5218–1, SEL–67–039. Stanford, California.

Nemnonov, S. A., Zyryanov, V. G. and Volkov, V. F. (1966). *Phys. Met. Metallogr.* **22**, 54.

Phillips, J. C. (1965). *Phys. Rev.* **137**, A1835.

Rooke, G. A. (1968). *J. Phys.* C. **1**, 767.

Spicer, W. E. and Berglund, C. N. (1964). *Rev. Sci. Instrum.* **35**, 1665.

Spicer, W. E. (1966a). *J. Appl. Phys.* **37**, 947.

Spicer, W. E. (1966b). *In* "Optical Properties and the Electronic Structure of Metals and Alloys", ed. by F. Abelès, p. 296. North-Holland Publishing Company, Amsterdam.

Wooten, F., Huen, T. and Stuart, R. N. (1966). *In* "Optical Properties and the Electronic Structure of Metals and Alloys", ed. by F. Abelès, p. 333. North-Holland Publishing Company, Amsterdam.

Part 3

BAND STRUCTURE AND THE THEORY OF SOFT X-RAY EMISSION

Preface

The Theory of Soft X-Ray Emission

The initial objective of measurements on the intensity $I(E)$ of the soft x-ray emission from metals, as a function of energy E, was to obtain the density of states in the valence band from which electrons make transitions to vacant core-levels. While it had been clear from the outset that transitions to an inner s-state from only p-components of the energy-band wavefunctions had finite probability, it was nevertheless thought for a long time that the density of such p-states would be faithfully reflected in the measured intensity $I(E)$. It has now become much clearer (see for example the papers in this Part by Harrison and by Stott and March) that not only is the density of states involved but also, and very intimately, the transition probabilities from the valence (or conduction) band to the core levels.

In one sense this is a disappointing outcome, for no longer can we believe that we are directly observing the density of states when we measure $I(E)$; but in another sense it gives us a much more powerful test of our theories, for the wavefunctions are involved in a quite fundamental manner. It is even tempting to ask whether we could, at least in principle, extract wavefunctions from the $I(E)$ curves. This would be doomed to failure in the case of metals with complicated Fermi-surfaces, for there, we surely could not map-out three dimensional wavefunctions from a function I of a single variable E. If the Fermi surface is spherical there is some hope of getting information on the wavefunctions, but it is premature at present to decide to what extent such an approach might succeed. So we are thrown back to attempting to use computations based on band theory, and immediately we meet the difficulties that available methods focus attention primarily on the calculation of the dispersion relations $E(k)$, and that few authors to date have adequately tabulated information on wavefunctions. But if this were the only consideration, then it would only be a matter of time before, by extensive electronic computation, we could break through this bottleneck and predict the spectra from first principles.

In fact, there are other fundamental considerations which make the problem at once more difficult theoretically and more fascinating to the physicist. These concern the very definition of the quantities we work with in soft x-ray emission, in terms of the one-electron approximation and the stationary-state wavefunctions obtained from it. With interacting electrons, there arises the question of whether we can even define one-electron states. The answer appears to be that in metals we can do so in the neighbourhood of the Fermi surface, where the lifetime of one-electron states becomes infinite as we approach the Fermi level. But it is certainly not true deep down in the valence bands of metals. Furthermore, a one-electron description takes no account of collective effects, such as plasmons, but fortunately these can be incorporated at a second step, without too much trouble.

The above considerations, on electron–electron interaction, apply generally to metals and are not specific to the problem of soft x-ray emission. However, in soft x-ray emission there is a further effect; this is the electron–hole interaction, which must be given careful consideration. Here, as for electron–electron interactions, the problem involves lifetime effects. For example, within the lifetime of the hole, do the valence electrons respond to its presence? Again it seems that the plasma oscillations are involved, because the time taken for electrons to rush in, to screen out a long-range electrostatic perturbation, must be of the order of 2π divided by the plasma frequency. While there are a few metals in which the lifetime of the hole and the period of the plasma oscillations are roughly comparable, it is true that in most cases the valence or conduction electrons have time to adjust themselves, and to pile-up around the atom with the vacancy in its core, in order to screen out its field. It is then clear that the measured intensity does not directly reflect the unperturbed valence-band states, but in accordance with the uncertainty principle, the very measurement of the soft x-ray spectrum has changed the system.

Is it permissible then, to any useful approximation in the theory of soft x-ray emission, to employ band theory to calculate valence-band wavefunctions with neglect of the hole? The situation is currently one of considerable interest, and in the opinion of the present writer remains somewhat open. For example, the problem of the anomalous spectrum for metallic lithium, which has been with us for many years, has prompted Goodings (1965) and more recently Allotey (1967) to suggest that in this case the spectrum is dominated by the electron–hole interaction. On the other hand, an alternative interpretation for lithium spectrum is possible in terms of band structure (Stott and March, 1966 and also this Part, p. 283); while for other metals Rooke (see p. 3), and other workers (Part 1), have shown that many detailed features in several metals can be understood without appeal to the hole.

As I remarked above, the hole in many cases must affect the valence-band wavefunctions, and the basic question is surely whether the hole introduces appreciable energy-dependent effects into the intensity $I(E)$. Certainly it must introduce energy-shifts and wavefunction enhancement near the hole; but do these vary appreciably through the band? Apart from first-principles calculations, of the kind referred to above, it may be fruitful to compare and contrast the theory of the second major technique for obtaining information on electron states in energy-bands, photoelectric emission, with that of the soft x-ray problem. Some comparison of the methods, including ion-neutralization electron emission, has been given in the foregoing discussion in Part 2 by Fabian. Unfortunately, in the photoemission experiment we may need a detailed understanding of what goes on at the metal surface, and this is a major complication.

On the alloy problem, a whole body of information is accumulating (see for example the review in Part 2 by Curry) and some progress in the theory is reported in the papers presented here (Harrison, p. 227; Stott, p. 303). The attractive impurity-centre has some similiarity to the electron–hole interaction, and careful use of alloy systems might again help to throw light on the electron–hole problem; although satellite-emission experiments, of the kind reported by Catterall and Trotter (1958), seem still to afford the most direct approach that we have for studying the effect of the hole. The influence of virtual and real bound-states around impurity centres in metals, is going to be one of the interesting problems in the soft x-ray field in the future.

Finally, a conference such as the one that led to this book, which had a nice blend of theory and experiment, and was sufficiently informal for controversial topics to be debated without inhibition, was most timely; the present writer has little doubt that now after a period in which the soft x-ray problem seemed to be in the wilderness, we are going to see a good deal of fundamental progress in the field over the next few years and—no doubt—a lot of surprises!

References

Allotey, F. K. (1967). *Phys. Rev.* **157**, 467.
Catterall, J. and Trotter, J. (1968). *Phil. Mag.* (8), **3**, 1424.
Goodings, D. A. (1965). *Proc. Phys. Soc.* **86**, 75.
Stott, M. J. and March, N.H. (1966). *Physics Letters* **23**, 408.

Electronic Structure and Soft X-Ray Spectra

WALTER A. HARRISON

Division of Applied Physics, Stanford University, Stanford, California, U.S.A.

ABSTRACT

Recent concepts in the theory of metals are applied to the problem of soft x-ray emission. The pseudopotential is introduced and the emission spectra are evaluated to successive orders in the pseudopotential. For simple metals, we see that structure due to variations in the oscillator strength is more prominent than that due to variations in the density of states.

Dilute alloys of simple metals, with broad valence-bands, are discussed in terms of phase shifts. It is found that in the dilute limit, emission from both constituents should sample the same common bandwidth. Application of the same approach to alloys that include noble or transition metals, suggests that structure due to the narrow *d*-bands seen by each constituent should be characteristic of the corresponding structure in the spectrum for the constituent as a pure metal. Modification of this behaviour in concentrated alloys is discussed.

Possible deviations from the simple description, due to the presence of the core hole, are considered. It is indicated that bound states are not expected, but a resonant *p*-state may arise in elements in the lithium row of the periodic table. Fermi-gas projections are considered and appear to give significant scaling of the intensity in simple metals, but not appreciably to affect the structure unless resonant states are present.

1. INTRODUCTION

I should like to discuss here some concepts which have recently become prominent in the theory of metals and which have particular applicability to our understanding of soft x-ray spectroscopy. It seems to me important to distinguish such a discussion from an *interpretation* of soft x-ray data. For an interpretation to be meaningful it must necessarily include a critical evaluation of the experiments and their significance. I am certainly no expert in this field.

In addition I have a general concern for a current trend in solid state physics, which also inhibits any tendency I might have to attempt a direct interpretation of the data. The pseudopotential method which I shall be discussing has made it possible now to attempt a direct calculation of experimental curves. However, calculation of a phenomenon such as soft x-ray emission inevitably involves a number of critical and questionable numerical approximations. Nevertheless, since no one is interested in the

227

details of numerical calculations, these remain unknown and frequently unspecified. The result of such a calculation is then a theoretical curve which could never be reproduced by another theoretician. It is even difficult for the theorist performing such a calculation to know what is the significance of his result. The reader is presented with a calculated curve, which he can compare with experiment but which he cannot assess in any other way. If the curve agrees with experiment he is tempted to say that the phenomena is understood and to forget both the experiment and the theory. Failure to agree with experiment tends to discredit the calculator. This of course provides an unhealthy pressure on the theorist to try several approximations but publish only that one which leads to the best agreement with the experiment in question. It is not clear to me that such attempts to interpret experiments serve any useful purpose.

An alternative approach which I have attempted here, is to proceed to the treatment of x-ray spectra by successive approximations. The approximation at each stage is easily stated and is simply a limitation in the order in an expansion parameter (the pseudopotential) to which the calculation is carried at that stage. Each stage is of course a successive step in the calculation of an experimental curve. However, it is easy to understand what is introduced and to see clearly what effects are being considered and therefore which aspects of the problem are important in determining the experimental curves. Hopefully, this in turn will lead to a clearer physical picture of which aspects of the problem are essential in determining the experimental results, and therefore which concepts may be useful in understanding experimental data.

The major part of my discussion will be based upon a self-consistent field approximation. Unfortunately the programme has not been carried far enough to determine to what extent the data are understandable in terms of this approximation alone. The analysis does suggest that the details of the spectra reflect, most directly, variations in the oscillator strength associated with the energy bands and not the density of states as is frequently envisaged.

Within this self-consistent-field framework we shall discuss the pseudopotential, which has been successful in describing many aspects of the properties of simple metals. We shall also discuss resonant states, arising from atomic d-states in noble and transition metals, and shall discuss the relevance of these concepts to the spectra of alloys. Finally we shall discuss the possibility of important effects arising that are beyond this simple self-consistent-field picture.

2. THE PSEUDOPOTENTIAL

We begin with a discussion of the pseudopotential method (Harrison, 1966) and first we shall see precisely what is meant by the pseudopotential.

This is most easily understood in terms of a calculation of the energy-band structure, though we shall see that a study of x-ray spectra is best made by avoiding the energy-band calculation itself.

Within the self-consistent-field approximation (this may be the Hartree or Hartree–Fock approximation) we seek the electronic eigenvalues determined by the time-independent Schrödinger equation,

$$(T+V)\psi = E\psi \tag{1}$$

T is the kinetic energy operator, $-\hbar^2 \partial^2/\partial r^2$; V is the self-consistent potential seen by an electron and is a function of position alone.

There are two types of solutions to this equation that may be clearly distinguished in the case of a simple (non-transition) metal. These are the core states, with energies E_α and eigenfunctions ψ_α, where α represents the principal and angular-momentum quantum numbers. These are the low-lying states of the free atoms and are essentially unchanged in going to the metal. In addition there are valence-band states ψ_k, with eigenvalues E_k, which correspond to the valence states on the free atom but are considerably modified in going to the metal. The core states are of course known from atomic-structure calculations, and the band-structure problem is the determination of the valence-band states.

The usual method for calculating the valence-band states, particularly in the age of electronic computers, is to make an expansion of the eigenstate ψ_k in terms of a complete set of functions such as plane waves; that is, we write the eigenstate as

$$\psi = \sum_k a_k |k\rangle \tag{2}$$

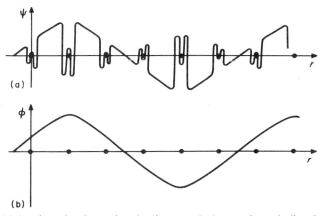

FIG. 1. (a) A valence-band wavefunction in a metal, shown schematically, plotted along a line of atoms in the crystal. The envelope of the wavefunction is approximately sinusoidal but is modulated by atomic-like structure at the atomic cores. (b) The corresponding pseudo-wavefunction, which is identical to the true wavefunction outside the cores but in which the atomic modulation has been removed. Thus the pseudo-wavefunction is quite free-electron-like.

where the a_k are coefficients, and where a normalized plane wave of wave-number k has been written as a ket. This expansion may be substituted in the eigenvalue equation, Eq. (1), multiplied on the left by the complex conjugate of a plane wave of wavenumber $k+q$, and integrated over the normalization volume. This converts the differential equation into a set of linear algebraic equations in the coefficients a_k.

The number of equations and the number of unknown coefficients depends on the number of expansion functions required in the expansion Eq. (2). In this respect it turns out that plane waves are a very poor set of basis functions and a very large number are required, perhaps two or three hundred. It is not difficult to see why this is so. The valence-band wavefunctions are shown schematically in Fig. 1(a). They tend to be rather smooth between atoms but to show sharp atomic-like oscillations in the vicinity of the metal ion. A very large number of plane waves are required to reproduce these sharp oscillations.

The essence of the various band-calculation techniques that have been developed is to seek a set of basis functions, each of which duplicates these oscillations and therefore permits the expansion of the wavefunction in a very limited number of terms. One of the earliest of such approaches is the orthogonalized-plane-wave method due to Herring (1940). He observed that the core eigenstates are known initially and that the valence eigenstates must be orthogonal to them. Thus, we may orthogonalize the individual plane-waves to the known core states and use these *orthogonalized plane-waves* as expansion functions. This approach has turned out to be useful. In most materials the valence-electron eigenstates can be rather accurately described in terms of a limited number, perhaps 20, of orthogonalized plane-waves. Since we shall base our pseudopotentials on the orthogonalized-plane-wave method, we should state the approach somewhat more explicitly.

A single orthogonalized-plane-wave is written

$$\begin{aligned} \text{OPW}_k &= |k\rangle - \sum_\alpha |\alpha\rangle\langle\alpha|k\rangle \\ &= (1 - \sum_\alpha |\alpha\rangle\langle\alpha|)\,|k\rangle \end{aligned} \tag{3}$$

The ket $|\alpha\rangle$ is a core wavefunction. The expression $\langle\alpha|k\rangle$ is an overlap integral of the complex conjugate of the core wavefunction and the normalized plane-wave $|k\rangle$. In the second expression we have written the ortho-gonalization terms as an operator. The meaning of the expression is precisely the same and implies that the same integration is to be performed. We may immediately verify that this OPW is orthogonal to every core state, by multiplying on the left by a particular core state and integrating over all volume. The first and second terms in Eq. (3) then cancel, assuming only that the core eigenstates on different atoms do not overlap; this is a very good approximation in the simple metals. We may now rewrite the expansion of Eq. (2), replacing the plane waves by orthogonalized plane-waves. In doing this we note that the a_k are simply numbers, and may be commuted through the orthogonalization operator. In addition the orthogonalization operator is independent of k and the summation may also be interchanged with that operator. Thus our expression for the wavefunction may be written

$$\psi = (1 - \sum_\alpha |\alpha\rangle\langle\alpha|) \sum_k a_k |k\rangle \qquad (4)$$

If we were now to substitute this form in the Schrödinger equation we could multiply on the left by a plane wave, integrate over the normalization volume, and proceed as before.

However, in the pseudopotential method we do not do this. We call the function $\sum_k a_k |k\rangle$ the *pseudo-wavefunction* Φ, and seek an equation for Φ. If we find the pseudo-wavefunction we can of course obtain the true wavefunction simply by making it orthogonal to the known core-states, using the procedure given in Eq. (4) which now becomes $\psi = (1 - \sum_\alpha |\alpha\rangle\langle\alpha|)\Phi$.

To obtain an equation for Φ we substitute this form for ψ in the Schrödinger equation, Eq. (1). This leads to a rather simple form since the Hamiltonian, $T + V$, operating on a core wavefunction $|\alpha\rangle$ leads to the core eigenvalue E_α times the same function. We bring all orthogonalization terms to the left-hand side, and obtain simply

$$T\Phi + W\Phi = E\Phi \qquad (5)$$

where W is called the *pseudopotential* and is given by

$$W\Phi = V\Phi + \sum_\alpha (E - E_\alpha)|\alpha\rangle\langle\alpha|\Phi\rangle \qquad (6)$$

Equation (5) is called the pseudopotential equation, and we note that in deriving it we have made no approximation on the initial Schrödinger equation. If we were to solve this equation exactly, we should obtain precisely the correct energy eigenvalue, and could obtain the true wavefunction simply by orthogonalizing the pseudo-wavefunction obtained to core states. We note, further, that the pseudopotential equation is of just the same form as the initial Schrödinger equation but with the potential replaced by a pseudopotential. The virtue of the pseudopotential equation arises from the fact that the pseudopotential must, in some sense, be small. We know that it must be small because the pseudo-wavefunction is expandable in a small number of plane waves and therefore rather smooth. At the same time the pseudo-wavefunction is equal to the true wavefunction outside the cores, since the expression $\sum_\alpha |\alpha\rangle\langle\alpha|$ is zero there (see Eq. 4). Thus the pseudo-wavefunction will be as shown schematically in Fig. 1(b). We may also more directly see that the pseudopotential is small by noting that the quantity $E - E_\alpha$ is always positive and that the corresponding positive term will tend to cancel the negative attractive true-potential, V.

More precisely, we should say that the virtue of the pseudopotential is that its Fourier components become small at large wavenumbers. This corresponds to the fact that only a small number of plane waves are needed in the expansion of the pseudo-wavefunction. We may see this in detail from calculated pseudopotentials, which in fact have Fourier components at small wavenumbers that are comparable to the corresponding Fourier components of the true potential, but which fall rapidly to zero as the wavenumber becomes large.

We might ask: what is the origin of this repulsive term in the pseudo-potential? It is frequently stated that it arises from the Pauli principle; that is, that the valence electrons are excluded from the core because of the presence of the core electrons. Thus the Pauli principle is thought of as excluding the valence electrons from the core and of effectively providing a repulsive potential which cancels the strong attractive coulomb potential. This cannot be the essence of the effect, which remains even if the core states are unoccupied. The exclusion of the electron from the core is in fact a classical effect. An energetic electron is accelerated as it passes near an ion and travels very rapidly through the core region. Thus, even classically, a particle of positive kinetic energy spends less time in the region of an attractive potential than elsewhere. This classical effect is the origin of the repulsive term in the pseudopotential which causes the exclusion of the valence electrons from the region of the cores.

Having arrived at the pseudopotential equation, Eq. (5), there are two ways to proceed. One is to seek a very accurate calculation of the energy eigenvalue, based upon Eq. (5). This is entirely equivalent to a traditional orthogonalized-plane-wave calculation of the energy bands. If we wish to make such a calculation less accurately we may also find the pseudopotential equation a convenient vehicle for making approximations. We may, for example, replace the complicated repulsive term in the pseudopotential by a simple repulsive potential. However, the real advance in the pseudopotential method, as applied to simple metals, is to take advantage of the smallness of the pseudopotential and use Eq. (5) to compute the *properties* of the metal. Therefore our second way is to proceed directly to the calculation of the property, including the pseudopotential as a perturbation. To do this it is essential that the band-calculation is not performed first. We shall use this latter approach in studying the x-ray spectra.

3. X-RAY SPECTRA OF SIMPLE METALS

To be specific we shall consider the *emission* spectra of the simple metal. The intensity of this emission is of course proportional to an oscillator strength and must be summed over all valence-band states which may be emptied in a transition to a core state. Thus we write the intensity at a given energy $\hbar\omega$ in the form

$$I(\hbar\omega) \propto \sum_k |\langle \psi_{core}|\boldsymbol{p}\cdot\boldsymbol{A}|\psi_k\rangle|^2 \delta(\hbar\omega - E_k + E_{core}) \qquad (7)$$

where $\boldsymbol{p}$ is the momentum operator and $\boldsymbol{A}$ is the vector potential for the x-ray in question. We note that there is no density of states in this expression; however, if the matrix elements were dependent only upon energy, we could replace the sum over wavenumber, by an integral over energy with a factor of the density of states. As we shall see this is not the case when the density of states is not constant.

The core wavefunctions are well known, and the difficulty in calculating the intensity is entirely in the evaluation of the valence-band wavefunctions, ψ_k, and in the determination of the energy of these states. In the pseudo-potential method we shall compute both the states and their energies by perturbation theory, regarding the pseudopotential as small. The pseudo-potential in the simple metals turns out ordinarily to be of the order of one-tenth of the Fermi energy, and therefore this is a suitable parameter for our expansion. It is important to notice that in this calculation it is the true wavefunction that enters the matrix element and not the pseudo-wavefunction; however, the true wavefunction is directly obtainable from the pseudo-wavefunction using Eq. (4), so this is not a fundamental problem.

We can immediately see what effects will arise at each order in our calculation. At zero-order the pseudopotential may be ignored altogether. The pseudo-wavefunctions become plane waves and the energy is simply given by $\hbar^2 k^2/2m$. The result will nevertheless differ from the result of a free-electron calculation, because of the orthogonalization terms which distinguish the true wavefunction from the pseudo-wavefunction. We shall see that the corresponding corrections to the x-ray intensity are significant.

In first-order the pseudo-wavefunction will be modified by the addition of other plane waves. These will enter the intensity through a modified oscillator-strength, which will be a large correction to the free-electron-like description. In addition the energies of the states will be modified by a term $\langle k|W|k\rangle$. Because of the operator nature of the pseudopotential, this term is not independent of k and will therefore give a slight shift in the density of states. This shift is of only minor importance and we shall not consider it here.

In second-order there will be an additional modification of the oscillator strength but this will be small in comparison to the first-order change and we shall not include it. In addition, there will be the first significant change in the energies of the states, and therefore a modification of the density of states. This is the effect which it is frequently imagined causes the main structure in the x-ray spectra, but we see here that it arises only in second-order.

We may now consider the various effects in more detail. Again in zero-order the eigenstates are given by single orthogonalized plane-waves and the energy E_k is simply $\hbar^2 k^2/2m$. It is the oscillator strength, or matrix element, which is of particular interest. Since the core is well localized, the vector potential may be regarded as a constant and taken outside the matrix element. In addition we may write p as $\hbar \mathbf{V}/i$. There are two terms in the single orthogonalized plane wave, the plane-wave and the orthogonalization term. Thus the matrix element takes the form

$$\langle \psi_{\text{core}} | p\, A | \psi_k \rangle \;=\; -i\hbar A \cdot (\langle \psi_{\text{core}} | \mathbf{V} | k \rangle - \sum_{\alpha} \langle \psi_{\text{core}} | \mathbf{V} | \alpha \rangle \langle \alpha | k \rangle) \qquad (8)$$

Now in the first term on the right-hand side of Eq. (8) we note that the gradient operator, operating on the plane-wave $|k\rangle$, yields simply $ik|k\rangle$. The ik may be taken out of the matrix element, and we are left with an overlap integral of the same form as the one that entered the determination of the pseudopotential itself. The matrix element $\langle \psi_{\text{core}} | \mathbf{V} | \alpha \rangle$ may also be written in a form which is easy to evaluate from the tabulated core-wavefunctions, using an identity given for example by Schiff (1949). Thus all of the expressions in the matrix element are readily evaluated from the tabulated core-wavefunction, though the angular averages may be complicated. In summing over core states, we must sum over the three orientations of the p-function and must sum over the three possible polarization directions of the vector potential associated with the x-ray. These summations may be performed analytically and exactly, and we shall not discuss them further. However, we shall note the general form of the two terms in the matrix element.

Here we must distinguish between x-ray emission associated with a transition to a core s-state and that associated with a transition to a core p-state. For a core s-state, the overlap $\langle \psi_{\text{core}} | k \rangle$ is approximately though not exactly independent of k. Thus for transitions to s states the first term in Eq. (8) is approximately proportional to k. In the second term there are matrix elements $\langle \psi_{\text{core}} | \mathbf{V} | \alpha \rangle$, only for α corresponding to a p-state. This matrix element is constant but the overlap $\langle \alpha | k \rangle$ is approximately proportional to k for a p-state. Thus both terms in Eq. (8) are approximately proportional to k, and therefore the oscillator strength for this case is approximately proportional to k; the intensity contains two factors of k from the oscillator strength and one from the density of states, to give an intensity proportional to k^3 or to energy raised to the power $3/2$. This is the familiar result obtained by writing the valence-band states as Bloch functions, with a periodic factor independent of k. However, detailed numerical calculation leads to overlaps which do not precisely have this simple form, and the intensity is found to rise less rapidly than $E^{3/2}$. Such a curve for sodium, computed with appropriate geometrical summations, is shown in Fig. 2. This result is more accurate than that obtained with the usual assumption of a simple form of Bloch function.

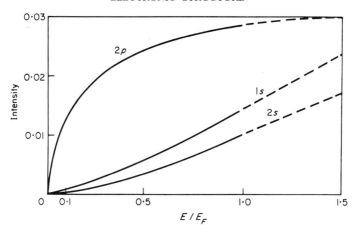

FIG. 2. The calculated intensity for sodium from a calculation to zero-order in the pseudo-potential (in arbitrary units, but the same units for each curve). The solid line represents x-ray emission; the dashed line x-ray absorption. The $1s$, $2s$, and $2p$ designate the core state to which transitions occur.

If the core state is a p-state, the first term in the matrix element of Eq. (8) contains a factor of k from the gradient and a second factor of k from the overlap of the core p-state with the plane wave. If this was the only term in the oscillator strength, which would be the case for a free-electron approximation, it would lead to an intensity proportional to $E^{5/2}$. However, the second term gives matrix elements with α corresponding to a core s-state; the corresponding overlap is a constant and therefore this term gives an approximately constant contribution to the oscillator strength. This leads to the familiar result of an intensity proportional to $E^{1/2}$. It is interesting to note that this result depends directly upon the orthogonalization terms in our approach. However, we note that with increasing wavenumber there is increasing cancellation between the first and second terms, and again an intensity which increases less rapidly with wavenumber than that obtained from the simple theory. This also is illustrated for sodium in Fig. 2. The deviation from an $E^{1/2}$ dependence is not so conspicuous in the energy range of interest in sodium, but the calculation has been carried to higher energies in aluminium with a result shown in Fig. 3. Here we see a conspicuous deviation from the $E^{1/2}$ dependence.

We now proceed to modifications of the intensity, arising in the first-order in the pseudopotential. Since these are simply small corrections to our zero-order curves, it is appropriate to compute correction factors which are of order 1 and which are to be applied to the zero-order curves. Within the framework of the perturbation theory used, it is proper to add the effects due to each of the Brillouin-zone planes. Each of these corrections can be calculated exactly. The only difficulty is a divergence of the corrections just

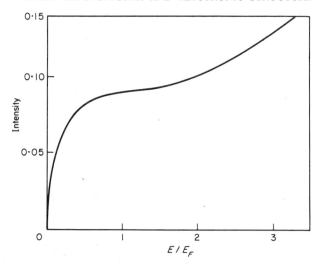

FIG. 3. Soft x-ray absorption and emission intensity for transitions involving the core 2p-state in aluminium, calculated to zero-order in the pseudopotential. Deviations from $E^{1/2}$ arise from orthogonalization terms.

at the zone faces. This is the familiar divergence of perturbation theory, arising from vanishing energy denominators at zone faces.

In calculating the first-order corrections, it is helpful to notice that the orthogonalization terms are intrinsically first-order, although we included them in our zero-order calculation above. The first-order nature of ortho-gonalization terms may be seen theoretically from the form of the pseudo-potential, and is numerically reasonable since the orthogonalization hole, $\sum_{\alpha} \langle k|\alpha\rangle\langle\alpha|k\rangle$, is of the order of 0·1. In any case, in evaluating the first-order corrections we may replace the true valence-band wavefunction in Eq. (7) by the pseudo-wavefunction. The pseudo-wavefunction, to first-order, con-tains the zero-order plane-wave and a first-order correction obtained in first-order perturbation theory. Since we may treat each Brillouin-zone face by itself, we need include only the correction for a single Brillouin-zone face. If that Brillouin-zone face is the plane bisecting the lattice wavenumber (the reciprocal-lattice vector) q, the first-order pseudo-wavefunction is given by

$$\Phi = |k\rangle + |k+q\rangle\frac{\langle k+q|W|k\rangle}{\epsilon_k - \epsilon_{k+q}} \tag{9}$$

where the ϵ_k are the zero-order energies, $\hbar^2 k^2/2m$. To a rather good approximation, $\langle k+q|W|k\rangle$ can be taken to be independent of k and is simply 1/2 the band-gap on the corresponding Brillouin-zone plane. In any case it is directly obtainable from pseudopotential calculations.

Substituting this form for the wavefunction in Eq. (7), we may perform the necessary geometrical summations and directly obtain a correction factor giving the modification of the intensity due to these first-order terms. The corrections are found to be small enough ($\sim 3\%$ for each of the [111] Brillouin-zone planes in aluminium) to justify the use of perturbation theory. However, when the effects for all eight of these planes are added, they become quite appreciable as indicated in Fig. 4. Note that we have omitted the region of the curve near the divergence.

Of particular interest, is the comparison of these first-order correction factors with the corresponding correction factors arising from modifications in the density of states. As we have indicated, these density-of-state correction factors are second-order in the pseudopotential. Their calculation in the pseudopotential method is entirely equivalent to the familiar calculation of the shift in density of states of the free-electron approximation, but of course the matrix element of the free-electron potential is replaced by a matrix element of the pseudopotential. This is the same matrix element that entered our calculation of the first-order correction. These density-of-states corrections are shown also in Fig. 4. As we expected, the corrections due to oscillator-strength modifications are much more pronounced than those due to density-of-states fluctuations, and to a good approximation we could regard the density of states as free-electron-like and consider the modifications to the intensity as arising only from the modified oscillator-strengths. Of course there are van-Hove singularities in the intensity due to the density-of-states corrections, but these are small features.

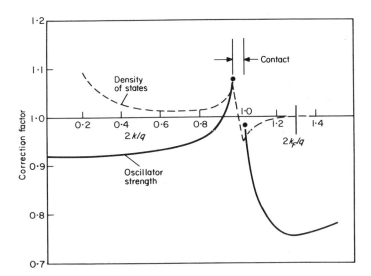

FIG. 4. First-order and second-order correction factors for the 1s-absorption intensity in aluminium, arising from the eight [111] Brillouin-zone faces. The solid line is the first-order correction and may be attributed to variations in the oscillator strength. The dashed curve is the second-order correction and may be associated with the density of states. The abscissa is twice the wavenumber of the valence-band electron, divided by the [111]-lattice wavenumber.

There is a second important distinction to be made between first-order and second-order corrections: the matrix element of the pseudopotential corresponding to a particular Brillouin-zone face, may be either positive or negative, depending upon whether the odd or even states lie lowest at the Brillouin-zone face. Wilson (1936), thinking in terms of matrix elements of the true potential, concluded naturally that the s-state would always be lower and considered only negative matrix-elements. In fact for aluminum, and for most simple metals, the pseudopotential matrix-elements are found to be positive. The modifications in oscillator-strength corrections, because they are proportional to the matrix elements of the pseudopotential, change sign with a change in sign of the pseudopotential. On the other hand, the density-of-states corrections, being proportional to the square of the pseudo-potential matrix-elements, are independent of the sign of these matrix elements.

4. ALLOYS

We shall discuss x-ray spectra in the alloys of simple metals, but in such a way that it will lead naturally to a discussion of the transition metals.

For an alloy A–B, the soft x-ray emission observed for atom A, and that observed for atom B, form spectra (i.e. component spectra) for which interpretation is quite easy but the theory is difficult. There are two simple experimental possibilities: either, each atom shows a spectrum very much the same as for that atom when pure, or, both types of atom show the same spectrum (i.e. the component spectra are the same as each other). In either case it is quite easy to propose a model which will explain the experimental result. In contrast, we seek here to deduce from first principles whether either of the two results is to be expected. We do not quite succeed in answering this question for the interesting case of concentrated alloys, but can make definitive statements with respect to extremely dilute alloys.

In particular, we consider the simple situation of a spherical crystal of pure metal in which we are to substitute a single impurity-atom at the centre. In the pure metal we may neglect the weak pseudopotential, as a first approximation, and construct free-electron pseudo-wavefunctions. Because we shall wish eventually to introduce an impurity, and for this reason have chosen a spherical boundary, it is appropriate to construct spherical states rather than plane-wave states. The angular momentum, with respect to the centre of the crystal, will be quantized and for simplicity we shall consider only the states of s-symmetry. The form of these spherically symmetric states is well known and is given by

$$\psi_k = \frac{A \sin kr}{kr} \tag{10}$$

A is a normalization constant which will not interest us, and k is a radial

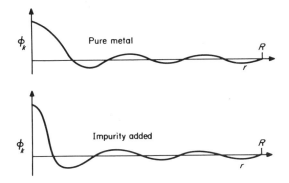

FIG. 5. A schematic representation of the radial s-function in a spherical metal of radius R. These are the $n = 6$ states, with six nodes. Other states may be constructed for all other integral values of n.

wavenumber. If we require vanishing boundary-conditions on the surface of the crystal, radius R, the restriction on k is simply $kR = n\pi$, where n is a positive integer. The corresponding state for $n = 6$ is illustrated in Fig. 5. These states for successive n will be very closely spaced in energy, with approximately as many states as there are atomic distances in R over an energy range of the order of 10 ev. If we consider the weak pseudopotential added for the pure material, the energies will be shifted very slightly and the true wavefunction will be orthogonalized to the cores on each atom.

Let us replace the central atom with an impurity. The potential at the center of the crystal, and thus the curvature of the pseudo-wavefunction there, are now modified. At large distances the wavefunction will approach

$$\psi'_k = \frac{A' \sin(k'r - \delta)}{k'r} \tag{11}$$

δ is called the phase shift and can be directly calculated from the difference in pseudopotential between the impurity and the parent atoms. The modified pseudo-wavefunction is also shown in Fig. 5. We must again satisfy the vanishing boundary-condition at R; this condition now becomes $k'R - \delta = n\pi$. The modified wavenumber becomes

$$k = \frac{(n\pi + \delta)}{R} \tag{12}$$

Now if δ is greater than π, this would introduce an additional node in the pseudo-wavefunction near the impurity atom. However, the pseudo-potential has enabled us to construct pseudo-wavefunctions which do not have such nodes. Pseudopotentials are weak, and we can be quite certain that the difference in pseudopotential between a simple impurity and a simple solute will not introduce extra nodes, and that therefore the phase shifts are less than π. This can readily be verified by evaluating phase shifts from the pseudopotential in, for example, aluminum or magnesium.

A shift in wavenumber or in energy for each state may be directly computed from Eq. (12). The important point is that, if the phase shifts are less than π, no wavenumber and therefore no energy is shifted as far as the neighbouring unperturbed state. As we have indicated, the levels are extremely closely spaced in a macroscopic crystal and therefore the shifts in each of the eigenvalues are extremely tiny. Such shifts are shown schematically in Fig. 6. It

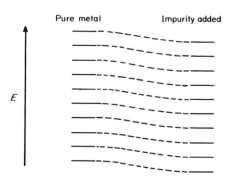

FIG. 6. A schematic representation of the shift in valence-band energy levels in a pure metal, as an impurity is added.

is possible in principle for one or more states to drop significantly from the bottom of the band; that is, it is possible for bound states to be formed. However, this would require phase shifts greater than π and it seems quite certain that these do not occur.

It is important to recognize that the states we have computed are the eigenstates of the alloy system within the self-consistent-field framework. Electron transitions must occur between these eigenstates and the core eigenstates, no matter whether these be transitions in the central impurity-atom or transitions to core states in the solvent atoms. Soft x-ray emission for either atom must see precisely the same valence-band energy eigenvalues. We have found here that when a single impurity is added the shift in the energy of these states is completely negligible. This must remain true with a finite, but very small, concentration of impurities. If the solvent is magnesium, with a bandwidth of 7 ev, it is inconceivable that an aluminum impurity could see a valence bandwidth equal to the 12 ev that would be present in pure aluminum. Of course in a concentrated alloy it is conceivable that the small shifts, which we have discussed here, could add up to a large shift corresponding to the difference in the two bandwidths. We shall briefly return to this question later; however, it is clear that this cannot be the case in the dilute limit.

Although the same states, and therefore the same bandwidth, are seen by both the impurity atom and solvent atom in soft x-ray emission, we may still ask whether emission in one atom will preferentially see one part of the valence band, while emission in another atom sees a different part of the band. This question may be answered in detail by considering the eigenstates themselves. To date such consideration has not been given, but we may readily answer a closely related question which arises as follows.

In the pure metals electrons in each energy range are equally shared by all atoms in the crystal, including the central atom. With the introduction of an impurity, and a corresponding modification of the wavefunctions, there is a redistribution of the charge density, and we ask to what extent the electrons of different energies are localized in the region of the impurity or are pushed into the solvent metal. We may in fact do better than this, and ask to what extent electrons of each energy and each angular momentum are shifted. It can readily be shown that the electrons of any angular momentum and energy are localized within the central atom to an extent proportional to $\partial\delta/\partial k$, where δ is the wavenumber-dependent phase shift for the angular momentum in question. This result appears as an intermediate step in the derivation of the Friedel sum-rule (Kittel, 1963). In the simple metals the phase shifts vary rather smoothly with energy and any shift in electron density to the impurity atom tends to be uniformly distributed over the valence band. This is particularly true for s-electrons, while the localization of p-electrons tends to occur higher in the bands. Much the same was true of the states in the pure metal. We conclude that no major effects are expected from variations in the oscillator strengths from the impurity atoms in the simple metals.

All of this discussion of alloys has omitted band-structure effects such as those we discussed for the pure elements. Further, we might ask whether the corresponding structure will show up in the alloy system. However, we recall that the structure is associated with Bragg-reflection planes and therefore specifically associated with interference effects among various atoms. Thus any structure in the alloys must be associated with the composite system and not with the pure material, at least if consideration is restricted to the dilute alloys.

We may now go on to discuss a transition-metal or noble-metal impurity inserted in a simple metal. In particular we shall consider a noble metal, such as copper, dissolved in a simple-metal solvent such as aluminium. We may proceed, just as before, to discuss the phase shifts for each angular momentum. The situation remains much the same as in the simple metals for states of s and p angular momentum. For these angular momenta the phase shifts are again small and vary slowly with energy, and the same conclusions obtain. However, the behaviour is profoundly different for electron states of d-symmetry. The atomic d-states of copper lie right in the middle of the copper valence-band and when copper is dissolved in aluminium they similarly lie in the midst of the aluminium valence-band. The attractive potential of the copper ion and the repulsive centrifugal potential are very nearly sufficient to bind the electron to the copper ion; that is, an integration of the Schrödinger equation for a d-state, at an energy near the atomic copper d-state energy, leads to an atomic-like function which decays exponentially to small values within a copper atomic cell. However, since the energy lies above the valence-band minimum, this state does not continue to decay exponentially at large distances but retains a very small oscillating tail analogous to the much larger tail indicated in Fig. 5. This state is profoundly modified from the free-electron d-state, and perturbation theory has broken down completely. This nearly-bound d-state is called a resonance level. It is precisely the same type of state that is occupied by an α-particle in an α-active nucleus. Similarly the α-particle is very nearly bound but lies at an energy above the rest-energy outside the nucleus, and therefore α-decay can occur, although the probability of decay per unit time is quite small. In such a resonance the phase shifts are very nearly zero (or very nearly an integral multiple of π) except at energies very close to the energy of resonance. Near resonance the phase shifts are given by

$$\tan \delta \approx \Gamma/(E - E_d) \qquad (13)$$

Γ is the energy-width of the resonance and is small compared to the valence bandwidth. E_d is the resonant energy, or the energy of the atomic d-state. We see that the phase shift increases by π over a very narrow range of energy. By considering a diagram such as Fig. 6, we see that the states are all shifted by an amount smaller than the distance to the next state. However, due to the π phase-shift, one additional state is introduced in a region of width Γ. In the case of a d-state there are really five states of a given spin introduced in this region, corresponding to the five different components of angular momentum along the axis of quantization. Counting both spins, ten states are introduced in a narrow energy-range due to the introduction of the copper atom. These of course correspond to the ten atomic d-states of a free copper atom, or to the ten d-states per atom in pure copper metal.

We may go on further to ask to what extent these extra d-states are localized within the copper impurity-atom. As before, the degree of localization is proportional to $\partial\delta/\partial k$; however, in this case we have seen that the phase-shifts change very rapidly over a narrow energy-range. Thus the localization is very high at energies near resonance, and negligible at other energies. We conclude that the extra ten d-states introduced by the copper atom are in fact strongly localized at that atom, and soft x-ray emission from the impurity will show strong evidence of these states while soft x-ray emission from the solvent will not show them appreciably.

Our conclusion, with respect to the soft x-ray emission from alloys with narrow d-bands, is just the reverse of the conclusion we came to in the case of simple metals. In simple metals, with broad valence-bands, we found that the alloy contained a common band-structure and that the electrons tended to be shared equally among impurity and solvent atoms. In the case of noble metals or transition metals we see that structure for the pure metal, that arises from their narrow d-bands, will tend to carry over to the alloy and will be observed in soft x-ray emission from the impurity, while the solvent-atom spectra will tend to be unaffected by the addition of the impurity. Again these considerations have been restricted to dilute impurities.

We should next go on to consider how these conclusions may be modified in concentrated alloys. It seems clear that the argument as applied to d-bands in the noble metals would carry over as more and more noble-metal atoms were added. Each added noble-metal atom produces a strongly localized set of states at the energy corresponding to a narrow band in the pure noble-metal, but leaves the states in the solvent material virtually unchanged. Thus, as more and more noble-metal atoms are added each finds itself in an environment which is little modified by the other noble-metal impurities, and its behaviour is therefore similar. Even in the limit, where the alloy would consist of two types of noble-metal or transition-metal atoms, we should expect each to show a soft x-ray structure with respect to its d-states that is similar to that in the pure metal. Inevitably there will be modification or broadening of the corresponding absorption bands, due to overlap of localized states on the different atoms, but the picture remains essentially the same. This conclusion seems to be largely consistent with the results of the x-ray emission studies of noble-metal and transition-metal alloys presented in Part 2 of this Volume.

With the alloys of simple metals the situation is much less clear. We may expect accumulative shifts in the levels as additional impurity-atoms are added. Also, of course, if we add impurities of higher valence than the solvent, we add with each impurity additional electrons and therefore cause the occupation of additional band states. It seems most probable that the shifts would be rather smoothly varying over the valence band and that the bands would remain free-electron-like throughout the alloy composition range. It is known, for example, that the bands are free-electron-like at both the pure-magnesium and pure-aluminium ends of the aluminium–magnesium alloys. We should therefore expect that the observed bandwidth would grow continuously, as aluminium was added, from the 7 ev bandwidth for magnesium to 12 ev for aluminium. We should further expect both constituents to show soft x-ray emission, at each concentration, corresponding to the same bandwidth.

This conclusion must remain somewhat speculative and clearly depends upon the manner in which the atoms are arranged. If the alloy consisted of clusters of aluminium and clusters of magnesium atoms, then the atoms in each cluster would tend to see a local bandwidth. In terms of our phase-shift analysis, we should say that clusters of atoms may form bound states though isolated atoms may not. The results for this alloy, described in this Volume,

p. 174, suggest a situation resembling the cluster model rather than the model we arrived at from consideration of dilute alloys. If these data are telling us what they appear to, it would mean either that the alloy consists of clusters of pure elements, which seems unlikely, or that even in the random alloy random clustering has the effect of leading to a result comparable to a truly clustered alloy. This is obviously a combined statistical and quantum-mechanical problem which to date has not been seriously addressed.

5. FAILURE OF THE SELF-CONSISTENT-FIELD APPROXIMATION

All of our discussion has been based on a picture in which the electrons make transitions between simple one-electron states and simple core-states. The interaction between electrons has been included only in terms of an average charge-density. Unquestionably this picture leaves out effects which are observable in the spectra. In particular there are plasmon satellites which arise from transitions accompanied by the emission of a plasma oscillation. These plasmon satellites are discussed at some length in Part 4. They are small effects, however, which do not change very much the picture we have presented. In addition, M. P. Weiner and H. Ehrenreich (to be published) have discussed the importance of the virtual emission and re-absorption of plasmons during transition. Such effects, as well as the bound excitons proposed by G. Mahan (to be published), may well distort the magnitude of the intensity observed at any energy. Because of such effects we cannot expect a detailed spectrum computed on the basis of the self-consistent-field approximation to fit precisely that measured. There is also the Auger tail to the spectrum which arises from a mechanism which we have not considered.

However, we are interested here in any effects which could fundamentally change the structure of the emission spectra through an invalidation of the description of valence-band states that we have used. It would seem that any such effect must arise from the presence of the empty core-state prior to emission, since our description of states is extensively confirmed in the theory of the properties of metals, such as transport properties of metals, without core holes.

One such modification was proposed by Friedel (1952) who indicated that the extra positive-charge, associated with the empty core-state, could give rise to a bound state dropping from the bottom of the valence band. Assuming that the extra potential was given by a simple screened coulomb potential, and treating the electrons as free, he in fact found such bound states. However, this binding of a state is no longer believed to occur. We may readily see that the tendency is inhibited if we consider the modification in the pseudopotential rather than in the potential itself. Consider the form of the pseudopotential as given in Eq. (6) but now written as

$$W = V + \sum_{\alpha} (E - E_{\alpha})|\alpha\rangle\langle\alpha| \tag{14}$$

It is true that the addition of a core hole will introduce a screened coulomb-potential localized near the core, and therefore will cause a sizable drop in V. That is, just the drop considered by Friedel. At the same time, however, this potential-well, at that site, will lower the core energies E_{α} for all cores at that site by an amount approximately equal to the depth of the well. Then, to the extent that the core states are a complete set for the expansion of functions within the core, the change in E_{α} will exactly cancel the lowering of V. This cancellation can never be complete, but the change in the repulsive term in the pseudopotential will go far to weaken the change in the pseudo-potential, and it seems most likely that this will prevent the formation of the bound state. We conclude that such fundamental changes in the valence bands due to the core hole are unlikely.

A second fundamental modification in the states due to the core hole has been discussed by Allotey (1967). He indicated that though no bound states are expected, a resonant p-state might arise within the valence band. This resonant p-state would be like the resonant d-states that we disussed in copper. He proposed this resonance to be the origin of the large hump in the soft x-ray emission of lithium. In fact the argument which we gave above, for the weakening of the pseudopotential due to the change in core eigen-values, does not apply to p-states in lithium since there are no core p-states. On the other hand, in sodium, which has a core p-state, we should expect the weakening of the change in pseudopotential as we discussed before, and should certainly not expect a p-resonance even if one were to occur in lithium. Allotey's calculations in fact indicated that such a p-resonance can occur in lithium but not in sodium. It is not clear how reliable these calculations are but the corresponding result is certainly conceivable. Thus, such an effect would be a special case restricted to the second row of the periodic table.

There is a second many-particle effect arising from the empty core-state; it was discussed earlier by Friedel (1952) but has only recently been solved. This is the question of *Fermi-gas projections*. In calculating the intensity, as in Eq. (7), we have written a matrix element of $p \cdot A$ between one-electron states. In fact, of course, even in the self-consistent-field approximation, we should use the many-electron states, product states, or Slater determinants. The assumption required in using the one-electron expression, of Eq. (7), is that all of the other occupied states (aside from the valence-band state which is to be emptied and the core state which is to be filled) remain essentially unchanged during the transition, and that the overlap between these other states (the Fermi-gas projection) simply gives a factor of unity.

Since in the transition a localized core-state becomes filled, there will be some change in the remaining states. In particular, in the presence of the core hole, there will be a non-zero phase shift in all of the valence-band states before transition and no phase shift afterwards. When no bound state is formed we may show that the overlap for each valence-band state differs from unity by a term of the order of the reciprocal of the number of atoms, $1/N$. However, the product of $\sim N$ such overlaps enters the Fermi-gas projection and it is not clear whether the result is near one or not. Friedel suggested that it was, but he was unable to evaluate the overlap explicitly. Very recently, Anderson (1967, with unpublished communication) addressed this problem in a different context and found that the total overlap is approximately given by

$$\frac{1}{N^{[\pi^{-2}\sum_l (2l+1)\delta_l^2]}}$$

The δ_l are characteristic phase-shifts for the various angular-momentum quantum numbers.

This is a most remarkable result. Clearly the exponent is positive and non-zero. Thus the Fermi-gas projections are of the order of one divided by the number of atoms present raised to a non-zero power. In the limit of a large number of particles, which is what we ordinarily consider in solid state physics, the expression approaches zero. We conclude that these Fermi-gas projections profoundly modify the transition probabilities.

However, we must look at the exponent more carefully. The $l = 0$ phase-shifts tend to be the largest. If they lead to a δ_0 of the order of $\pi/10$, we see that the exponent is of the order 0·01. Then even with 10^{22} particles present the denominator becomes approximately $10^{1/5}$, which is of order unity. The effect may be significant in ordinary materials but at the same time it is not overwhelming, as we might guess from a casual glance at the high-N limit. Furthermore, the Fermi-gas projections will tend to be roughly the same for transitions to all states in the valence band and therefore can, at most, scale the absolute intensities. This conclusion depends on our knowledge that in the simple metals the phase shifts tend to be rather small and to drop rapidly with increasing angular momentum.

On the other hand, if there exist p-state resonances in a metal, such as in lithium, the phase shifts will be π above resonance and the exponent becomes near unity for the state $l = 1$, i.e. for an appreciable fraction of the states. In this case the Fermi-gas projections could become small and the transitions would be greatly inhibited. However, we should expect transitions from the resonant p-states to the core to be less inhibited since in these the filling of the core hole is accompanied by the emptying of a rather well-localized state. Thus, the difference in phase shifts between the states before and after absorption becomes smaller and the Fermi-gas projections become nearer to unity. This would tend to enhance the prominence of the resonance peak in the intensity if such resonances do occur. Such an enhancement might also be expected for absorption by the resonant d-state in noble and transition metals.

We conclude that most of the features of the soft x-ray emission should be understandable in terms of the simple one-electron band-structure point of view. We expect that effects beyond this one-electron picture may cause scaling of the emission bands and even appreciable distortion. They can also, as is well known, give rise to the Auger tail and to plasmon satellites. Finally these effects may in some special cases, such as lithium, cause a profound modification in the observed intensity if resonance states occur such as those proposed by Allotey. Of course we have not in any sense considered all possible many-electron effects and it is possible that other effects not considered here may become important in special systems.

5. CONCLUSION

If there is any central message in the various considerations given here, it must be that soft x-ray spectra do not give a direct measure of the density of valence-band states in the metal as at times has been supposed. In this context it is important that the experimentalist takes care to present his results in terms of the physical quantities that are measured, and not in terms of the density of states which some theorists might hope he could measure. It is appropriate, where possible, to reduce the data by removing extraneous effects which are understood to some degree, such as the Auger tail and the plasmon satellites. On the other hand, it is not appropriate to attempt to remove from the experiments the dependence on the oscillator strength in order to obtain a density of states. In fact, it is more nearly the oscillator-strengths which are measured than the density of states, and the two are in any case strongly intertwined.

To the extent that many-electron effects can be eliminated, the experiments do lead us to the summation, Eq. (7), of the oscillator-strengths over the states in a narrow energy-range. It is then the theoretician's job to understand or predict *this* function of energy, and not the density of states which may have conceptual simplicity but which has little direct bearing on observable experimental quantities.

Acknowledgement

This work was supported by the Advanced Research Project Agency, through the Centre for Materials Research at Stanford University.

References

Allotey, F. K. (1967). *Phys. Rev.* **157**, 467.
Anderson, P. W. (1967). *Phys. Rev. Letters* **18**, 1049.
Friedel, J. (1952). *Phil. Mag.* **43**, 1115.

Harrison, W. A. (1966). *In* "Pseudopotentials in the Theory of Metals". Benjamin, New York.

Herring, C. (1940). *Phys. Rev.* **57**, 1169.

Kittel, C. (1963). *In* "Quantum Theory of Solids", p. 343. Wiley, New York.

Schiff, L. I. (1949). *In* "Quantum Mechanics", p. 247. McGraw-Hill, New York.

Wilson, A. H. (1936). *In* "The Theory of Metals", 1st Ed., p. 140. Cambridge University Press.

Density of States in Simple Metals and the Soft **X**-Ray Spectrum

N. W. ASHCROFT

Laboratory of Atomic and Solid State Physics, Cornell University, Ithaca, New York, U.S.A.

ABSTRACT

The general one-electron pseudopotential for simple metals is developed and discussed in terms of its inherent energy-dependence and of the subsequent effect on the density of states. Analysis of the soft x-ray emission process is most easily accomplished in terms of a pseudopotential for the crystal whose matrix elements separate into the product of a structure factor and an individual ionic-pseudo-potential. The effect on the latter, of removing a core state, changes the pseudo-potential by approximately the amount that the actual potential is altered. Possible effects of an assembly of these modified pseudopotentials on the observed emission-spectrum (via the band structure) are discussed in terms of a local model-pseudopotential method; the latter is also used to calculate the density of states in indium.

1. INTRODUCTION

It is well known that upon formation of the metallic state the atomic valence wavefunction undergoes substantial changes outside the atomic-core region but inside is relatively unchanged. Consequently, the metal valence wavefunction in the core preserved the oscillatory behaviour there; a characteristic of the atomic state which results from the requirements of orthogonality to the lower core-states. The valence wavefunction for the crystalline solid conforms to the Bloch-type

$$\psi_k(r) = e^{ik \cdot r} u_k(r) \tag{1}$$

where $u_k(r)$ has the periodicity of the lattice and, as we have indicated, must vary sharply near each atomic core. If we attempt to expand the wave-function (1) as a Fourier series over reciprocal lattice vectors G,

$$\psi_k(r) = \sum_G C_G e^{i(k+G) \cdot r} \tag{2}$$

then, because of the rapid variation of the wavefunctions near each lattice site, we expect a very poorly convergent sum. Values of $(k+G)$ such that

249

$\dfrac{\hbar^2}{2m}(k+G)^2$ is of the order of an x-ray energy must be included. At first sight it is somewhat contradictory to find that analysis of Fermi-surface data leads to the opposite conclusion; i.e. that the wavefunctions of valence electrons in simple metals may be expressed as the sum of a very few plane-waves; in fact a single plane-wave is a good representation throughout most of the zone, which follows from the experimental finding that the Fermi surfaces in these metals are remarkably spherical.

The resolution of this paradox relies on a correct interpretation of the data. It is clear that the difficulty is in accounting for the oscillations in the wavefunction, which are evident in the core regions. The suggestion by Phillips and Kleinman (1959) and Kleinman and Phillips (1960) that the valence electrons can actually be regarded as moving in a different effective potential, with associated nearly-plane-wave-like states, gave considerable impetus to the understanding of the nearly-free-electron behaviour of metals in a one-electron picture. Their argument is based on the concept of eliminating (from the outset) the effect of the core states. For the crystal Hamiltonian we write

$$H = T + V$$

where V is the total self-consistent potential seen by the valence electrons. If $\psi_k(r)$ are the valence states with energies ϵ_v then

$$H\psi_k(r) = \epsilon_v \psi_k(r) \qquad (3)$$

Interstitially we have an approximately constant potential and consequently a plane-wave-like solution of (3). To approximate the core solution of $\psi_k(r)$, Phillips and Kleinman suggest writing it as a summation of the wavefunctions appropriate to the tightly bound core-states of H. Thus after orthogonalizing valence and core states

$$\psi_k(r) = \phi_k(r) - \sum_c (\psi_c, \phi_k)\psi_c$$

where the ψ_c are the Bloch core-states (i.e. $H\psi_c = E_c\psi_c$), and $\phi_k(r)$ is a smooth wavefunction. Substituting in (3), and rewriting in terms of ϕ_k, we find

$$H_{ps}\phi_k = \epsilon_v \phi_k$$

where

$$H_{ps} = H + V_R = T + (V + V_R).$$

We write

$$V_{ps} = V + V_R = V + \sum_c (\epsilon_v - E_c)|c\rangle\langle c|$$

$$= V + \sum_{n\alpha} (\epsilon_v - E_n)|\alpha n\rangle\langle n\alpha| \qquad (5)$$

where $|n\alpha\rangle = \phi_n(r - r_\alpha)$ is a tight-binding atomic-like state situated on the lattice site r_α. This latter approximation is appropriate to simple metals. Taking diagonal matrix-elements of (5), it is clear that the second term in (5) substantially cancels the largely attractive potential V so that $V + V_R$ tends to be small, and perturbation theory can be carried out using a plane-wave expansion for ϕ_k;

i.e.
$$\phi_k = \sum_G a_G e^{i(k+G)\cdot r} \tag{6}$$

By construction, the oscillations have been removed from $\phi_k(r)$ using the transformation (4), hence we expect a much more quickly convergent sum (6), in agreement with the interpretation of the Fermi-surface data in terms of the nearly-free-electron model.

Now the quantity $V + V_R = V_{ps}$ is not uniquely defined by the requirement that the pseudo-wavefunction is a synthesis of the real valence-state with a summation over the core-space states. In fact, consider any Hamiltonian H_{ps} written as follows:
$$H_{ps} = H + P_C \Xi$$
where P_C is a projection operator onto the entire core space, and Ξ is an arbitrary linear operator, or sum of linear operators with the same primitive translations as H. We can show that H_{ps} gives the same valence energy value as H; suppose
$$H_{ps}|\phi_k\rangle = \bar{\epsilon}_v|\phi_k\rangle$$
$$= H|\phi_k\rangle + \sum_c |\psi_c\rangle\langle\psi_c|\Xi|\phi_k\rangle$$
we take the scalar product with $\langle\psi_k|$, and find
$$\epsilon_v\langle\psi_k|\phi_k\rangle = \bar{\epsilon}_v\langle\psi_k|\phi_k\rangle$$
Hence, if $\langle\psi_k|$ and $|\phi_k\rangle$ are not orthogonal (which they are not, by construction), then this Hamiltonian produces the same valence energies. Our most general pseudopotential[†] may therefore be taken as
$$V_{ps} = V + P_C\Xi$$

Now for the real potential V, which in the single-particle picture is local, we always find that for plane-wave states $|k\rangle$
$$\langle k'|V|k\rangle = \rho_{k'-k}v_{k'-k} \tag{7}$$

[†] It is obvious that non-diagonal forms
$$V + P_C\Xi = V + \sum_{c,c'} |c\rangle\langle c'|\Xi$$
also give the same valence energies. Use of this pseudopotential does not modify the conclusions based on the diagonal form.

where, the $v(r)$ are individual electron–ion potentials from which V is composed, i.e. $V = \sum_\alpha v(r-r_\alpha)$, and $\rho_{k'-k} = \sum_\alpha e^{-i(k'-k)\cdot r_\alpha}$ is a density fluctuation of the ions with wave vector $k-k'$. For a periodic lattice $\rho_{k'-k} = N\delta_{k'-k,G}$, where G is a vector in the reciprocal lattice and N is the number of ions. Similarly let us consider

$$\langle k'|P_c\Xi|k\rangle = \sum_{ck''}\langle k'|\psi_c\rangle\langle\psi_c|k''\rangle\langle k''|\Xi|k\rangle \tag{8}$$

where we have introduced a complete set $|k''\rangle$. Under certain conditions we may write this in the form of (7). If, for instance, Ξ happens to be diagonal in the core states of the original crystal Hamiltonian, then we can separate (8) as

$$\langle k'|P_c\Xi|k\rangle = \rho_{k'-k}v_{k',k}$$

An example is the Phillips–Kleinman form (PK), where we have $\Xi = (\epsilon_v - H)$ giving

$$\langle k'|P_c(\epsilon_v - H)|k\rangle = \rho_{k'-k}v^{PK}_{k'-k}$$

with

$$v^{PK}_{k',k} = \sum_n \langle k'|n\rangle\langle n|k\rangle(\epsilon_v - E_n)$$

which may be regarded as the matrix element of a single-ion pseudopotential. This follows in the case of a regular solid composed of ions with tightly bound core-states, because we can write $P_C = \sum_{\alpha n}|\alpha n\rangle\langle n\alpha|$.

Then

$$\langle k|n\alpha\rangle = e^{-ik\cdot r_\alpha}\langle k|n\rangle$$

and in general

$$\langle k'|P_c\Xi|k\rangle = \sum_\alpha\sum_n\sum_{k''}\langle k'|n\rangle\langle n|k''\rangle\langle k''|\Xi|k\rangle e^{i(k''-k')\cdot r_\alpha} \tag{9}$$

When k' and k differ by G (a reciprocal lattice vector), the matrix element of the local potential is simply V_G which is related to the usual band gaps;

i.e.

$$\langle k'|V|k\rangle = v_G \qquad (k' = k+G)$$

since, for simple lattices, $\rho_G = N$.
For the general pseudopotential (9) we find

$$\langle k'|P_c\Xi|k\rangle = \sum_n\sum_{k''}\langle k'|n\rangle\langle n|k''\rangle\langle k''|\Xi|k\rangle\sum_\alpha e^{i(k''-k-G)\cdot r_\alpha}$$

$$= \sum_n\sum_{k''}\langle k+G|n\rangle\langle n|k''\rangle\langle k''|\Xi|k\rangle N\delta_{k''-k-G,G'}$$

$$= \sum_{G'}\sum_n\langle k+G|n\rangle\langle n|k+G'\rangle\langle k+G'|\Xi|k\rangle N$$

which involves matrix elements evaluated for states with wave vectors separated not only by the particular G but by all the reciprocal-lattice vectors.

The so-called Austin-pseudopotential (Austin *et al.*, 1962), which is known to lead to a smooth pseudo-wavefunction, is obtained by setting $\Xi = -V$, for which

$$\langle k + G'|V|k\rangle = v_{G'}$$

and

$$\langle k'|P_C(-V)|k\rangle = N \sum_{G'} \sum_n \langle k + G|n\rangle\langle n|k + G'\rangle v_{G'}$$

$$(k' = k + G)$$

Again this is not simply of the form $\rho_G V_G$ involving a single Fourier-component; moreover, for an arbitrary Ξ which satisfies the pseudopotential transformation, we *cannot* always write

$$\langle k'|V_{\text{ps}}|k\rangle = \rho_{k'-k}V_{k',k}$$

where $V_{k',k}$ is a pseudopotential associated with a single ion. However, the Phillips–Kleinman form, while *a priori* not yielding the smoothest wave-function, does lead to a simple structure-factor separation, and this is convenient for a discussion of the soft x-ray emission process.[†]

2. ENERGY DEPENDENCE

We now wish to discuss the energy dependence of the pseudopotential, $V_{\text{ps}} = V + P_C \Xi$, and the effect on the density of states. If we take, as pseudo-valence wavefunctions, a sum

$$\phi_k = \sum_G C_G e^{i(k + G)\cdot r}$$

[†] As another example, if we choose the operator Ξ to be $\lambda_c H$ with arbitrary coefficients, $\lambda_c = A_c/E_c$, then V_{ps} takes the form

$$V_{\text{ps}} = V + \sum_c |c\rangle\langle c|\lambda_c H = V + \sum_{n\alpha} |n\alpha\rangle\langle \alpha n|A_n$$

The last step follows from the usual tight-binding approximation made above. For this potential we find

$$\langle k'|V_{\text{ps}}|k\rangle = \rho_{k'-k}(v_{(k'-k)} + \sum_n \langle k'|n\rangle\langle n|k\rangle A_n)$$

which therefore leads us to consider individual ion pseudopotentials

$$v_{\text{ps}} = v + \sum_n |n\rangle\langle n|A_n$$

Heine and Abarenkov (1964), in an approximation to this form, make the usual identification of the cancellation of the real potential v by the second term of this equation to be occurring mostly within the core region; i.e. in $r < R_M$ say. They propose a model potential

$$v_M = \sum_n |n\rangle\langle n|A_n \quad \text{for } r < R_M \qquad \text{and} \qquad v_M = -Ze^2/r \quad \text{for } r > R_M$$

where the A_n are now specifically calculated from an application of the same form to known atomic spectroscopic properties.

then this leads us to a secular equation

$$\begin{vmatrix} k^2 - (\epsilon_v + \langle k|V + P_c\Xi|k\rangle) & \langle k|V + P_c\Xi|k - G\rangle \cdots \cdots \\ \langle k - G|V + P_c\Xi|k\rangle & (k - G)^2 - (\epsilon_v + \langle k - G|V + P_c\Xi|k - G\rangle) \\ \vdots & \cdots \cdots \end{vmatrix} = 0$$

We can choose here to set the zero of energy by writing

$$E = \epsilon_v + \langle k|V + P_c\Xi|k\rangle \qquad (\hbar^2/2m = 1)$$

and we define

$$\Delta_{k,G} = \langle k - G|P_c\Xi|k - G\rangle - \langle k|P_c\Xi|k\rangle$$

and

$$V_{k,k-G} = \langle k|P_c\Xi|k - G\rangle$$

so that the equation is

$$\begin{vmatrix} k^2 - E & V_G + V_{k,k-G} \cdots \cdots \\ V_G + V_{k-G,k} & (k - G)^2 - E - \Delta_{k,G} \\ \vdots & \cdots \cdots \end{vmatrix} = 0$$

which normally is simply approximated, in the pseudopotential interpolation scheme, by

$$\begin{vmatrix} k^2 - E & U_G \cdots \cdots \\ U_G: & (k - G)^2 - E \\ \vdots & \cdots \cdots \end{vmatrix} = 0$$

However, note that, although the $\Delta_{k,G}$ are assumed to be zero, the fact that E is not the real valence-energy leads to a correction to the density-of-states function. A direct evaluation of the number of states enclosed by two nearby surfaces of constant energy in k space, defined by E and $E + \Delta E$ in the secular equation, simply involves a calculation of the volume change $\Delta \mathscr{V}$; and gives

$$N(\epsilon) = \frac{1}{4\pi^3} \left|\frac{\Delta \mathscr{V}}{\Delta E}\right| \cdot \frac{dE}{d\epsilon}$$

$$= N(E)\left(1 + \frac{d}{d\epsilon}\langle k|P_c\Xi|k\rangle\right)$$

which may therefore lead to a general scaling of the band. Energy dependence also appears in the U_G: hence if the U_G are determined at one particular energy (say ϵ_F) they will not necessarily be the same as another. A direct consequence is that in the pseudopotential approximation the band gaps will vary across the appropriate zone faces.

For the particular case of the Phillips–Kleinman potential we find

$$V_{k,k-G} = \sum_n (\epsilon_v - E_n)\langle k|n\rangle\langle n|k - G\rangle = V^*_{k-G,k}$$

and the PK potential is Hermitian within degenerate valence-states. Further we see that

$$\Delta_{k,G} = \sum_n (\epsilon_v - E_n)[\langle k-G|n\rangle\langle n|k-G\rangle - \langle k|n\rangle\langle n|k\rangle]$$

so that to a good approximation $\Delta_{k,G}$ is negligible when k satisfies the Bragg conditions. Thus the secular equation appropriate to a weak potential is suited (and indeed has more validity) to problems involving elastic or near-elastic scattering of electrons. The factor scaling $N(E)$ is given by

$$1 + \frac{\mathrm{d}}{\mathrm{d}\epsilon_v} \sum_n (\epsilon_v - E_n)\langle k|n\rangle\langle n|k\rangle$$

$$\doteqdot 1 + \sum_n \langle k|n\rangle\langle n|k\rangle + \sum_n (\epsilon_v - E_n)\frac{\mathrm{d}}{\mathrm{d}k}|\langle k|n\rangle|^2 \frac{\mathrm{d}k}{\mathrm{d}\epsilon_v}$$

which may be estimated knowing the various overlaps involved. It is well known, from constructing the normalizations of OPWs, that the second term may be as large as 0·1. We shall see in the next section that the overlaps for metals with tightly bound cores are very slowly varying functions of k for $k \langle k_F$. We also expect them to decrease with increasing k, hence the third summation is small and partially cancels the effect of the first summation.

3. PSEUDOPOTENTIAL DESCRIPTION OF SOFT X-RAY EMISSION

The conventional description of the soft x-ray emission process portrays a "prepared" hole state in an inner core-shell of the metallic ion, which is subsequently filled by an electron making a transition from the valence band and giving up a photon. In view of the fact that at the "prepared" ion the potential seen by the valence electrons may be substantially different we must examine the mechanism a little more closely. We know for example that the valence electrons respond to the new potential (and may indeed contribute to it): what is less clear is the characteristic time τ required for the polarization charges to be set up around the ionized site. We can broadly distinguish between two classes of emission; namely, "fast" in which the transition rate is larger than the inverse of τ, and "slow" in which it is smaller.

A rough estimate of the rate at which polarization charge clouds are established is simply $\epsilon/\hbar \approx 5 \times 10^{16}/r_s \sec^{-1}$ where ϵ is a typical energy of interaction within the electron gas (say $\epsilon \approx 27/r_s$ ev) and r_s is the usual electron-spacing parameter. In the case of fast emission (i.e. for rates $> 5 \times 10^{16}/r_s \sec^{-1}$, which might occur with K-shell emission for example), the hole is created (giving the "prepared" state) but is filled again by one electron from the valence band before the rest of the electrons in the band respond to the new field. The hole that is left in the valence band disperses fairly slowly, either by Auger-type transitions or by "bubbling" up to the top of the Fermi distribution under the action of the same interactions that establish the polarization clouds around the ionized site. The dispersion is therefore also slow on the scale of the transition rate, which means that the emitted photon can in principle be partly re-absorbed with the hole, subsequently re-emitted, and so on. For a single-hole state in the valence band, this process is not likely to affect significantly the lifetime of the photon. A more important broadening process is the excitation by the photon of the higher core-level electrons into the vacant states of the valence band (e.g. a K-emission photon is broadened by excitation and reabsorption of L-shell electrons into band states above ϵ_F.) A simple calculation (see Appendix A) shows the lifetime to be comparable to the observed broadening. We shall not go into the details of the processes involved but the observation can be made that, because of the broadening, no significant sharp cutoff should appear at the Fermi-energy.

In "slow" emission the polarization of the valence band electrons is essentially accomplished before the emission process occurs. In contrast with "fast" emission, the hole state in the core is now appropriate to a solution of a Hamiltonian in which the Hartree–Fock field incorporates the redistribution of the valence electrons. In addition, the transition rate requires a knowledge of the wavefunctions of the valence-band state and core state, both evaluated for the new configuration. Expressed as a final-state problem, the observed intensity must depend on the density of hole-states in the valence band weighted by this transition probability. We remark that in addition to modifications from Auger transitions, the intensity may also require correction for lifetime broadening arising from the virtual excitations of *any* core states that lie above the one being filled.

Many observed spectra fall into the category of "slow" emission and it is this class that we shall examine by the pseudopotential method. It is clear that in "fast" emission, different approximations apply. There is no reason, therefore, to suppose that the intensities appropriate to a particular valence band should strictly reflect the same structure in both classes of emission.

Let us use a pseudopotential for which we may make the simple structure-factor separation (e.g. the P-K type). Then the total pseudopotential is written

$$V = \sum_{\alpha \neq i} v(\mathbf{r} - \mathbf{r}_\alpha) + \omega(\mathbf{r} - \mathbf{r}_i)$$

where $\omega(\mathbf{r} - \mathbf{r}_i)$ is the potential seen by a valence electron as contributed by the ionized site. Rewriting this as

$$V = \sum_{\alpha} v(\mathbf{r} - \mathbf{r}_\alpha) + \Delta v(\mathbf{r} - \mathbf{r}_i)$$

where,

$$\Delta v = \omega(\mathbf{r} - \mathbf{r}_i) - v(\mathbf{r} - \mathbf{r}_i)$$

then

$$\langle k'|V|k \rangle = \rho_{k'-k} v_{k'-k} + e^{-i(k'-k)\cdot r_i} \Delta v_{k',k} \tag{10}$$

We may, of course, following the steps of Section 2, carry out the pseudo-potential transformation on the Hamiltonian appropriate to the inclusion of the ionized site and obtain $H^i_{ps}\phi' = \epsilon_{v'}\phi'$; where $\epsilon_{v'}$ refers to the valence-state energies in the presence of the ionized sites. Here

$$H^i_{ps} = \sum_{n'}\sum_{\alpha \neq i} (\epsilon_{v'} - E_{n'})|n'\alpha\rangle\langle\alpha n'| + \sum_{n''} (\epsilon_{v'} - E_{n''})|n''i\rangle\langle in''|$$

In the tight-binding approximation for the core states the primes, on the n and the v in the first term, indicate the new valence and core states evaluated with a hole in the core of one atom. For "fast" emission we expect the valence states to be virtually unchanged from the original states. For "slow" emissions there may be small changes in all core states.

In the second term, however, we expect the Hartree–Fock field in the ionized atom to be substantially changed, and the core states and energies will be different from those in the undisturbed ions even for "fast" emission. Furthermore, if the states are determined self-consistently (i.e. including effects of the new valence states), then for "slow" emission where we include the effects of polarization in the valence electrons to the new field, the states n'' and energies $E_{n''}$, required for the evaluation of transition probabilities, will not be the same as those evaluated for "fast" emission.

It is important to know the effect of the additional potential Δv at the ionized site, and to know how it behaves when viewed as a pseudopotential. The individual pseudopotential for the site is

$$v^i_{ps} = v^i + \sum_{n''} (\epsilon_{v'} - E_{n''})|n''\rangle\langle n''|$$

where the v^i is now the Hartree–Fock or self-consistant potential for the atom, with an additional hole-state in one of the core shells. It clearly takes the form of a screened version of $-\dfrac{(Z+1)e^2}{r}$ outside the core region, and is more complicated inside but obviously different from the situation appropriate to no hole-states.

We wish to compare this with $v_{ps} = v + \sum_{n} (\epsilon_v - E_n)|n\rangle\langle n|$ and shall do so by considering the diagonal matrix-elements, i.e.

$$\langle k|v^i_{ps}|k \rangle = \langle k|v^i|k \rangle + \sum_{n''} (\epsilon_{v'} - E_{n''})\langle k|n''\rangle\langle n|k\rangle$$

and

$$\langle k|v_{ps}|k \rangle = \langle k|v|k \rangle + \sum_{n} (\epsilon_v - E_n)\langle k|n\rangle\langle n|k\rangle$$

where $0 < k < k_F$ is a reasonable range of wave vectors.

To arrive at a meaningful comparison, we first assume ϵ_v to be essentially the same as $\epsilon_{v'}$ (this is reasonable for the factor $\epsilon_v - E_n$, relative to the core energies concerned). We should like to know how the core energies change when a core-state electron has been removed. This question has been discussed by Bagus (1965) for several elements including sodium, which we consider here as an example. The details pertaining to the change in the pseudopotential are given in Appendix B, but the salient points are as follows. The factors $(\epsilon_v - E_n)$ substantially increase through the states n as might be anticipated. However, the overlaps $\langle k|n \rangle$ *decrease* (since the core states become more compact) and do so in a roughly compensating manner. As a result, the pseudopotential at the ionized site increases by the amount by which the *real* potential changes, i.e. in its unscreened form, from $\dfrac{-Ze^2}{r}$ to $\dfrac{-(Z+1)e^2}{r}$ outside the ion core, and by a substantial amount inside the core. Solutions to the same problem, with an extra charge distributed over a sphere *outside* the ion, have also been given by Sacks (1961). Changes in the core-state functions and levels are small and, using this is a model for inclusion of the polarization cloud, our basic conclusion about a strong additional pseudopotential existing at the ionized site is unaltered.

If we now look at the scattering in terms of the expected phase-shifts δ_l, the ionized site looks asymptotically like an impurity of valence $(Z+1)$. In the Born approximation the quantities $\tan \delta_l$ are integrals of the Fourier transform of the scattering potential, and if the screening remains essentially fixed each $\tan \delta_l$ suffers a change by a factor $\dfrac{Z+1}{Z}$; a result which satisfies the Friedel rule (Friedel, 1952) provided the δ_l are all small.

4. STRUCTURE IN THE EMISSION SPECTRUM

We now wish to use a simple model to analyse the expected structure in soft x-ray emission in terms of the ideas expressed above, but also incorporating one further physical condition. We refer to the fact that soft x-ray emission, due to the large absorption coefficients found in metals, principally occurs at the metal–vacuum interface. However, it is the surface layers there that are exposed to the primary electron-beam, the latter normally being intense in order to achieve high counting-rates of the x-ray emission after subsequent diffraction by the analysing crystal. It follows that at any instant of time an atom emitting a soft x-ray photon which penetrates the surface does so in the presence of a considerable number of ionized neighbours. The precise number of ionized sites in the surface layers is difficult to estimate, but it is not a negligible fraction of the number of ions within a characteristic absorption distance. Parratt's opinion (personal communication) is that for some metals this may run as high as 50%, depending on the type of state

(K, L, etc.) and on the experimental conditions. Consequently, it is difficult to see how the measured intensity can produce the combined transition-rate and density-of-states factor for the bulk of the sample. We can indicate this difficulty by analysing the problem in terms of a local model-pseudopotential, as follows.

As in most of the nearly-free-electron (NFE) calculations throughout this text, it is convenient to take the valence pseudo-wavefunctions as plane waves. This state, being the smoothest approximation to the actual pseudo-wavefunction, has as a natural counterpart the "Austin" form for the pseudopotential; that is,

$$V_{ps} = (1 - P_C)V \qquad (11)$$

If the core states were complete by themselves this would lead to ideal cancellation. As it is, they are not complete over the space of the whole crystal, but according to the usual pseudopotential argument they form a reasonably complete set over the volume of the core space. Therefore, using the cancellation principle outlined earlier, and Eq. (11), we may regard the potential to be effectively eliminated within the core volume and use a simple model-potential of the type

$$v(r) = 0, \quad r < R_{core}$$
$$v(r) = -Ze^2/r, \quad r > R_{core} \qquad (12)$$

where R_{core} (defining the region of optimum cancellation) must be of the order of the usual ionic radius. Regarding Eq. (12) as a basic single-ion pseudopotential, we find its screened Fourier transform to be

$$u(x) = (\tfrac{2}{3}\epsilon_F)\left[\frac{-\lambda^2 \cos Sx}{x^2 + \lambda^2 f(x)}\right] \qquad (13)$$

where $\lambda^2 = (\pi a_0 k_F)^{-1}$, $x = k/2k_F$, $S = 2k_F R_{core}$; and $f(x)$ is the usual Lindhard-function for linear screening, modified by an exchange correction whose precise form is not essential here. Actual calculations of metallic properties, using Eq. (13), have been reasonably successful and Fermi-surface studies have shown R_{core} to be indeed close to the ionic radius. Plots of pseudopotentials obtained using Eq. (13) are shown in Fig. 1, for Al, Pb, and In.

Let us suppose, now, that some of the sites are ionized. In terms of the potential (12) two changes must be made: firstly, as we have seen, the states inside the core become more compact, hence we expect R_{core} to decrease; secondly, outside we have $-(Z+1)e^2/r$ for the potential. The second change increases the scale of $u(x)$, around $x \approx 1$, by a factor $(Z+1)/Z$; whereas the first, assuming the valence-electron density is fixed, moves the first node of (13), say x_0, outwards. Now the conventional semi-band gaps u_G are just ordinates on the curve (13), evaluated at $x_G = G/2k_F$. Hence if $x_G < x_0$ the two shifts reinforce each other, whereas if $x_G > x_0$ they partially cancel.

Within the surface layers, the appropriate band-structure is determined by a weight of the pseudopotentials for the ionized sites, the proportions being determined through a structure factor appropriate to the fraction of ionized sites (see eq. (10)). Metals whose principal reciprocal-lattice vectors x_G satisfy $x_G > x_0$ include Na, K, and Al, whereas those with $x_G < x_0$ include In and Li

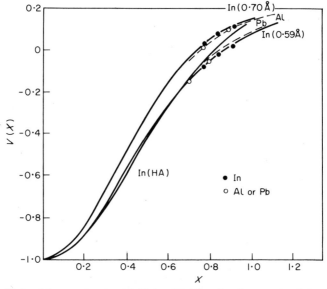

FIG. 1. Plots of the pseudopotential $V(x) = U(x)/\frac{2}{3}\epsilon_F$ given by equation (13), with values of R_{core} adjusted to make the curves pass through the points determined by Fermi-surface-distortion data (Al, $R_{\text{core}} = 0.59\text{Å}$; Pb, $R_{\text{core}} = 0.58\text{Å}$; In, $R_{\text{core}} = 0.58\text{Å}$). We also show the sensitivity of the curves to the value of R_{core} by plotting the In curve with a value $R_{\text{core}} = 0.70\text{Å}$.

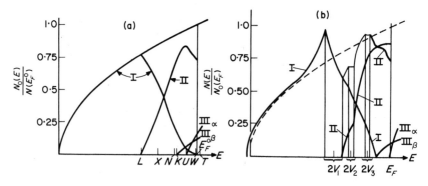

FIG. 2. (a) The contribution to the free-electron density of states from the first, second, and third Brillouin-zones. The latter is further subdivided into contributions from α arms (III$_\alpha$), and β arms (III$_\beta$). All the contributions are normalized to the total free-electron density of states at the Fermi energy. (b) Approximate density-of-states function for In obtained from the curves in (a), as outlined in the text.

(Ashcroft, 1968). In the band structures we expect the perturbations from the ionized sites to be small for the former group of metals, whereas from the latter they may be significant, especially if, as we indicate, the region of metal being probed contains a large proportion of ionized sites. Hence the following model-calculation of the density of states for In may require substantial correction when interpreted as an intensity.

R_{core} for In has been determined from the available Fermi-surface data (O'Sullivan et al., 1967; Ashcroft and Lawrence, to be published). The components $V_{11\alpha}$, $V_{2\alpha00}$ are both negative, and V_{200} is weakly positive. In the NFE picture, we evaluate the density of states

$$N(E) = \frac{1}{4\pi^3} \int_{S(E)} \frac{dS}{|\nabla_k E_k|}$$

zone by zone, by first approximating $\nabla_k E$ by a constant and evaluating the fractional surface area in each Brillouin-zone. This calculation is a straight-forward exercise in geometry and the result is shown in Fig. 2(a). To correct for the fact that $|\nabla_k E|$ is not constant, we use the result, first obtained by Jones (1937), that near a Brillouin-zone plane with free electron energy $E_{G/2} = \frac{\hbar^2}{2m} \frac{G^2}{4}$, the density of states to first order in $|V_G|$ at $E^{\pm} = E_{G/2} \pm |V_G|$ is simply

$$N(E^{\pm}) = N_0(E^{\pm}) \left[1 \mp \frac{|V_G|}{2E_{G/2}} \right]$$

By approximating the secular equation, resulting from a second-order determinant, a smooth interpolation away from the plane was also obtained by Jones and hence by subtraction the net effect of a single-zone plane. When the effects of the nearest set of zone planes is summed the resulting density-of-states curve is as shown in Fig. 2.

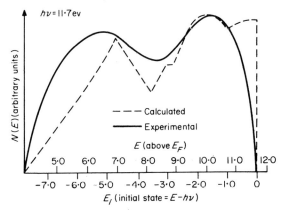

FIG. 3. Spectral intensity of photo-emitted electrons from In, compared with the curve (b) in Fig. 2 multiplied by a transmission function for the surface.

If we wish to make a comparison with experiment (with the usual assumption of a constant transition-rate) the only data available are those of Noreland (1964). These are measurements of L emission from In, and the energies involved are >3 kev. Very little structure is seen although the bandwidth for the L_3 emission is approximately correct. However, some structure is seen in the optical density of states as determined from photo-emission data, and for the purposes of comparison we show in Fig. 3 the curve of Fig. 2(b) multiplied by a transmission function and plotted along with the data of Koyama et al. (1967, 1968). We note that the gross structure, resulting mainly from the fairly large value of $V_{11\alpha}$, coincides reasonably well. Some slight scaling of the energy E is required to make the two curves match in the position of the main peaks. This we interpret as the factor $(1 + \dfrac{\mathrm{d}}{\mathrm{d}\epsilon}\langle k|P_C\Xi|k\rangle)$ mentioned earlier; its actual value corresponds to an effective mass $\dfrac{m^*}{m} \doteq 0.85$, although this number must be regarded as approximate because the structure itself depends on the magnitude of the band gaps which, as we have indicated, are functions of the energy E.

5. RECAPITULATION

The nearly-free-electron model, with its obvious simplifications resulting from the use of plane-wave-like valence states, can be established in terms of a pseudo-Hamiltonian in which the effective potential is weak. The benefit of this small expansion parameter comes at the expense of extra non-local and energy-dependent terms in the formalism. These contribute to the density of states, both in a scaling of the overall bandwidth and also in a dependence of the conventional band-gaps on valence-state energy. For simple metals, estimates of the density-of-states function appropriate to a given model-pseudopotential can now be made, but a meaningful comparison with experimental intensities can only be achieved after, (1) allowing for the energy dependence of the transition probability, and (2) including the potential changes that occur at the ionized sites. We have not discussed (1) here; the calculation of this energy-dependence will depend on the effect of (2), and we simply remark that an accurate assessment of the number of ionized sites in the absorption layer of the sample is necessary, in order to invert the measured intensity and hence deduce information regarding the form of the bands in the metal.

Acknowledgements

This work was supported by the National Science Foundation. I am very grateful for helpful and stimulating discussions with Professor J. W. Wilkins, and Professors L. G. Parratt and J. A. Krumhansl.

References

Ashcroft, N. W. (1968). *J. Phys. C.* (*Proc. Phys. Soc.*), **1**, 232.

Austin, B. J., Heine, V. and Sham, L. J. (1962). *Phys. Rev.* **127**, 276.

Bagus, P. S. (1965). *Phys. Rev.* **139**, A619.

Friedel, J. (1952). *Phil. Mag.* **43**, 1115.

Heine, V. and Abarenkov, I. (1964). *Phil. Mag.* **9**, 451.

Jones, H. (1937). *Proc. Phys. Soc.* **49**, 250.

Kleinman, L. and Phillips, J. C. (1960). *Phys. Rev.* **116**, 880; **117**, 460.

Koyama, R., Spicer, W. E., Ashcroft, N. W. and Lawrence, W. E. (1967). *Phys. Rev. Letters* **22**, 1284.

Noreland, E. (1964). *Ark. Fys.* **26**, 361.

O'Sullivan, W. J., Schirber, J. E. and Anderson, J. R. (1967). *Solid State Commun.* **5**, 525.

Phillips, J. C. and Kleinman, L. (1959). *Phys. Rev.* **116**, 287.

Sachs, L. M. (1961). *Phys. Rev.* **124**, 1283.

Slater, J. C. (1960). *In* "Quantum Theory of Atomic Structure", Vol. 1. McGraw-Hill, New York.

APPENDIX A

The following calculations were stimulated by discussions with Prof. J. W. Wilkins. We imagine an emission photon of energy $\hbar\omega$ appropriate to an inner core-state to suffer life-time broadening by a process involving the excitation of a higher core-state electron (with wave function ψ_n and energy E_n) into a vacant band-state with wavefunction $\psi_p \sim e^{i\boldsymbol{p}\cdot\boldsymbol{r}}$.

Averaging over the various polarizations of the photon, we may write for the lifetime

$$\frac{1}{\tau} = \frac{2\pi}{\hbar} \sum_p (1-n_p)\frac{1}{3}\left|\left\langle \psi_p \left| \frac{e}{mc}\boldsymbol{A}\cdot\nabla \right| \psi_n \right\rangle\right|^2 \delta(\hbar\omega + E_n - E_p)$$

where n_p is the Fermi distribution function. If $\boldsymbol{q}$ is the wave vector of the photon then the above equation becomes

$$\frac{1}{\tau} = \frac{2\pi}{\hbar} \sum_p (1-n_p)(\boldsymbol{p}+\boldsymbol{q})^2 |\psi_n (\boldsymbol{p}+\boldsymbol{q})|^2 A^2 \delta(\hbar\omega + E_n - E_p)$$

where $\psi_n(\boldsymbol{p}+\boldsymbol{q})$ is the Fourier transform of the core state. We note in passing that $q \approx 0.5 \times 10^8 \bar{E}$ cm^{-1} where $\bar{E}$ is the energy of the photon ($\hbar\omega$) measured in kev (hence the transitions are generally "non-vertical"). Converting the $\sum_p$ to an integral, we find

$$\frac{1}{\tau} = \frac{2\pi}{\hbar}\frac{1}{\hbar^3}\frac{(\hbar\omega)^{3/2}}{4\pi^2}(2m)^{5/2}\left(\frac{e}{cm}\right)^2\left(\frac{\hbar c^2}{\omega}\right)M^2$$

where the factor $\dfrac{\hbar c^2}{\omega}$ is the standard term arising from the quantization of the photon field with vector potential A. We have taken $M^2 = |\psi_n(\boldsymbol{p}+\boldsymbol{q})|^2$

to be a constant, which is probably satisfactory in a calculation of this level. We have also neglected E_n in relation to $\hbar\omega$. Finally, after a little rewriting

$$\frac{1}{\tau} \doteq \frac{4\sqrt{2}}{3} \left(\frac{\hbar\omega}{ma_0 2}\right)^{1/2} M^2$$

As an example, $\sum_n |\psi_n(0)|^2$ for Al is about 0·1, most of the contributions coming from the $n = 2$ shell. For high wave-vectors p, the overlap will be much smaller but the fact remains that for a 1-kev photon the factor $\left(\dfrac{\hbar\omega}{ma_0 2}\right)^{1/2}$ is of order 10^{17}/sec which means that the broadening will be significant even for overlap sums of the order of 10^{-3}–10^{-4}.

APPENDIX B

Here we consider as a particular example the change in core-level energies and core-state overlaps when an electron is removed from an inner shell of a sodium ion. According to the self-consistent-field calculations by Bagus the shifts for holes in the K and L shells are as follows:

State shifts	Position of hole	
$(A - U)$	K (%)	L (%)
1s	$-5\cdot06$ (12%)	$-1\cdot10$ ($\sim 10\%$)
2s	$-1\cdot15$ (38%)	$-0\cdot86$ ($\sim 28\%$)
2p	$-1\cdot12$ ($\sim 60\%$)	$-0\cdot94$ ($\sim 50\%$)

Clearly the changes are large and far from constant. They are all negative which indicates, as expected, that $\langle k|v^i|k\rangle$ is less than $\langle k|v|k\rangle$.

We now consider changes in the overlaps. To do so we construct "Slater" wavefunctions, using the "Slater" rules (J. C. Slater, 1960. "Quantum Theory of Atomic Structure", p. 369) to arrive at the appropriate effective-charges for the core-state wavefunctions. For Na we require 1s-, 2s-, and 2p-states, where for example

$$\psi_{1,0} = N_1 e^{-Z_1 r}$$

$$\psi_{2,0} = N_2 (r e^{-Z_2 r} - \alpha e^{-Z_1 r})$$

for the radial parts. The nomalization condition easily determines N_1 and N_2 and α for these and the p-states. Evaluating the Fourier transforms of these functions, we find that they are essentially independent of k for the range $0 < k < k_F$. This result can be easily anticipated by knowing the range of the states in r-space. On the other hand, the overlaps are sensitive to Z_2 (the effective charge for the $n = 2$ shell) and this, of course, changes when a core electron is removed (as also does Z_1 slightly). For a 1s-hole in Na, we find a 15% reduction in the overlap for the 2s state. The 30% change in the square, therefore, balances the change in $(\epsilon_v - E_{2s})$ and a similar pattern emerges for the other states.

The Band Structure of the Hexagonal Close-packed Metals Sc, Ti, Y and Zr[†]

S. L. ALTMANN and C. J. BRADLEY

Department of Metallurgy, University of Oxford, England

ABSTRACT

The band structures, density-of-states curves and Fermi surfaces are presented for the hexagonal close-packed configurations of Sc, Ti, Y and Zr, as calculated by the cellular method. A study is made of the effect of the potential field on the band structure and Fermi surfaces of Ti. Heat capacities and bandwidths are discussed. Comments are given of the validity of the methods of band-structure calculation.

1. INTRODUCTION

We present in this paper the band structures, Fermi surfaces and density-of-states curves for the hexagonal close-packed metals Sc, Ti, Y and Zr, as computed by the cellular method. A full description of the method, as well as a study of the errors involved in it, has been given in a previous paper (Altmann and Bradley, 1965b).

The calculations are of an *ab-initio* nature, the only experimental data required being the lattice constants of the metals. These are shown in Table I.

TABLE I. Lattice constants

Metal	a (A.U.)	c/a	Reference
Sc	6·25	1·59	Pearson, 1958
Ti	5·575	1·59	McQuillan and McQuillan, 1956
Y	6·88	1·57	Pearson, 1958
Zr	6·117	1·59	Lustman and Kerze, 1955

2. POTENTIAL FIELDS

The most serious problem in the calculation of the band structure of a transition metal is that of the choice of an appropriate metal field. It is now well known that quite reasonable changes of the potential field can alter

[†] This paper is, in part, reprinted with permission of the Institute of Physics and the Physical Society (from Altmann and Bradley, 1967).

265

the spacing of the energy eigenvalues by factors of 2 or 3. Mattheiss (1964), for instance, on using the augmented-plane-wave method, obtained such factors when comparing results for two vanadium fields, derived from the atomic configurations $3d^3\,4s^2$ and $3d^4\,4s^1$ respectively. Admittedly, the order of the levels is unaltered in this case and most of the features of the Fermi surface would presumably be preserved. Nevertheless, this is so only because vanadium has a favourable band structure with well-separated bands. There is no reason whatever to expect this to be the case for other transition metals.

In view of uncertainties of the order of 0·3 ryd (1 ryd = 13·6 ev) in the energy levels arising from the choice of field, we did not consider it would have been consistent to introduce a number of refinements in the field which, however desirable on physical grounds, would nevertheless affect the eigen-values by a few hundredths of a rydberg. Instead, we tried to achieve some overall consistency over a region of the periodic table as follows.

For a given metal, say A, with n valency-electrons, we started from a Hartree–Fock self-consistent field for the ion A^{m+} (m generally equal to n, but in some cases taken to be one of the various integers $m \leqslant n$). We added to this field the contribution due to a uniform charge distribution of $m-1$ electrons over half the unit cell (since there are two atoms per unit cell in the hexagonal close-packed lattice).

The whole of such a field must be considered as a parameter of the calcula-tions and its goodness can be gauged in two distinct ways. The first is, of course, to compare the resulting band structure with whatever experimental evidence is available for a given metal, such as bandwidths or de Haas–van Alphen periods. This will provide support for the results obtained from a given field. Secondly, when no such experimental evidence is available for a given metal, we can apply a consistency criterion over a region of the periodic table, since we expect to preserve a certain element of rigidity of the band structure in going from Sc to Ti and from Y to Zr.

Thus, from three fields used for Ti, only one (Ti3) gave a bandwidth in reasonable agreement with experiment (see Section 6). For Sc, where experi-mental evidence is lacking, we observe good consistency of the bands with that of the best field for Ti. Likewise, the field used for Zr gives a bandwidth, heat capacity and Fermi surface in reasonable agreement with experiment and that for Y is consistant with the Zr band structure. On these grounds, we are reasonably satisfied that the results we present for these metals are of some significance.

We show in Fig. 1 the potential fields used for Sc, Y and Zr, derived from those of the ions Sc^{3+}, Y^{3+} and Zr^{4+} respectively. In Fig. 2 we display four fields used for Ti. The number n given in the figure indicates that the field derives from the Ti^{n+} ion. We also investigated a muffin-tin field (designated MT) constructed from a field obtained by L. F. Mattheiss and kindly pro-vided by him. Although the basic field is the same as that used for Ti by Mattheiss (1964), the choice of muffin tin is different so that our results

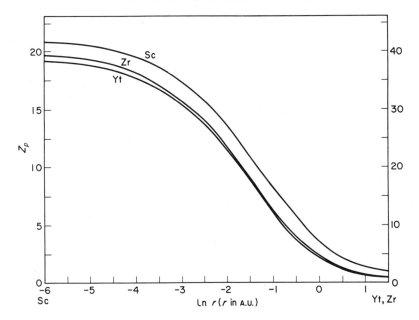

FIG. 1. The potential field Z_p for Sc, Y, Zr; the scale on the right is for Y and Zr.

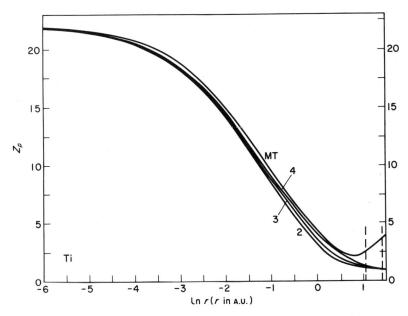

FIG. 2. The potential fields for Ti; the vertical lines mark the shortest and longest radii of the Wigner–Seitz cell.

cannot be directly compared with his. However, it is interesting to consider this field for the following reason. For the other fields used the potential is not spherically symmetric outside the inscribed sphere, and this might introduce some errors of convergence in the cellular method, which are not present when a muffin-tin potential is used. However, we computed all the energy eigenvalues for the first half-dozen bands at all points of symmetry with the muffin-tin potential and we did not find any difference with regard to the convergence of the expansions. In fact, when the lowest level at Γ is taken as the zero of energy we found that, with two or three exceptions, all the levels for MT coincided with those for Ti4 within 0·03 ryd, and that there is no serious difference in the Fermi surface obtained for those two fields. Since the difference for MT and Ti4 is not significant, the results for MT are not presented in this paper.

All the self-consistent fields of the ions were kindly computed for us by Dr. D. F. Mayers.

3. ENERGY BANDS

The energy bands computed are shown in Figs. 3–8. For Sc, Ti4, Y and Zr five points (including the ends) were computed along all the lines of symmetry shown, except KM for which three points were computed. Broken lines indicate interpolated bands. For Ti2 and Ti3 only points of symmetry were computed. In every case, the notation is that of Altmann and Bradley (1965a).

Mattheiss (1964) computed only the ΓK direction for Ti. His bands agree very well qualitatively with those for Ti3, except that the levels ΓE'_ and ΓA''_{2+} are interchanged. The quantitative agreement is also quite good as comparison with his bands will show. The only other set of bands published

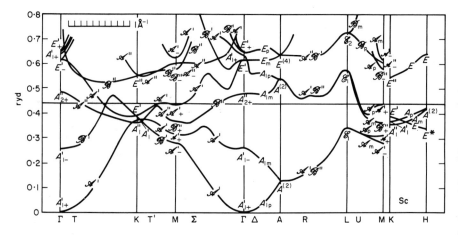

Fig. 3. Bands for Sc; the value of ΓA_{1+} is -0.607 ryd.

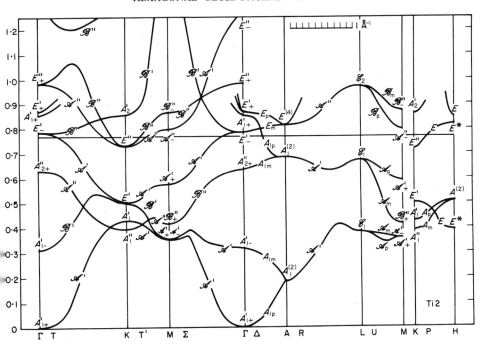

FIG. 4. Bands for Ti2; the value of ΓA_{1+} is -0.390 ryd.

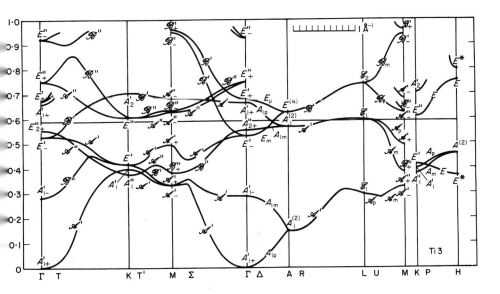

FIG. 5. Bands for Ti3; the value of ΓA_{1+} is -0.634 ryd.

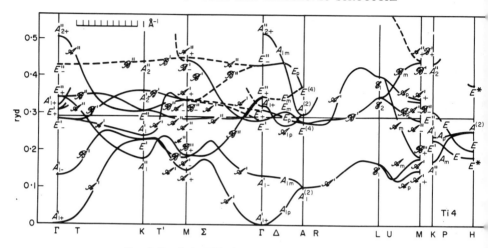

FIG. 6. Bands for Ti4; the value of ΓA_{1+} is -0.856 ryd.

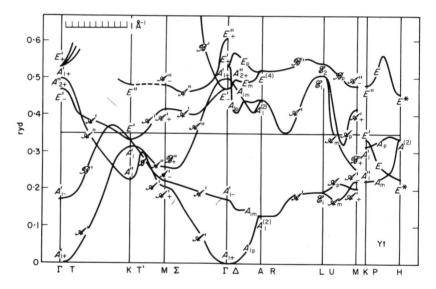

FIG. 7. Bands for Y; the value of ΓA_{1+} is -0.411 ryd.

for these metals is that for Zr computed by Loucks (1967). The agreement both qualitatively and quantitatively is reasonable for the first three or four bands along the ΓK and ΓM directions but the dip of the $T\mathscr{B}''$ curve does not appear in Loucks's calculation. We must admit, however, that this line was the very first that we computed in 1962, and the standard of accuracy then achieved was not quite as good as in later calculations. Indeed, for all later calculations along the line ΓK we used more terms in the wave function and obtained better convergence. On the other hand, there is a crossing in the bands obtained by Loucks that appears unlikely to be genuine. This is the one between the band that joins the lowest level at Γ and the doubly degenerate level at K (which must be $T\mathscr{A}'$), and the band that starts from the level immediately below at K. This latter level is most unlikely to be anything else except A_1' so that the band coming from it must be $T\mathscr{A}'$ and cannot cross a $T\mathscr{A}'$ band. Loucks has treated all the values of k as general ones and cannot identify directly the symmetry of his levels, which may explain this difficulty.

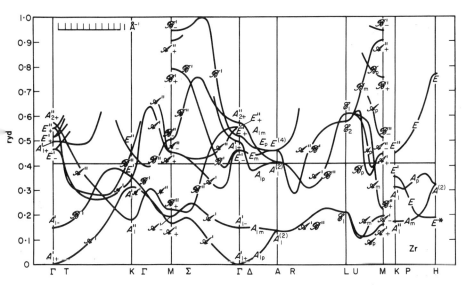

FIG. 8. Bands for Zr; the value of ΓA_{1+} is -0.537 ryd.

We describe briefly in Table II the nature of the levels below the Fermi surface at the points of symmetry. It can be seen that the main s-contribution appears at Γ and that, as the edge of the Fermi surface is approached, these levels turn into d-states. Also, as expected, hybridization is considerable and even the f-contribution is not negligible. The fact that the Ti wavefunctions are more d-like is a consequence of the field Ti4 used. From Fig. 5 it can be seen that Ti3 is more s-like.

TABLE II. Nature of the wave functions

	Sc	Ti4	Y	Zr
ΓA_{1+}	S	S	S	S
ΓA_{1-}	ds	ds	ds	ds
ΓE_{-}	D	D	D	D
ΓA_{2+}	P	P	Pf	Pf
KA_1''	D	D	Df	D
KA_1'	dp	Dp	dp	dp
KE'	dsp	Dps	dsp	sdp
$M\mathscr{A}_+'$	dps	D	Ds	Ds
$M\mathscr{A}_-'$	sdp	dsp	dsp	sdp
$M\mathscr{B}_+''$	Ðp	Dp	dp	dp
$M\mathscr{A}_+'$	pds	dps	pds	pds
$AA_1^{(2)}$	sp	sdp	sdp	sdp
$AA_1^{(2)}$	dp	dps	dp	dp
$L\mathscr{E}_1$	dps	Ds	ds	ds

The contributions are listed in order of magnitude and where contributions are greater than 80% the appropriate letter is written in capitals. Contributions are listed only if they are greater than 5%.

4. THE FERMI SURFACES

In order to display the sticking together of the n and $n+1$ bands on the *AHL* plane we represent, as in the original of this paper (Altmann and Bradley, 1967), all the Fermi surfaces in the double Brillouin-zone that we discussed in a previous publication (1964).

We reproduce here, for scandium, sections of the Fermi surface (Fig. 9) and a perspective (Fig. 10). The basic feature of the surface is a sinuous tube of holes along ΓA and a pillar of electrons along KH multiply connected *ad infinitum*, from which extensions protrude towards the central axis of the cell as well as towards *M*.

Since there is no experimental evidence on this metal as yet, it is premature to discuss orbits in detail. Estimates for them are not difficult to obtain from the figures.

For titanium, Fig. 11 illustrates the drastic changes in the Fermi surface brought about by changes in the field. As discussed in Section 2, the most reliable field is Ti3 (see also Section 6). Thus we expect the Fermi surface to consist essentially of a globular pocket of holes in the third and fourth bands around *L* and a globular pocket of electrons around Γ in the fifth and sixth bands. These are exactly the major features of our Fermi surface for Zr (Altmann and Bradley, 1964), although in this metal the pocket of the electrons around Γ has a narrow hollow tube at Γ itself along ΓA.

We present and discuss the Fermi-surface results for yttrium in the original paper (1967) and for zirconium in our earlier publication (1964). Our results reproduce the same general features as obtained by Loucks (1966),

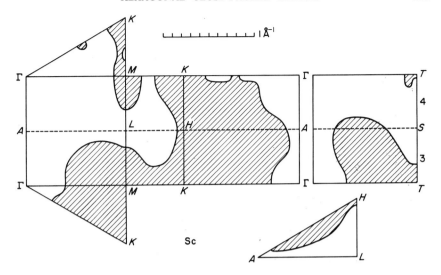

FIG. 9. Sections of the Fermi surface of Sc. S is the midpoint of LH. The hatched areas contain electrons.

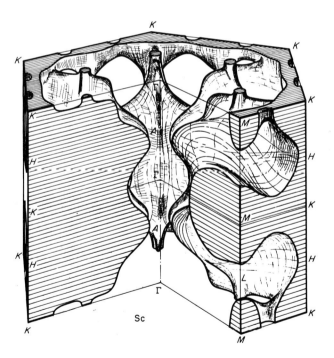

FIG. 10. The Fermi surface of Sc (third and fourth bands). The solid object shown contains electrons.

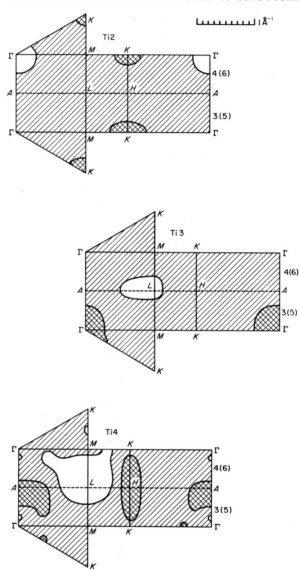

FIG. 11. The Fermi surface of Ti. The empty areas contain holes and the hatched areas electrons. Areas with double hatching indicate pockets of electrons in the fifth and sixth bands.

while minor differences are indicated in the two papers cited, together with some brief discussion of the substantial dependence to be expected of computed Fermi surfaces on the relative position of the eigenvalues for the bands. It can be concluded that better experimental results will help greatly in interpreting the calculations.

5. THE DENSITY OF STATES

Figs. 12 and 13 display the density of states for the various fields used. It will be noticed that the curves for the pairs Sc, Ti3 and Y, Zr show a reasonable degree of rigidity.

The curve for Zr differs from that given in a preliminary communication

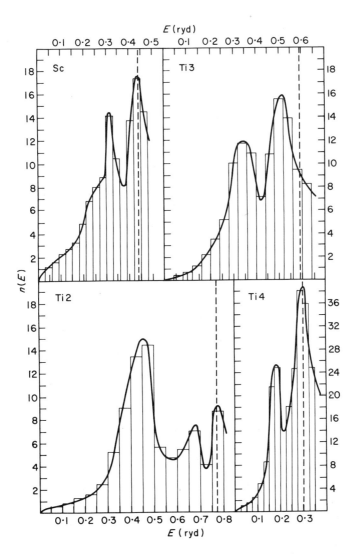

FIG. 12. The density of states $n(E)$ for Sc and Ti. $n(E)$ is given in electron states of both spins per rydberg per atom. The histograms are calculated and the curves are plausible interpolations that have the same area as the histogram. The broken lines indicate the Fermi energies. The change of scale for Ti4 should be noticed.

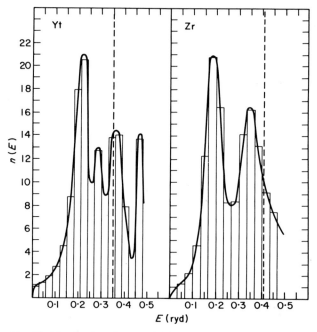

FIG. 13. The density of states $n(E)$ for Y and Zr (see legend to Fig. 12).

(Altmann and Bradley, 1962)[†] which explains an apparent discrepancy pointed out by Loucks. It can be seen that, as expected by this author, the Fermi energy cuts the curve near the sharp minimum.

6. HEAT CAPACITIES AND BANDWIDTHS

Table III collects our main results on heat capacities and bandwidths. It can be seen that, after phonon enhancement is allowed for, the agreement is quite fair for Ti3 and Zr. In fact, the values of $n(E_F)$ "exp." are probably fairly good estimates although (M. A. Jensen, unpublished) they are likely to be overestimated by 30% or so. This would make the agreement even better for these metals.

On the other hand, when the Fermi surface hits a peak of the density-of-states curve, as in Sc and Y the agreement is poor (although the agreement with Loucks for Y is very good). However, the values of $n(E_F)$ "exp." for these metals are not as reliable as those for Ti and Zr and they must be considered as no more than rough guesses (M. A. Jensen, unpublished). It appears, nevertheless, unlikely that they could go down to our value unless

---

[†] The new curve was shown at the Bristol Solid State Conference of January 1964 and was widely circulated.

some new effect such as a spin-fluctuation enhancement could be operative.

It must be appreciated, on the other hand, that we are dealing here with narrow peaks and that quite small changes in the bands could raise the peaks considerably.

TABLE III. Heat capacities and bandwidths

	$n(E_F)$ Calc.	$n(E_F)$ "Exp."	γ (10^{-4} cal mole^{-1} deg K^{-2}) Calc	γ (10^{-4} cal mole^{-1} deg K^{-2}) Exp.	E_F (ryd)	δ_s (ryd)	Bandwidth Estimate (ryd)	Bandwidth Estimate (ev)	Bandwidth Exp. (ev)
Sc	17·6	33·6	7·4	25·8	0·44	0·11	0·33	4·5	
Ti2	9·1	14·4	3·8	8·15	0·77	0·11	0·66	8·9	6·0[a]
Ti3	9·2		3·9		0·59	0·11	0·48	6·5	
Ti4	38·6		16·2		0·30	0·10	0·20	2·7	
Y	14·3	31	6·0	24·1	0·35	0·11	0·24	3·3	
Y (Loucks)	26·8		11·2		0·337				
Zr	10·3	12·7	4·3	6·95	0·40	0·12	0·29	3·9	4·0[b]
Zr (Loucks)	9·1		3·8		0·387				

[a] Skinner, Bullen and Johnston, 1954.   [b] Shaw and Jossem, 1959.

$n(E_F)$, the density of states at the Fermi energy, is given in electron states of both spins per rydberg per atom. The calculated value is that obtained directly from the density-of-states curves. The value headed "Exp." is a rough estimate of the value of $n(E_F)$ which, after allowing for phonon enhancement, would reproduce the experimental value of γ. These values come from Jensen and Maita (1966) except the one for Y (M. A. Jensen, unpublished).

All the values of γ come from Gschneidner (1964).

The calculated values of γ are not corrected for phonon enhancement. They are obtained from the calculated values of $n(E_F)$ of the first column.

δ_s is an estimate of the width of the s band, obtained by working out the width of that part of the first band in the ΓM direction for which the wavefunction contains an s-contribution of 80% or more. The values for Ti2 and Ti3 are guesses since the wavefunctions were not computed for these fields.

The agreement between our Fermi energies and those of Loucks for Y and Zr is as good as it could be. We have estimated the bandwidths by subtracting the correction δ_s defined in the table, and the agreements shown for Ti3 and Zr are very good.

It is clear from the table that Ti3 is the only one of the titanium fields that gives reasonable density of states and bandwidth.

Acknowledgement

This work was supported by a grant from the United Kingdom Atomic Energy Authority, which is gratefully acknowledged. We are very grateful to Dr. D. F. Mayers for the generous provision of the atomic fields required and to Professor L. Fox, Director of the Oxford University Computing Laboratory, for making available its facilities to us. Drs. J. E. Jeacocke and A. P. Cracknell have helped in the preparation and running of the programmes. Useful correspondence with Professor M. A. Jensen is acknowledged, as well as communication and discussion with Professor T. Loucks. We are grateful to Mr. D. Codd for his help with the illustrations.

References

Altmann, S. L. and Bradley, C. J. (1962). *Phys. Letters* **1**, 336.

Altmann, S. L. and Bradley, C. J. (1964). *Phys. Rev.* **135**, A1253.

Altmann, S. L. and Bradley, C. J. (1965a). *Rev. Mod. Phys.* **37**, 33.

Altmann, S. L. and Bradley, C. J. (1965b). *Proc. Phys. Soc.* **86**, 915.

Altmann, S. L. and Bradley, C. J. (1967). *Proc. Phys. Soc.* **92**, 764.

Gschneidner, K. A. (1964). *Solid St. Phys.* **16**, 275.

Jensen, M. A. and Maita, J. P. (1966). *Phys. Rev.* **149**, 409.

Loucks, T. L. (1966). *Phys. Rev.* **144**, 504.

Loucks, T. L. (1967). *Phys. Rev.* **159**, 544.

Lustman, B. and Kerze, F. (1955). "The Metallurgy of Zirconium". McGraw-Hill, New York.

McQuillan, A. D. and McQuillan, M. K. (1956). "Titanium". Butterworths, London.

Mattheiss, L. F. (1964). *Phys. Rev.* **134** , A970.

Pearson, W. B. (1958). "Handbook of Lattice Spacings and Structures of Metals". Pergamon Press, New York.

Shaw, C. H. and Jossem, E. L. (1959). *Rep.* 471, Ohio State University Research Foundation.

Skinner, H. W. B., Bullen, T. G. and Johnston, J. E. (1954). *Phil. Mag.* **45**, 1070.

Thorsen, A. C. and Joseph, A. S. (1963). *Phys. Rev.* **131**, 2078.

Comment

The Validity of Band-structure Calculations

Simon L. Altmann

I present here some general comments on the validity of methods of computation and on the use of the results of band-structure calculations; and I have added an epilogue regarding the interpretation of the results described in the foregoing paper in the light of new experimental data for these metals presented elsewhere in this Volume.

Three major methods have been used in the past two decades for the accurate calculation of energy eigenvalues in transition metals; namely the Augmented-plane-wave (APW) method, the Green-function method and the version of the cellular method proposed by the present author. A period of fundamental discussion has ensued regarding the relative merits of the methods. If objective, such discussion is certainly useful; but in practice it has obscured the most important problem in the whole subject, that of choosing the crystal potential-field.

The APW and Green-function methods are now accepted as equivalent and accurate on the strength of the agreement of some copper eigenvalues which were computed, with exactly the same potential-field using both methods, by Segall and Burdick. However, an important difficulty in the Green-function method has been pointed out recently by Ballinger and Marshall (1967), who found that spurious eigenvalues appear for points of symmetry in k-space if the spherical harmonics used in the expansions are not symmetry-adapted. Such eigenvalues pertain, in fact, to wavefunctions that admit of forbidden harmonics. It was pointed out by the present author that, by continuity, these spurious levels should give rise to spurious bands, for which no clear criterion of rejection exists since symmetry cannot be invoked. The existence of such bands has now been confirmed by R. A. Ballinger (unpublished), and this means that some careful re-examination of the method is required.

Although the cellular method has been somewhat maligned in the past, it now appears to be quite satisfactory. The calculations reported here were carried out with an out-dated programme which was not very accurate. However, an excellent check was obtained with it of the eigenvalues for aluminium obtained by Segall (1961), using the same potential field. For most eigenvalues the agreement is 0·01 ryd or better, which is within the accuracy of the programme. A new programme for a larger computer has now been developed and all the empty lattice errors, for the first six or seven bands at all points of symmetry, are smaller than 10^{-4} ryd. The same accuracy is obtained for the first two bands at general values of k, but errors of the order of 0·01 ryd appear for the sixth and seventh band. Changes in the sets of points used introduce errors of 10^{-4} ryd.

When a muffin-tin potential is used, the cellular method is conceptually as satisfactory as other methods. On the other hand, for some metals one may have to make a choice between a poor physical model that can be mathematically exactly computed (the muffin-tin model) and a good physical model which might yield eigenvalues that are mathematically less exact. Since it is now abundantly clear that the overwhelming problem is the choice of field, it may sometimes be useful to employ a better model of the field even at the slight expense of some "mathematical" accuracy. The cellular method offers the unique advantage of permitting either course of action.

The main purpose of the foregoing paper is to stress the dependence of the computed band-structure on the choice of field, and to present enough data in a manner such that the computed results can be reasonably adjusted to interpret the experimental results as these become available. In the past, too many calculations have been presented which were obtained with a single potential-field, and no reliance can be placed upon them. Until some major advance is made, on the question of the choice of metal field, the only way in which detailed band-structure calculations can help in the theory of the transition metals is by providing enough data on the field-dependence of the results to permit their adjustment to the experimental results. Careful attention must also be paid to the consistency of the results throughout some region of the periodic table.

An example of this approach arose at the conference that led to this Volume. When the foregoing paper was written the only results available for the band-width of titanium were those obtained by Skinner, which suggested a value of 6 ev. Holliday presents results (this Volume, see p. 111) that support a smaller value of some 4 ev. If this value is accepted, the correct picture of the band structure of titanium would be obtained by slightly modifying the results that we found with the field Ti3, shifting them in the direction of those of the Ti4 field. From Figs. 5 and 6 it can be seen that the level $\Gamma A_{2+}''$ is then likely to move from Ti3 across the Fermi surface, so that the pocket of electrons at Γ in the fifth band of Ti3 will move towards A, this being the only major change to be expected in the Fermi surface, with respect to that given for Ti3 in Fig. 11. Since the value of the Fermi energy for Ti4 is so low, a comparatively small change from Ti3 would bring about agreement with Holliday's suggested value. In doing this, the important point is that we can speculate with some degree of certainty on the more probable shape of the Fermi surface to be expected. Also a collateral effect of the change in question would be to raise $n(E_F)$, the density of states at the Fermi level, from the value given in Table 3; thus ameliorating the agreement with experiment. Corresponding small changes would also be required in the bands for scandium, in order to restore the balance between the two metals. It should be noticed that the agreement between the bandwidth measured by Holliday for zirconium (see p. 111) and our own value is very good.

I hope that this result indicates how band-structure calculations can be used to correlate experimental findings while at the same time strongly supporting the remarks made by Harrison in his opening discussion to Part 3, p. 228. Results of calculations of a single metal with a single field should be used with extreme care.

References

Ballinger, R. A. and Marshall, C. A. W. (1967). *Proc. Phys. Soc.* **91**, 203.
Segall, B. (1961). *Phys. Rev.* **124**, 1797.

Soft X-Ray Emission Spectrum and Momentum Eigenfunction for Metallic Lithium

M. J. STOTT and N. H. MARCH

Department of Physics, The University of Sheffield, England

ABSTRACT

An attempt is made to calculate the intensity of soft x-ray emission of metallic lithium by expanding the Bloch wavefunctions in terms of the momentum eigenfunction $v(p)$. The spectrum is shown to depend sensitively on the form of this function and a preliminary calculation is carried out using a model in which $v(p)$ is independent of the direction in k-space. Though this cannot be quantitatively correct, comparisons of $v(p)$ with existing calculated wavefunctions show that peaks in v in different directions in k-space occur at very nearly the same energy, and that the calculated values explain the maximum in the observed x-ray intensity.

As an additional result of the calculations, we obtain the general features of the Wannier-function for lithium; this is found to extend to a large distance, and to oscillate with a relatively small exponential decay.

Finally, a brief examination of a dilute lithium–beryllium alloy is carried out to ascertain the validity of the rigid-band model. We consider that this model should be a good approximation in this case, though not for lithium–magnesium.

1. INTRODUCTION

The form of the soft x-ray emission spectrum of lithium has presented considerable difficulty for some time; the intensity profile is abnormal, showing a maximum to the low-energy side of the cut-off that corresponds to the Fermi level. Interest in the problem has recently revived and, in particular, Goodings (1965) interpreted this peak in the $I(E)$ curve (i.e. the curve of emission intensity against energy E) as indicating that the hole in the K shell dominates the problem. He explained the peak in terms of the distortion of the wavefunctions caused by the hole, but since Catterall and Trotter (1958) have shown experimentally that the form of the emission band is hardly altered for satellite emission (for which there are two holes present), we conclude that Gooding's explanation cannot be correct and that the peak must be explained solely in terms of undistorted wavefunctions. Our work suggests that it is the form of the wavefunctions in the occupied part of the valence band that dominates the shape of the intensity curve, which is in accordance with the conclusion of Catterall and Trotter that the hole hardly affects the shape of the $I(E)$ curve.

The arrangement of the paper is as follows. In Section 2 we show how the momentum eigenfunction $v(p)$ is related to the intensity $I(E)$ of soft x-ray emission and discuss the theoretical considerations governing the form of $v(p)$; the experimental results are also briefly summarized. In Section 3 a crude spherical model for the momentum eigenfunction is used to illustrate how a maximum in the transition probability can arise, and the dependence of $I(E)$ on $v(p)$, together with the corresponding transition probability and density of states is discussed for realistic $v(p)$ compatible with band-theory studies. We stress that we calculate both the transition probability and the density of states consistently from a wavefunction. The Wannier function for the valence band is investigated and its main features are established from the properties of $v(p)$. In Section 4 some discussion of the influence of the hole in the K shell is given, along with a discussion of the related problem of a dilute lithium–beryllium alloy. Finally, we discuss the reason for lithium occupying a unique place in the alkali series, the momentum eigenfunction for sodium being sketched for comparison.

2. THE SOFT X-RAY EMISSION SPECTRUM IN TERMS OF THE MOMENTUM EIGENFUNCTION

A. Relation of the Momentum Eigenfunction to the Emission Intensity

Instead of using Bloch wavefunctions $\psi_k(r)$ for the valence electrons in metals, we shall work with the momentum eigenfunction v_{kK_n} defined by

$$\psi_k(r) = \sum_{K_n} v_{kK_n} e^{i(k+K_n)\cdot r} \tag{1}$$

where the vectors K_n describe the reciprocal lattice, and Eq. (1) simply expresses Bloch's theorem. Clearly $|v(k+K_n)|^2$ gives the probability that the 'Bloch' electron with wave-vector k will be found with momentum $k+K_n$ (in units $\hbar = 1$). Actually v_{kK_n} is a function solely of momentum $p = k+K_n$, and hence the whole valence band may be described by a single function $v(p)$, the momentum eigenfunction of the band.

We now turn to the main purpose of this paper: the development of the theory of soft x-ray emission in terms of the momentum eigenfunction. We can first say that the probability of a transition from a "Bloch" state $\psi_k(r)$ to a core state $\phi_c(r)$ is determined by the usual expression:

$$\frac{\left| \int \phi_c^*(r)\nabla_r\psi_k(r)dr \right|^2}{\int \phi_c^*(r)\phi_c(r)dr \int \psi_k^*(r)\psi_k(r)dr} \tag{2}$$

The numerator may be written as

$$\int dr dr' \phi_c^*(r)\phi_c(r')\nabla_r\nabla_{r'}\psi_k(r)\psi_k^*(r') \tag{3}$$

and if we write the Bloch wave $\psi_k(r)$ in terms of momentum eigenfunctions through (1), then we find, after a simple calculation, that the transition probability is proportional to

$$T(k) = \frac{\sum_{K_n}\sum_{K_m} \phi_c^*(k+K_n)\phi_c(k+K_m)(k+K_n)\cdot(k+K_m)v(k+K_n)v^*(k+K_m)}{\sum_{K_n}|\phi_c(k+K_n)|^2 \sum_{K_m}|v(k+K_m)|^2} \tag{4}$$

where $\phi_c(p)$ is the momentum wavefunction of the final state. This is the result we shall use for our calculations.

Finally, the group velocity $\dfrac{dE}{dk}$ can be calculated via

$$\nabla_k E(k) = \frac{1}{i}\frac{\int \psi_k^*(r)\nabla_r\psi_k(r)dr}{\int |\psi_k(r)|^2 dr} \tag{5}$$

Thus, without knowing the potential, $E(k)$ can be found from the momentum eigenfunction, since

$$\nabla_k E(k) = \frac{\sum_{K_n}|v(k+K_n)|^2(k+K_n)}{\sum_{K_n}|v(k+K_n)|^2} \tag{6}$$

or equivalently (cf. Eq. (8) below)

$$\nabla_k E(k) = k + \frac{\sum_{K_n}|v(k+K_n)|^2 K_n}{\sum_{K_n}|v(k+K_n)|^2} \tag{7}$$

For a non-local potential, the result (5) is not exact; the correction is explicitly derived in the Appendix, and estimates of the effect of non-locality made using the formula given there suggest that it plays a minor role in the case of lithium, using a crude Bohm–Pines correction for exchange and correlation (cf. Heine, 1957).

B. Form of the Momentum Eigenfunction for Lithium

$v(k+K_n)$ must satisfy the following general conditions: (i) it must be continuous in the argument $k+K_n$; (ii) it must be normalized in the sense that, for all k,

$$\sum_{K_n}|v(k+K_n)|^2 = 1 \tag{8}$$

(iii) it must be orthogonal to the core states; (iv) it must have the symmetry of the reciprocal lattice.

We consider now the form of $v(k+K_n)$ specific to metallic lithium. In earlier work by Donovan and March (1956), on the form of the Compton profile for lithium, the wavefunction at the bottom of the band ($k = 0$) was discussed in detail. While the physical consequences of this work have recently been experimentally confirmed by Cooper, Leake and Weiss (1965), the "sphere" approximation employed by these investigators led to some irregularities in the high-momentum components, and R. E. Borland (unpublished) has kindly recalculated the results for the Seitz wavefunction $u_0(r)$ avoiding the sphere approximation. We shall base our analysis on these results, which are shown in Table 1 for $k_f = 0.589$ A.U.

TABLE I. Momentum eigenfunction at $k = 0$

$|K_n|$	Number of vectors of length $|K_n|$	v_{0,K_n}
0	1	-0.9477
1.353	12	$+0.0617$
1.914	6	$+0.0414$
2.344	24	$+0.0289$
2.706	12	$+0.0210$
3.026	24	$+0.0160$
3.315	8	$+0.0126$
3.580	48	$+0.0102$
3.828	6	$+0.0085$
4.060	36	$+0.0071$
4.279	24	$+0.0060$
4.488	24	$+0.0051$
4.688	24	$+0.0044$
4.879	72	$+0.0038$
5.241	48	$+0.0029$
5.413	12	$+0.0026$
5.580	48	$+0.0023$
5.742	30	$+0.0020$

These results, valid for $k = 0$, can in principle be extended to higher values of k by writing the approximate Bloch wavefunction $\psi_k(r)$ as

$$\psi_k(r) = u_0(r)e^{ik \cdot r} \qquad (9)$$

This yields a form for $v(k + K_n)$ that is flat in every Brillouin-zone, with a value equal to that at $k = 0$. Hence there are discontinuities at all the zone boundaries, and condition (i) above is violated. The form is shown schematically in Fig. 1(a).

Certain related attempts to expand the Bloch wavefunction around $k = 0$ are worth noting at this stage. The early work of Bardeen (1938) leads to the form shown in Fig. 1(b), while later work of Silverman and Kohn (1950) corresponds to the improved form of Fig. 1(c). This theory gives the "Seitz" values at the K_n, but it also gives the correct non-zero slope. However, continuity of $v(p)$ is violated.

The best calculation available to-date is that of Brown and Krumhansl

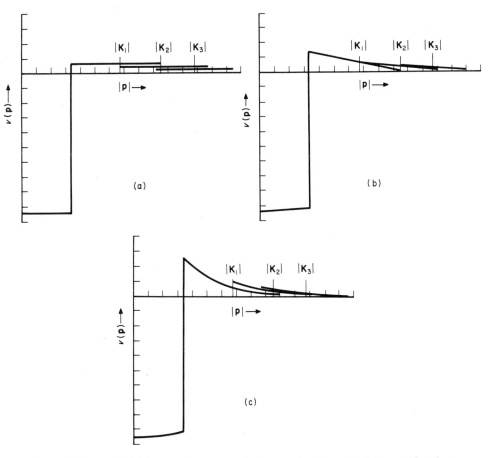

FIG. 1. Forms of $v(p)$ from various approximate wavefunctions. (a) Seitz wavefunction. (b) Bardeen wavefunction. (c) Silverman–Kohn wavefunction.

(1958). These workers calculated $v(k + K_n)$ by a generalization of the ortho-gonalized-plane-wave method, and we lean heavily on their results in our discussion below.

The important features regarding the momentum eigenfunction[†], that emerge from these band-theory studies, are: (i) $v(p)$ is rather flat inside the Fermi surface, with a value varying between -0.95 and -0.90; (ii) $v(p)$ changes sign at $p = K_1$, the first reciprocal lattice vector; (iii) $v(p)$ rises going inwards from K_1 (from the calculations of Brown and Krumhansl, and also of Silverman and Kohn, cf. Fig. 1(c); (iv) $v(p)$ falls off like the atomic momentum-wavefunction for 2s-electrons at large momenta ($p > 5$ A.U.), at the reciprocal lattice points; (v) from (i), (ii) and (iv), it follows that $v(p)$ must have a maximum inside the second Brillouin zone, whichever direction

[†] A preliminary note (Stott and March, 1966) has appeared giving the general form of $v(p)$.

we take from the origin of momentum space; (vi) from dispersion-relation calculations (Ham, 1962), and also from experiment (see Melngailis and De-Benedetti, 1966), it is known that $E(k)$ is accurately spherical within the Fermi surface; any $v(p)$ chosen must obviously be consistent with this, from the group-velocity formula (7).

We must stress at this stage that, while band-theory calculations tell us the position and height of the maximum in the (100) direction in k-space,[†] uncertainty regarding the detailed shape of $v(p)$ remains. We shall show that the surfaces corresponding to the zero and maximal values of $v(p)$ are crucial in determining the form of the transition probability. However, before turning to this, we shall briefly summarize the experimental work on lithium.

width of the core state is to broaden the intensity profile. The resulting

C. Experimental Results on Lithium

The experimental results of Bedo and Tomboulian (1958) for the intensity of soft x-ray emission are shown in Fig. 2. In this case the effect of a non-zero width of the core state is to broaden the intensity profile. The resulting intensity is given by

$$I(E) = \int_{-\infty}^{\infty} \frac{I'(\epsilon)}{(E-\epsilon)^2 + \lambda^2} \, d\epsilon \tag{10}$$

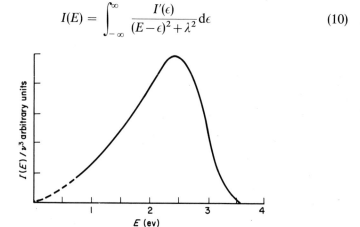

FIG. 2. Experimentally determined intensity of soft x-ray emission from metallic lithium.

where $I'(E)$ is the intensity calculated assuming a discrete core-level, and the energy profile of the core states is given by a Lorentzian of half-width λ. The inversion of (10), to give $I'(E)$ in terms of $I(E)$, will not significantly alter the general shape of the curve if a suitable half-width, $\lambda \simeq 0.1$ ev, is inserted. It was therefore not considered necessary to unfold the experi-

---

[†] We are very grateful to Dr. L. Pincherle and Mr. Masharrafa for communicating some recent results which add significantly to the data of Brown and Krumhansl.

mental intensity-curve. We shall now try to incorporate some of the main features of $v(p)$ discussed into a simple model, in order to illustrate how such an unusual form of $I(E)$ can occur for lithium.

3. MODEL FOR OBTAINING THE MOMENTUM EIGENFUNCTION

A. Spherical and Finite-range Model for $v(p)$

We wish to focus attention on the properties of the transition probability $T(k)$ given by Eq. (4), and in particular on its relation to the nodal surface and maxima in $v(p)$. Due to the complexity of $v(p)$ in metallic lithium, we shall first choose a very crude form that is spherical and cuts off at a finite momentum. The form chosen is shown in Fig. 3; where, the flat region is somewhat arbitrarily taken to extend to $\frac{1}{2}|K_1|$ with $v = -1$, the maximum height is chosen as 0·2 A.U., and the range is fixed at ~ 3 A.U. The distance of the peak from the origin is denoted by A, and the node then occurs at a distance $s = \dfrac{1}{6}\left[5A + \dfrac{|K_1|}{2}\right]$ from the origin.

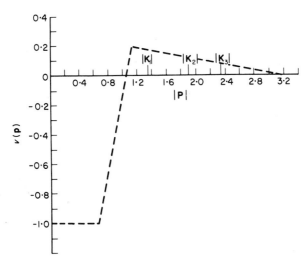

FIG. 3. Schematic model of $v(p)$ for valence band of metallic lithium.

Clearly a comment on the finite range chosen for $v(p)$ is needed. It is known that reproduction of the node in the 2s-like wavefunction at $k = 0$ requires many high-momentum components. We include these later, but it turns out that they do not dominate the energy dependence of the transition probability and we at first omit them to avoid complicating the main issue.

The model of Fig. 3 will not correctly normalize for all k, due to its

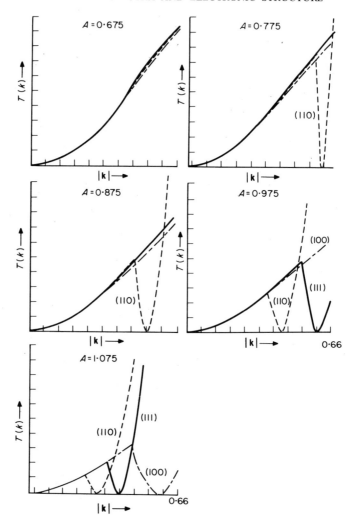

FIG. 4. Transition probability $T(k)$ along different directions in k-space for model of $v(p)$ given in Fig. 3. Different values of the position of the peak are considered.

spherical nature. However, this is approximately accounted for by the denominator on the right-hand side of Eq. (4).

The core wavefunction adopted throughout the paper is the Slater function

$$\phi_c(r) = \left(\frac{Z'^3}{\pi}\right)^{1/2} e^{-Z'r} \qquad (11)$$

with $Z' = 2.70$. The corresponding momentum eigenfunction required in evaluating (4) is

$$\phi_c(p) = \left(\frac{Z'^3}{\pi}\right)^{1/2} \frac{8\pi Z'}{(p^2 + Z'^2)^2} \tag{12}$$

For this model, $T(\mathbf{k})$ was calculated exactly as a function of $|\mathbf{k}|$ for the three directions (100), (110) and (111) and for various values of the parameter A; i.e. for various positions of the maximum (and therefore also the node) in $v(p)$.

The summations in (4) then lead to results shown in Figure 4. The main conclusions can be summarized by plotting the maximum in $T(\mathbf{k})$, for a particular direction, against the distance between the maximum in $v(p)$ in the chosen direction $\mathbf{K}$ and the nearest first reciprocal lattice point. For example, if θ is the angle between $\mathbf{k}$ and the appropriate vector $\mathbf{K}_1$, then this variable is

$$t = [|\mathbf{K}_1| \cos\theta - (A^2 - |\mathbf{K}_1|^2 \sin^2\theta)^{1/2}] \tag{13}$$

the significance of which is clear from Figure 5. The relation between the maximum in $T(\mathbf{k})$ and the variable t is shown for the three directions in Table II. There is an immediate correlation. For only one of the chosen values of A is there no maximum in $T(\mathbf{k})$ in any direction, with $|\mathbf{k}| < \frac{1}{2}|\mathbf{K}_1|$, and this is when the nodal and maximal surfaces occur at radius $\frac{1}{2}|\mathbf{K}_1|$.

For a given A, the maxima are seen to occur in this spherical model at different values of $\mathbf{k}$, and this is connected with the anisotropy in $E(\mathbf{k})$ inside the Fermi surface which is an inevitable consequence of a spherical $v(\mathbf{p})$. A more realistic model with angularity will be discussed below, and shown to remove this difficulty.

But already, from the present model, the maximum in $v(\mathbf{p})$ in the second zone is intimately connected with the maximum in the experimental soft x-ray intensity for lithium.

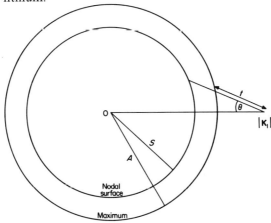

FIG. 5. Significance of the variable t of Eq. (13).

TABLE II. Position of calculated peak and distance from K_n

Value of A	Position of peak in $T(k)$	Position of zero	Distance t of peak from K_1	Distance of K_1 from zero
(110) direction				
0·775	0·58	0·62	0·58	0·59
0·875	0·48	0·53	0·48	0·51
0·975	0·37	0·44	0·38	0·43
1·075	0·27	0·33	0·27	0·34
(111) direction				
0·975	0·53	0·61	0·52	0·61
1·075	0·47	0·42	0·36	0·46
(100) direction				
1·075	0·46	0·61	0·46	0·63

B. Realistic Momentum Eigenfunction for Lithium

Even though the model outlined above shows how a peak in the transition probability can arise, we must clearly deal more realistically with the following points: (i) $v(p)$ must be long range—in particular, at the reciprocal lattice vectors, it must fall off like p^{-4} for large p; (ii) $v(p)$ must be continuous, because otherwise we should get undamped oscillations of the Wannier function, which only occur for free electrons (cf. the discussion in the next Section); (iii) $v(p)$, as discussed already, must have some appreciable angularity in the second zone, in order to give a highly spherical Fermi-surface.

These three points are met in the following manner. Firstly, at the reciprocal lattice points $v(p)$ is always given the values shown in Table I. Secondly, $v(p)$ is chosen to be continuous along the (110) direction, and is normalized; the normalization is always satisfied to within 2 or 3%. Thirdly, and most important, the model of the last Section shows that the distance t in Fig. 5, i.e. from reciprocal lattice points to the maximum of $v(p)$, must be practically constant over a substantial region of momentum space.

While we shall summarize the essential nature of $v(p)$ in lithium below, we have taken for explicit calculation a model which we now describe. We begin with the values of Table I, at $p = K_n$. For Brillouin-zones other than the first two we chose

$$v(p) = a_n + b_n k \cdot K_n + c_n (k \cdot K_n)^2 \qquad (14)$$

where k is measured from the nth reciprocal lattice point K_n and is restricted to lie within the zone, and a_n is the Wigner–Seitz value. b_n and c_n were chosen to give a smooth curve along the (110) direction.

In the zone centred on the shortest non-zero reciprocal lattice vector, a_1 is again taken as the Wigner–Seitz value, while b_1 and c_1 are chosen to be functions of $|k|$ such that $v(p)$ is continuous along (110). Thus, the flexibility in $v(p)$, which we exploit below, arises from the choice of $b_1(k)$ and $c_1(k)$. This choice is always guided by the results of the calculations made by Brown and Krumhansl, and leads to the required form of angularity, such that $E(k)$ is almost spherical within the Fermi surface.

Finally, in the first zone, b_0 and c_0 are chosen as zero, and $a_0(k)$ is used to give agreement with existing band-theory calculations out to the Fermi surface. It should be stressed that the different zones are related to one another always through normalization. The forms chosen were checked for orthogonality to the 1s-functions: this condition was always satisfied to satisfactory accuracy (i.e. to $\sim 5\%$ of the normalization integral; the essential point here is the presence of the nodal surface in $v(p)$).

We have chosen a family of curves for $v(p)$, which are shown in Fig. 6 along the (110) direction. With each of these functions A to D we have evaluated: (i) the intensity of soft x-ray emission, (ii) the density of states, (iii) the dispersion relation. The results for (i) and (ii) are displayed in Fig. 7 for the corresponding forms of $v(p)$ shown in Fig. 6, A to D.

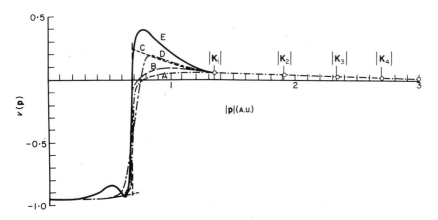

FIG. 6. Forms chosen for $v(p)$ along (110) direction.

Note that $I(E)$ has a maximum in cases A, B and D, but only in case of D does the density of states have a maximum.

On closer examination, it turns out that a whole family of $v(p)$ could lead to the same soft x-ray intensity (apart from a multiplying constant, which could only be obtained from a knowledge of absolute atom-intensities). Thus, it is not even in principle possible to read a unique $v(p)$ from the experiments. In any event, this could not be done from a single $I(E)$ curve since $v(p)$ is not spherical.

However, a momentum eigenfunction can be found that is compatible with all known requirements and will fit the soft x-ray data; it is presented in curve E of Fig. 6.[†] The transition probability and the density of states for this $v(p)$ are shown in Fig. 7, E. A maximum appears in the density of states, at roughly 0·5 ev below the Fermi level.

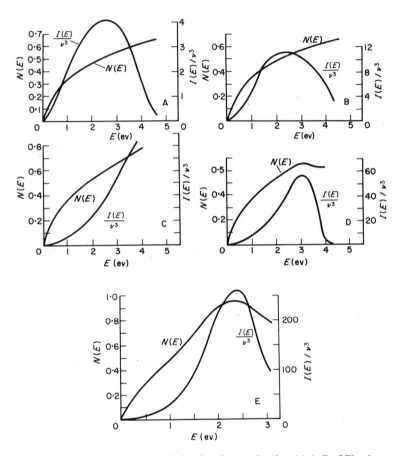

FIG. 7. Soft x-ray intensity and density of states for the $v(p)$ A–E of Fig. 6.

We wish to stress that there is no need to invoke such a maximum in the density of states to explain the soft x-ray results. The over-riding feature is the transition probability, and semi-quantitatively there is complete agreement that all realistic forms of $v(p)$ lead to this for Li. We believe that available evidence is against a maximum in $N(E)$ for Li, but that an experiment on a Li–Be alloy would settle the point (cf. Section below).

[†] The small spur in curve D that arises because we have satisfied the normalization condition (8) to high accuracy, is now exaggerated in curve E. We have considered the effect of removing the spur, and of explicitly introducing the normalization factor in Eq. (7); doing so does not alter the conclusion that there is a maximum in the density of states for these two cases.

It is also of interest to note in Fig. 8 the dispersion relations for the various choices of $v(p)$ of Fig. 6. From curves D and E for $v(p)$, we find E rising more rapidly than k^2 near k_f and this is the origin of the maximum in $N(E)$ for these two cases.

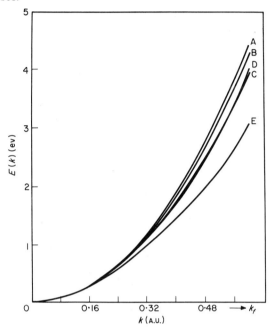

FIG. 8. Dispersion relations for forms of $v(p)$ A–E of Fig. 6.

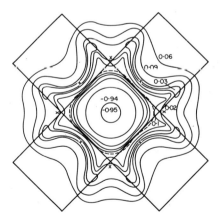

FIG. 9. Momentum eigenfunction $v(p)$ for valence band of lithium. Contours of equal v are shown in a section perpendicular to (010). The nodal surface is shown by dashed curve. Crosses indicate maxima in the radial symmetry directions.

Finally, we have attempted in Fig. 9 to draw together all of our information on $v(p)$ by plotting contours of equal v for a section of k-space perpendicular to the (010) direction. The nodal surface shown by the dotted line lies rather near to the zone boundary. The maxima along radial-symmetry directions are shown by the crosses. As we remarked earlier, some uncertainties remain in this picture, but there can be little doubt that the most essential features of $v(p)$ are revealed in this contour diagram.

C. Wannier Function for the Valence Band

The Wannier function $a(r)$ is related to $v(p)$ through

$$a(r) = \frac{\Omega^{1/2}}{(2\pi)^3} \int v(p) e^{ip \cdot r} dp \qquad (15)$$

where Ω is the volume of the unit cell.

We must stress that only the spherical average $a_0(r)$ of the Wannier function will be reported here. Undoubtedly, from $v(p)$ shown in Fig. 9, the Wannier function has appreciable angularity, which could be calculated

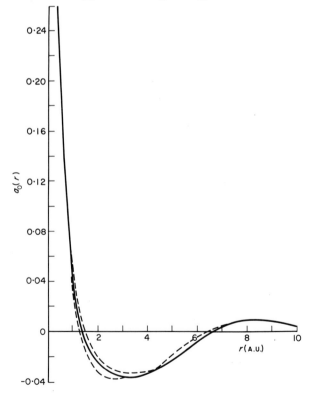

FIG. 10. Form of spherical average $a_0(r)$ of Wannier-function. Dashed curves indicate likely errors.

should it be required later. Again some uncertainty arises because the spherical average of $v(p)$ is not precisely defined in the region where it is rapidly rising through a node to its maximum. The likely errors that this will cause in the Wannier function have been estimated, and are indicated in the dashed curves of Fig. 10. As far as we are aware, this is the first realistic Wannier-function that has been obtained in a metal.

We see that $a_0(r)$ oscillates with distance, and it also turns out to be exponentially damped. This exponential damping is easily understood; it has the form $e^{-\alpha r}$ where α is a measure of the "breadth" of the sharp rise in $v(p)$ around $|K_1|/2$. Only for free electrons, where there is a sharp edge, will α be zero. Therefore the free-electron Wannier-function oscillates in an undamped manner and, for a simple cubic lattice, like $\dfrac{\sin x}{x} \dfrac{\sin y}{y} \dfrac{\sin z}{z}$.

4. DISCUSSION

A. Influence of the Hole in K Shell

It can be argued that so far we have ignored the influence of the hole in the K shell. It seems clear that, if this was a crucial feature of the calculation, it would lead to more high-momentum components than a simple band theory which neglects the hole. However, we have seen that the form of the intensity curve is very sensitive to the position of the maximum; Catterall and Trotter have measured the intensity for the satellite emission arising when an atom is doubly ionized, and no marked change in the intensity was found. We conclude that the position of the maximum in $v(p)$ cannot be appreciably altered by the hole.

However, we should stress that our argument is not that the electron states are unaffected by the hole but that in accordance with the above discussion the hole has an unimportant effect on the shape of the intensity vs. energy curve. This conclusion is supported by both the perturbative and variational estimates that we have made—although admittedly for free electrons. We consider our interpretation to be a more natural one than the alternative proposed by Goodings (1965), who focused all attention on the influence of the hole.

B. Band Structure of Lithium–Beryllium Alloy

In view of the possibility that the density of states for lithium may be anomalous, it might be interesting to measure the Pauli paramagnetism of Li–Be alloys, using electron spin resonance techniques. Unfortunately, the solubility of Be in Li is low and the experiment will necessitate high accuracy.

To verify that the rigid-band model is good for Li–Be, we show in Table III the level shift calculated using our wavefunction of curve E of Fig. 6. This shift is given in first order, by

$$\Delta E(k) = \int \psi_k^* V(r)\psi_k(r)dr \qquad (16)$$

where $V(r)$ is the solute potential.[†] In spite of appreciable k-dependence of the Bloch waves, $\Delta E(k)$ is independent of k to within 5 or 6%, and a rigid-band description seems very good. Thus, the question we raise about $N(E)$ would be settled by knowing whether the Pauli susceptibility of Li is increased or decreased by addition of Be.

Further calculations that we have made on Li–Mg, where the solubility of Mg in Li is much greater than that of Be, show that, because of the major difference in core structure between solute and solvent, the rigid-band model will unfortunately not be reliable.

TABLE III. Level shift per atomic % Be in Li–Be alloy

$|k|$	Level shift $\Delta E(k)$
0	−0·107
0·1	−0·107
0·2	−0·106
0·3	−0·106
0·4	−0·107
0·5	−0·110
0·6	−0·105

C. Soft X-Ray Spectra of Other Alkalis

It is clear from experiment that lithium occupies an anomalous position in the alkali series with regard to its soft x-ray spectra. The most obvious point to make is that Li has no p-electrons, and this affects its valence band relative to that of other alkalis because of the importance of core ortho-gonalization.

More specifically, we have plotted in Fig. 11 the momentum eigenfunctions for the 2s-electrons in Li and the 3s-electrons in Na, using a single ortho-gonalized plane-wave. The rough characteristics of $v(p)$ shown for the metal in Fig. 11 are already present in the 2s atomic momentum-function for Li (Fig. 12), although, of course, the first Brillouin-zone is dominated by free-electron behaviour and has nothing to do with the atomic correspondence. However, the slope of $v(p)$ at the first reciprocal lattice vector is negative for Li, and it appears for Na to be positive even though the peaks in the atomic functions occur at similar momentum.

[†] For the present purpose, a crude "screened" Coulomb-potential sufficed for the solute atom.

Furthermore, there appears to be a close connection between our findings and the work of G. M. Stocks *et al.* (1968) who find, from a pseudo-atom analysis for Li, an anomalous behaviour of the *p*-wave phase shift as a function of k near the Fermi level.

Thus there is no difficulty in qualitatively understanding why Li occupies a unique place in the series.

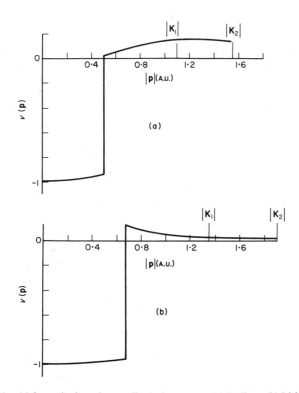

FIG. 11. $v(p)$ from single orthogonalized plane-wave. (a) Sodium. (b) Lithium.

5. CONCLUSION

The present theory attributes the anomalous shape of the soft x-ray intensity in Li entirely to band shape. Existing knowledge of the momentum eigenfunction demands that the transition probability $T(k)$ has a pronounced maximum. It is possible, although we do not think probable, that there is also a gentle maximum in the density of states near the Fermi level. We have proposed an experiment to settle this point. The form of the momentum eigenfunction for lithium appears now to be established in its main features, but a full band-theory calculation is needed to refine the picture shown in Fig. 9, and would be of obvious interest.

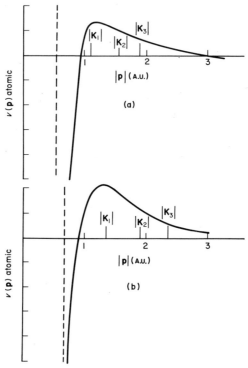

Fig. 12. Atomic momentum-eigenfunctions. (a) 3s-electron in sodium. (b) 2s-electron in lithium.

Acknowledgement

Professor Sir Neville F. Mott and Dr. V. Heine made helpful comments on our paper at an earlier stage, and suggested the possible importance of the angularity of $v(\mathbf{p})$.

One of us (M.J.S.) wishes to acknowledge the award by the Science Research Council of a Postgraduate Studentship. The later stages of the work were performed in part under contract with the Royal Radar Establishment, Malvern.

References

Bardeen, J. (1938). *J. Chem. Phys.* **6**, 367.
Bedo, D. E. and Tomboulian, D. H. (1958). *Phys. Rev.* **109**, 35.
Brown, E. and Krumhansl, J. A. (1958). *Phys. Rev.* **109**, 30.
Catterall, J. A. and Trotter, J. (1958). *Phil. Mag.* **3**, 1424.
Cooper, M., Leake, J. A. and Weiss, R. J. (1965). *Phil. Mag.* **12**, 797.
Donovan, B. and March, N. H. (1956). *Proc. Phys. Soc.* **B69**, 1249.
Goodings, D. A. (1965). *Proc. Phys. Soc.* **86**, 75.
Ham, F. S. (1962). *Phys. Rev.* **128**, 82.
Heine, V. (1957). *Proc. Roy. Soc.* **A240**, 340.
Melngailis, J. and De Benedetti, S. (1966). *Phys. Rev.* **145**, 400.
Silverman, R. A. and Kohn, W. (1950). *Phys. Rev.* **80**, 912.
Stocks, G. M., Young, W. H. and Meyer, A. (1968). *Phil. Mag.* in press.
Stott, M. J. and March, N. H. (1966). *Phys. Letters* **23**, 408.

APPENDIX. DERIVATION OF THE $\Delta E(k)$ RELATION FOR AN ENERGY-DEPENDENT POTENTIAL

Consider the Hamiltonian for a periodic lattice where the crystal potential has been replaced by some k-dependent potential (perhaps a pseudo-potential), or alternatively where electron correlations have been included; i.e.

$$H = -\frac{\nabla^2}{2} + V_k(r)$$

Writing the Schrödinger equation

$$\left[-\frac{\nabla^2}{2} + V_k(r)\right]\psi_k(r) = E(k)\psi_k(r) \tag{17}$$

where, $\psi_k(r) = u_k(r)e^{ik \cdot r}$, and $u_k(r)$ has the periodicity of the lattice, we have by substituting for $\psi_k(r)$ in (17),

$$\left[-\frac{\nabla^2}{2} + \frac{|k|^2}{2} - ik \cdot \nabla_r + V_k(r)\right]u_k(r) = E(k)u_k(r) \tag{18}$$

with the boundary condition that $u_k(r + R_n) = u_k(r)$.
Then

$$E(k) = \int_\Omega u_k^*(r)\left[-\frac{\nabla^2}{2} + \frac{|k|^2}{2} - ik \cdot \nabla_r + V_k(r)\right]u_k(r)dr \tag{19}$$

and by Feynman's theorem

$$\nabla_k E(k) = \int_\Omega u_k^*(r)\nabla_k\left[-\frac{\nabla^2}{2} + \frac{|k|^2}{2} - ik \cdot \nabla_r + V_k(r)\right]u_k(r)dr \tag{20}$$

where, on the right-hand side, ∇_k operates only on the term in brackets. Writing the second term in terms of the momentum eigenfunction, we have

$$\nabla_k[E(k)] = k + \sum_{K_n}|v(k + K_n)|^2 K_n + \int u_k^*(r)[\nabla_k V_k(r)]u_k(r)dr \tag{21}$$

The first two terms give the conventional group-velocity formula, while the third term is the correction which takes account of the non-local nature of the potential. As remarked in the text, this term appears to play a minor role in metallic Li, when we take the k-dependence of v from Bohm–Pines theory.

The Solvent-metal Soft X-Ray Emission from a Dilute Alloy

M. J. STOTT

Department of Physics, University of Sheffield, England

ABSTRACT

The profile of the soft x-ray emission from the solvent atoms in a dilute alloy depends largely on the density of states for the alloy and on the transition probability for the emission. The latter involves the behaviour of the alloy valence-electron wavefunctions in the region of the emitting solvent-atom only.

An expression for the average emission intensity for the solvent in some dilute binary-alloys has been derived by approximating the alloy valence-electron wavefunctions in the region of the emitting atom to the asymptotic form of the pure solvent-metal "Bloch" waves, scattered by the solute atoms.

Using phase shifts computed from model potentials, the lithium K-emission intensity from 5% and 10% lithium–magnesium alloys has been calculated. The results indicate that the premature peak in the pure lithium K-emission is diminished upon alloying with magnesium; this effect is due mainly to changes in the energy dependence of the transition probability, not to changes of the density of states.

The application of the theory to dilute aluminium–magnesium alloys is also discussed.

1. INTRODUCTION

A detailed calculation of the band structure of a disordered alloy starting from first principles—that is, solving for the density matrix or Green function of the system from the potential—will be difficult. There are no stringent boundary conditions which simplify calculation, as for the case of the perfect solid. When calculating the electronic properties of a perfect solid through the wavefunction, Bloch's theorem makes it sufficient to calculate the wavefunction within a unit cell only; the wavefunction throughout the whole periodic system is then obtained, apart from a phase factor. In the alloy problem, unless simplifying assumptions are made about the order, we must calculate the chosen quantity—either the density matrix or Green function—throughout the whole system.

Perhaps at first a more promising approach would be to relate the properties in question—such as the intensity of soft x-ray emission, the Knight shift, etc.—to the properties of the pure metals of which the alloy is constituted. The intensity of soft x-ray emission from the solvent atoms of a dilute alloy resembles closely the emission intensity from the pure solvent metal. This suggests that there is a simple relation between the solvent-atom soft x-ray emission from a dilute alloy and that from the pure solvent metal. It is with this relation that we are concerned here.

An expression has been derived for the solvent-atom emission from certain dilute alloys, which relates the solvent emission intensity to the pure solvent-metal transition probability and involves the scattering amplitude of the pure solvent-metal "Bloch" waves scattered by the solute or impurity atoms. Using this expression, an estimate of the solvent emission intensity from 5% and 10% lithium–magnesium alloys has been made.

2. THE DILUTE ALLOY PROBLEM

A. The Intensities of Soft X-Ray Emission from the Solute and Solvent Atoms

Consider an alloy A–B where the number of A atoms N_A is much greater than the number of B atoms, N_B. We assume that all the A and B atoms are situated on pure A-metal atom sites. The concentration C, of B atoms, is given by

$$C = N_B/(N_A + N_B)$$

When the concentration is small enough, we may consider a B atom to scatter Bloch waves independently of the other impurities. That is, to determine some matrix-element for the dilute alloy, we need only to calculate it for a single impurity in the A metal, and then multiply the change from the pure metal by C to obtain the alloy matrix-element correct to the order of the concentration.

The soft x-ray emission from such an alloy will be composed of two sets of emission bands; one set when the transition involves an A-atom core state, to be termed the "solvent" emission intensity, and the other set when a B-atom core state is involved, termed the "solute" emission intensity. we now discuss each of these in turn.

(i) *Solute-atom emission*

The core-state wavefunctions involved in the expression for the soft x-ray emission are localized functions, and so a calculation of the solute emission intensity will involve the alloy wavefunctions near to the impurity atom only.

The method of Koster and Slater (1954) provides a straightforward way of calculating the wavefunctions around an impurity in a solid. Here the perturbed Bloch functions are expanded in terms of the pure-solid Wannier functions. However, this method depends, for its tractability, on the localized nature of the pure-solid Wannier functions; the Wannier functions for the valence band in question must not extend much outside the unit cell. This may be a good approximation for semiconductors and insulators, where the tight-binding limit is more appropriate than the free-electron limit, but for lithium in particular, and the alkali metals in general, the Wannier function for the valence band is extended over many unit cells (Stott and March, this Volume, p. 296) and the Koster–Slater method is not suitable.

(ii) *Solvent-atom emission*

Consider now the emission of a photon from a solvent A-atom. In this case the expression for the emission intensity will involve knowledge of the alloy wavefunctions outside of the unit cell of the impurity B-atom. We shall require the Bloch waves of the pure metal and the scattered Bloch-waves outside the unit cell of the impurity.

B. The Asymptotic Form of the Scattered Bloch-waves

Under certain limiting conditions, the asymptotic form of the scattered Bloch-waves is a particularly simple form, which we shall now discuss.

A solute atom in a metal is considered to give rise to a perturbing potential $V_D(r)$ acting on the "Bloch" electrons. The electrons are then described by the Schrödinger equation,

$$\left[-\frac{\nabla^2}{2} + V_L(r) + V_D(r)\right] \Psi_k(r) = E(k)\Psi_k(r)$$

where $V_L(r)$ is the periodic potential of the pure metal. In the absence of the perturbing potential the solutions are Bloch waves, $\psi_k(r)$, for the particular band, with dispersion relation $E_0(k)$.

Under the limiting conditions: (i) that, the surfaces of constant energy in the occupied regions of k-space are spheres, i.e. $E_0(k)$ is a function only of $|k|$ for $k < k_f$; (ii) that there is no band-degeneracy, for example copper is excluded because of the overlapping s- and d-bands; (iii) that the perturbing potential is not strong enough to form a bound state below the bottom of the valence band; it will follow that the asymptotic form of $\Psi_k(r)$ corresponding to an incident Bloch-wave $\psi_k(r)$ and an outgoing scattered-wave, is given by

$$\Psi_k(r) \underset{r \to \infty}{=} \psi_k(r) + f(k, k')\frac{\psi_{k'}(r)}{|r|} \tag{1}$$

where the origin is taken to be at the solute atom, $f(k, k')$ is the scattering amplitude, and k' is the vector of magnitude $|k|$ in the r direction, $k' = |k|\dfrac{r}{|r|}$.

The asymptotic form of Eq. (1) should apply well to divalent or trivalent impurities in alkali metals, but it is uncertain whether it will apply to aluminium-based alloys. It is accepted that the Fermi surface of aluminium is highly spherical in an appropriate band-scheme, but clearly the constant-energy surfaces below the Fermi surface will deviate from spheres near the

Brillouin-zone boundaries and it remains to be seen how these deviations will affect the asymptotic form of the scattered Bloch-waves.

Detailed information on the Bloch scattering-amplitude is not available at the moment, so by analogy with the theory of the scattering of plane waves we take $f(k, k')$ to be a function only of $|k|$ and θ, the angle between k and k', and furthermore to be of the form

$$f_k(\theta) = \frac{1}{2ik} \sum_{l=0}^{\infty} (2l+1)(e^{i2\xi_l} - 1)P_l(\cos\theta) \qquad (2)$$

where the ξ_l are some phase shifts and the $P_l(x)$ are Legendre polynomials.

We now examine the expression for the solvent emission intensity from a dilute alloy.

3. THE SOFT X-RAY EMISSION INTENSITY

The total soft x-ray intensity emitted by a solvent atom situated at R_n, where the origin is taken to be at the solute atom, is given by

$$I(E) = \frac{16\pi^2 v^2}{3c^3} N(E) \langle |\int \phi^*(r - R_n)\nabla_r \Psi_k(r)dr|^2 \rangle$$

where, $\phi(r)$ is the core-electron wavefunction (for K-emission $\phi(r)$ will be a $1s$-like core-wavefunction), $N(E)$ is the density of states, the Dirac brackets indicate an average over the surface of constant energy $E = E(k)$, and ∇_r is the gradient operator. We neglect the effect of the hole in the core state on the intensity profile, and also the effect of many-body interactions which predominantly affect the energy dependence of the low-energy side of the intensity profile.

Since we are primarily concerned with the energy-dependence of the intensity, the effect of the slowly varying v^2-factor is neglected and the units of intensity are chosen such that

$$I(E) = N(E) \langle |\int \phi^*(r - R_n)\nabla_r \Psi_k(r)dr|^2 \rangle \qquad (3)$$

It is convenient to split up this expression for the intensity, thus

$$I(E) = N(E)T(k) \qquad (4)$$

where the density of states $N(E)$ is a quantity depending on the whole system, while the transition probability $T(k)$ depends only on the behaviour of the system near the emitting atom. Throughout, it is implied that $T(k)$ is related to the energy through the dispersion relation $E = E(k)$;

$$T(k) = \langle |\int \phi^*(r - R_n)\nabla_r \Psi_k(r)dr|^2 \rangle \qquad (5)$$

If the surfaces of constant energy in the occupied regions of k-space are spheres then $T(k)$ and $E(k)$ will be functions of $|k|$ only and can therefore be written $T(k)$ and $E(k)$.

We now consider the expression for the intensity emitted by the same A-atom in the pure A-metal, that is in the absence of the impurity.

$$I_0(E) = N_0(E)T_0(k) \tag{6}$$

where $N_0(E)$ is the pure A-metal density of states and $T_0(k)$ is the corresponding transition probability given by

$$T_0(k) = \langle |\int \phi^*(r - R_n)\nabla_r \Psi_k(r)dr|^2 \rangle \tag{7}$$

where the average is taken over the pure A-metal constant-energy surface, $E = E_0(k)$, which relates k to E.

We notice immediately that the pure-metal transition probability is independent of the position of the emitting atom. As the gradient operator is translationally invariant we have, through Bloch's theorem

$$\int \phi^*(r - R_n)\nabla_r \psi_K(r)dr = \int \phi^*(r)\nabla_r \psi_k(r)e^{ik \cdot R_n}dr$$

and hence

$$|\int \phi^*(r - R_n)\nabla_r \psi_k(r)dr|^2 = |\int \phi^*(r)\nabla_r \psi_k(r)dr|^2$$

which is independent of R_n. This result is to be expected because of the periodicity in the pure metal. When the periodicity is removed by the insertion of an impurity atom, we find that the transition probability depends on the position of the emitting atom relative to the impurity.

4. THE ALLOY TRANSITION-PROBABILITY

We now wish to examine the alloy transition-probability for a solvent atom far from an impurity, for the particular case in which the asymptotic form of the scattered Bloch-wave is given by Eq. (1).

$T(k)$ may be rewritten

$$T(k) = \langle \int \int \phi^*(r_0 - R_n)\phi(r_1 - R_n)\nabla_{r_0}\nabla_{r_1}\Psi_k(r_0)\Psi_k^*(r_1)dr_0 dr_1 \rangle$$
$$= \int \int \phi^*(r_0 - R_n)\phi(r_1 - R_n)\nabla_{r_0}\nabla_{r_1}\langle \Psi_k(r_0)\Psi_k^*(r_1)\rangle dr_0 dr_1 \tag{8}$$

Defining

$$\sigma_k(r_0, r_1) = \langle \Psi_k(r_0)\Psi_k^*(r_1)\rangle \tag{9}$$

(that is, $\sigma_k(r_0, r_1)$ is the product of the alloy wavefunctions at r_0, r_1 averaged over the sphere in k-space of radius k) we see the relationship between this work and that of Kohn and Vosko (1960) on the nuclear-resonance intensity in dilute alloys. The key quantity in the Kohn–Vosko calculations is $\sigma_k(r, r)$, the diagonal elements of the matrix (9). $\sigma_k(r, r)$, is the density of electrons in the alloy on the energy shell $E = E(k)$.

Considering only the K-emission intensity for which the core state involved is $1s$-like: for lithium K-emission, the wavefunction $\phi(r)$ has the form

$$\phi(r) \propto e^{-\alpha r}, \quad \alpha \simeq 2 \cdot 7 \text{ A.U.}$$

Clearly $\phi(r)$ is a very localized function, and in fact $\phi(r)$ has dropped to $1/e$ of its value at $r = 0$, by $r = 0 \cdot 37$ A.U., which is to be compared with the nearest-neighbour distance for lithium, $\sim 5 \cdot 7$ A.U.; thus $\sigma_k(r_0, r_1)$ will only give a contribution to $T(k)$ when $r_0 \simeq r_1 \simeq R_n$, and we need to calculate $\sigma_k(r_0, r_1)$ around the diagonal only.

A. The Calculation of the term $\sigma_k(r_0, r_1)$

With the view to calculating the alloy transition-probability $T(k)$, for a solvent atom far from an impurity, we form $\sigma_k(r_0, r_1)$ from the asymptotic form of $\Psi_k(r)$ given in Eq. (1).

$$\sigma_k(r_0, r_1) = \langle \psi_k(r_0)\psi_k^*(r_1) \rangle + \left\langle \frac{\psi_k(r_0)\psi_{k_1}^*(r_1)f_k^*(\theta_{12})}{|r_1|} \right\rangle +$$

$$\left\langle \frac{\psi_{k_0}(r_0)\psi_k^*(r_1)f_k(\theta_{02})}{|r_0|} \right\rangle +$$

$$\left\langle \frac{\psi_{k_1}^*(r_1)\psi_{k_0}(r_0)}{|r_1||r_0|}f_k^*(\theta_{12})f_k(\theta_{02}) \right\rangle \tag{10}$$

where $k_0 = |k|\dfrac{r_0}{|r_0|}$, and $k_1 = |k|\dfrac{r_1}{|r_1|}$; θ_{02} and θ_{12} are respectively the angles between k_0 and k, and k_1 and k. Then for large R_n and when r_0 and r_1 are very nearly equal to R_n, we have, after averaging over the constant-energy surfaces,

$$\sigma_k(r_0, r_1) = \langle \psi_k(r_0)\psi_k^*(r_1) \rangle + \frac{1}{2ik|r_0||r_1|}[\psi_{k_0}^*(r_0)\psi_{k_1}^*(r_1)f_k^*(\pi) - \text{c.c.}] + \text{terms}$$

$$\text{that fall off faster than } \frac{1}{R_n^2} \tag{11}$$

Now substituting this expression for $\sigma_k(r_0, r_1)$ into $T(k)$ given by (8), we have

$$T(k) = T_0(k)\left\{ 1 + \frac{1}{2ik}\left(f_k^*(\pi)\frac{e^{-i2kR_n}}{R_n^2} - f_k(\pi)\frac{e^{i2kR_n}}{R_n^2} \right) \right\} + \text{terms that fall off faster}$$

$$\text{than } \frac{1}{R_n^2} \tag{12}$$

where k is related to the energy through the alloy dispersion-relation. We see that we have related the alloy solvent-atom transition probability to that of the pure metal.

Note, that in expression (12) for $T(k)$, the transition probability for an

A-atom depends on the position of that atom relative to the impurity. This is a consequence of the Friedel electron-density oscillations; for example, if the emitting atom is situated at a node of the oscillation the probability of emission will be different from the probability if it is situated at a maximum.

5. CALCULATION OF THE AVERAGE INTENSITY WHEN A CONCENTRATION C OF IMPURITIES IS PRESENT

We have calculated $T(k)$ for a single emitting atom when there is a single impurity, and we wish now to extend this result to take account of a concentration C of impurities, and then to average over all possible emitting solvent atoms.

Following the method used by Rowlands (1962) in the study of Knight shifts in dilute alloys, we find that the average K-emission transition probability, when a concentration C of impurities is present, is given by

$$\bar{T}(k) \simeq T_0(k)\left\{1 + \frac{C}{2ik} f_k^*(\pi) \sum_{R_n \neq 0} \frac{e^{-i2kR_n}}{R_n^2} - f_k(\pi) \sum_{R_n \neq 0} \frac{e^{i2kR_n}}{R_n^2}\right)\right\} \quad (13)$$

where the summations are over all pure solvent-metal lattice sites, excluding the one at the origin.

To obtain this result we have to assume that the asymptotic form of the scattered Bloch-wave is a good approximation to the actual form, to as close as the nearest-neighbour distances. This approximation may be improved by including higher-order terms in the asymptotic form of $\Psi_k(r)$; for example, the next term to fall off as $1/r^2$ will involve derivatives of the scattering amplitude. Work is proceeding to find the correction terms to the leading term for $\bar{T}(k)$ given by (13).

In the expression (13) for the transition probability we have related $\bar{T}(k)$, the alloy transition-probability, to $T_0(k)$ the pure solvent-metal transition-probability; if an estimate of the dispersion relation for the pure solvent-metal is available then $T_0(k)$ can be obtained from experiment using relation (6) for the intensity of soft x-ray emission.

The average intensity $\bar{I}(E)$ for the alloy is given by

$$\bar{I}(E) = N(E)T_0(k)\left\{1 + \frac{C}{2ik}\left(f_k^*(\pi) \sum_{R_n \neq 0} \frac{e^{-2kR_n}}{R_n^2} - f_k(\pi) \sum_{R_n \neq 0} \frac{e^{i2kR_n}}{R_n^2}\right)\right\}$$

If $f_k^R(\pi)$ and $f_k^I(\pi)$ respectively denote the real and imaginary parts of the back scattering amplitude then $\bar{I}(E)$ becomes

$$\bar{I}(E) = N(E)T_0(k)\{1 - CF(k)\} \quad (14)$$

where

$$F(k) = \frac{f_k^R(\pi)}{k} \sum_{R_n \neq 0} \frac{\sin 2kR_n}{R_n^2} + \frac{f_k^I(\pi)}{k} \sum_{R_n \neq 0} \frac{\cos 2kR_n}{R_n^2}$$

and as we increase C, the concentration of impurities, the effect is to subtract $F(k)$ from the pure-metal transition probability $T_0(k)$.

To calculate $F(k)$ we need to evaluate the summations in $F(k)$ given by Eq. (15). These sums have been evaluated as a function of $|k| < k_f$ for the b.c.c. lattice appropriate to lithium. The procedure adopted, say for the sine summation, is to sum the quantity $\dfrac{\sin 2kR_n}{R_n^2} e^{-\alpha R_n}$ over the lattice sites within a sphere of radius 30 lattice spacings, to integrate over the rest of space, and then to let the convergence factor α tend to zero. Convergence is obtained for all except very small values of k, $k < 0.15$ A.U. The results are shown in Fig. 1, the functions

$$\sum_{R_n \neq 0} \frac{\sin 2kR_n}{kR_n^2} \quad \text{and} \quad \sum_{R_n \neq 0} \frac{\cos 2kR_n}{R_n^2}$$

vary considerably with k.

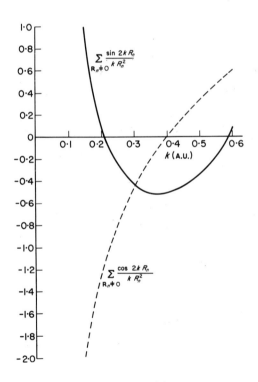

FIG. 1. Lattice summations as functions of $|k|$ for the b.c.c. lithium lattice.

6. DISCUSSION OF THE THEORETICAL RESULTS AND COMPARISON WITH EXPERIMENT

A. The Dispersion Relation and the Choice of Phase Shifts

To compare results with experiment we need an estimate of both the phase shifts and the dispersion relation.

In the case of a spherical dispersion-relation the density of states is given by

$$N(E) = \frac{4\pi k^2}{8\pi^3 \dfrac{dE}{dk}}$$

At present there is no satisfactory way of estimating the dispersion relation for a dilute alloy; there is no reason to expect that the rigid-band model, which predicts a constant energy-shift throughout the band, should hold. However, the energy-shift formula

$$E(k) - E_0(k) = \frac{\int \psi_k^*(r) V_D(r) \Psi_k(r) dr}{\int \psi_k^*(r) \Psi_k(r) dr}$$

suggests that the energy shift is simply related to the forward, Bloch scattering-amplitude and, in our approximation, to the phase shifts. Lloyd (1967) has obtained an expression for the density of states for a low density of plane-wave scattering centres. In the dilute alloy we have a low density of Bloch-wave scattering centres. In entire analogy with the result obtained by Lloyd, we take

$$E(k) - E_0(k) = -\frac{C2\pi}{\Omega} \sum_{l=0}^{\infty} (2l+1) \frac{\xi_l}{k} \qquad (16)$$

where Ω is the unit-cell volume.

From their work on the nuclear-resonance intensity, Kohn and Vosko estimated the first two phase-shifts to be inserted into the expression for the total electron-density, $\rho_{k_f}(r, r)$, by fitting ξ_0 and ξ_1 to the Friedel sum-rule and the experimental results for the residual resistivity of the alloys in question. This procedure will only give ξ_0 and ξ_1 at the Fermi level. We require the phase shifts as a function of k, up to $k = k_f$. To do this we use phase shifts obtained from the model potential of Meyer et al. (1967), which consists of the bare-ion potential screened by a shell of charge the radius of which is adjusted to satisfy charge-normalization through the Friedel sum-rule. Considerable success has been achieved with this model in predicting the transport properties of liquid metals (Young et al., 1967), of pure materials under pressure (Dickey et al., 1967a), and of dilute alloys (Dickey et al., 1967b).

We have evaluated the phase shifts for lithium and magnesium as a function of k from the Meyer model potentials, for an electron density corresponding to lithium, $k_f = 0.589$ A.U. The phase shifts to be inserted into the expressions for $F(k)$ and $E(k)$, from respectively Eq. (15) and Eq. (16), for a dilute lithium–magnesium alloy, have been taken to be the difference of the lithium and magnesium phase-shifts, $\xi_l = \xi_l$ (Mg) $- \xi_l$ (Li). The real and imaginary parts of the back scattering amplitude are shown in Fig. 2.

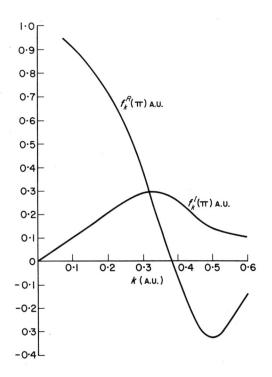

FIG. 2. The real and imaginary parts of the back scattering amplitude for lithium–magnesium as functions of $|k|$.

B. The Solvent Emission-intensity from a Dilute Lithium–Magnesium Alloy

In view of the current interest in the soft x-ray emission from pure lithium and also because in this case there are reasonably spherical constant-energy surfaces for $k < k_f$, the quantities $F(k)$, $E(k)$ and $N(E)$ have been calculated for a dilute lithium–magnesium alloy, using the phase shifts described above. For this case, the unperturbed dispersion-relation $E_0(k)$, for pure lithium, was taken to be the spherical average of the dispersion relation obtained by Ham (1962). The variation of the quantity $F(k)$ through the valence band is particularly interesting and is shown in Fig. 3. There is a sharp peak in $F(k)$ at an energy ~ 2.4 ev; this compares well with the energy of the premature peak in the pure-lithium emission intensity. The effect of adding magnesium atoms to lithium is to subtract the quantity $F(k)$ from the pure-lithium transition probability; that is, $F(k)$ tends to remove the peak in the pure-lithium emission intensity. The changes in the dispersion relation and in the density of states as we go to the alloy, have little effect on the energy-dependence of the intensity; it is the change in the transition probability which causes most of the alteration in the energy-dependence of $\bar{I}(E)$.

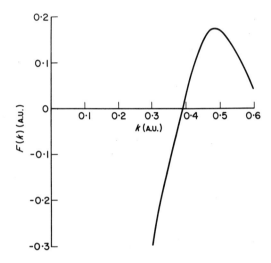

FIG. 3. The function $F(k)$ for lithium–magnesium alloys as a function of $|k|$.

The average lithium K-intensities have been calculated for a 5% and 10% dilution of magnesium in lithium and are shown together with the pure-lithium emission intensity obtained by Bedo and Tomboulian (1958) in Fig. 4. Unfortunately, there are no experimental alloy-results available for comparison. Catterall and Trotter (1959) have measured the lithium K-intensity from a 20% lithium–magnesium alloy, but this is a concentration of magnesium outside the scope of the present theory.

7. CONCLUSIONS

This paper has been concerned with the intensity of the solvent soft x-ray emission from some dilute alloys, and a calculation has been performed pertaining to the lithium K-intensity from dilute lithium–magnesium alloys. Unfortunately, there are no experimental results available for this alloy system in the small magnesium-concentration region. The results of the theoretical calculation indicate that changes of the energy-dependence of the pure-lithium emission intensity on alloying with magnesium are due

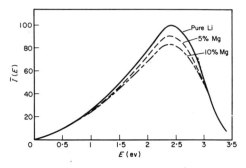

FIG. 4. The estimated lithium K-emission intensities for 5% and 10 %magnesium in lithium, and the pure lithium K-emission intensity obtained by Bedo and Tomboulian (1958).

largely to changes in transition probability. The important feature is the presence of the sharp maximum in the function $F(k)$, at an energy which is approximately that of the premature peak in the pure-lithium emission intensity-profile. This premature peak diminishes as the concentration of magnesium atoms is increased, and so the peak in the emission intensity tends to be removed. This supports the view (Stott and March, 1966) that the premature peak in the pure-lithium emission intensity is due to variations of transition probability through the occupied part of the valence band.

It would be useful to repeat the calculations described here, for aluminium–magnesium alloys. There are reliable experimental results for this system for low magnesium concentrations (Das Gupta and Wood, 1955); these indicate that the high-energy peak in the pure aluminium L-intensity decreases relative to the hump at lower energies upon alloying with magnesium. However, we are uncertain how the asymptotic form of the scattered Bloch-waves, given by (1), is altered when the constant-energy surfaces deviate from spheres near the Brillouin-zone boundaries. The present theory will not adequately deal with the shifts in energy of the points of fine-structure in the aluminium L-emission curve, but may shed light upon the gross changes on alloying.

Acknowledgement

This work was performed under a Ministry of Technology (R.R.E.) contract. The The author wishes to thank Mr. M. J. Stocks for computing the phase shifts, and Dr. W. H. Young for many useful discussions.

References

Bedo, D. E. and Tomboulian, D. H. (1958). *Phys. Rev.* **109**, 35.
Catterall, J. A. and Trotter, J. (1958). *Phil. Mag.* **3**, 1424.
Das Gupta, K. and Wood, E. (1955). *Phil. Mag.* **46**, 77.
Dickey, J. M., Meyer, A. and Young, W. H. (1967a). *Proc. Phys. Soc.* **92**, 460.

Dickey, J. M., Meyer, A. and Young, W. H. (1967b). *Phys. Rev.* **160**, 490.

Ham, F. S. (1962). *Phys. Rev.* **128**, 82.

Kohn, W. and Vosko, S. H. (1960). *Phys. Rev.* **119**, 912.

Koster, G. F. and Slater, J. C. (1954). *Phys. Rev.* **95**, 1167.

Lloyd, P. (1967). *Proc. Phys. Soc.* **90**, 207.

Meyer, A., Nestor, C. W. and Young, W. H. (1967). *Proc. Phys. Soc.* **92**, 446.

Rowland, T. J. (1962). *Phys. Rev.* **125**, 459.

Stott, M. J. and March, N. H. (1966). *Phys. Letters* **23**, 408.

Young, W. H., Meyer, A. and Kilby, G. E. (1967). *Phys. Rev.* **160**, 482.

Part 4

ELECTRON–ELECTRON INTERACTIONS IN SOFT X-RAY EMISSION

The Effect of Electron Interaction on Soft X-Ray Emission Spectra of Metals

A. J. GLICK[†]

Faculté des Sciences, Physique des Solides, Orsay, France

P. LONGE[‡]

Institut de Physique, Université de Liège, Belgium

S. M. BOSE[¶]

Department of Physics and Astronomy, University of Maryland, U.S.A.

ABSTRACT

The effect of electron–electron interactions on the soft x-ray emission-spectrum of sodium has been investigated using the methods of many-body theory. A first-order theory gives good results for the low-energy tail and the plasmon-satellite band, but breaks down in the parent band region. Here we discuss the reasons for the breakdown and the results obtained with a re-normalized theory. It was found that interactions have an important effect on the spectrum. They change its absolute magnitude by factors, and qualitatively introduce new structure which cannot be explained by a one-electron or density-of-states model. Among these effects in the intensity peak at the high-energy edge of the spectrum, which is found to be due to an infrared divergence of fermion–hole-pair excitations. A range of experiments is suggested to test the validity of the theory.

1. INTRODUCTION

In the interpretation of the shape of soft x-ray emission bands of metals, the conduction-electron (i.e. valence-electron) interactions assume an important role. An early article by Skinner (1940) described several mechanisms by which interactions could modify the emission spectrum. These were subsequently calculated by Landsberg (1949) and by Pirenne and Longe (1964). Here we describe the results of a more complete theory based on methods, which have recently been developed for treating many particle systems. The results suggest new structure in the spectrum of even the simple metals such as sodium.

[†] Permanent address: Department of Physics and Astronomy, University of Maryland, U.S.A.
[‡] Chercheur IISN, Belgium.
[¶] Present address: Catholic University, Washington, D.C., U.S.A.

We shall attempt to describe the general features of the theory and the origin of the structure introduced into the spectrum by interactions. Details of the calculations will appear elsewhere (P. Longe, A. J. Glick and S. M. Bose, to be published). The last section of the paper contains a summary of those features of the spectrum for which additional experimental information would be desirable. Our results are based on a study of the $L_{2,3}$-emission spectrum of sodium.

2. GENERAL FORMULATION AND FIRST-ORDER THEORY

The emission intensity as a function of frequency is proportional to $I(\omega)$, where

$$I(\omega) = \frac{\omega}{l} \sum_{i,f} |\langle f|\theta|i\rangle|^2 \, \delta(\omega + E_f - E_i) \tag{1}$$

where i and f are the exact eigenstates of the total Hamiltonian of the metal, l is the number of initial states, and θ is the operator coupling the metal to the radiation field. In the dipole approximation

$$\theta = \sum_{k=1}^{N} \mathbf{n} \cdot \mathbf{p}_k$$

Equation (1) can be rewritten in a convenient form by introducing a representation of the δ-function in terms of the real part ($\mathcal{R}$) of an integral

$$\delta(\omega + E_f - E_i) = \mathcal{R} \frac{1}{\pi} \int_0^\infty dt \, e^{-i(\omega + E_f - E_i)t}$$

After appropriately rearranging terms in (1) and introducing the Hamiltonian of the metal, one can apply closure to get

$$I(\omega) = \frac{\omega}{\pi l} \sum_i \mathcal{R} \int_0^\infty dt \, e^{-i\omega t} \langle i| \theta^\dagger(t) \, \theta(0)|i\rangle \tag{2}$$

where we have Heisenberg-type operators

$$\theta(t) = e^{iHt} \, \theta \, e^{-iHt}$$

(in this form there is no explicit reference to the final state, but in fact all possible final states will enter into any explicit calculation). The diagonal matrix-element appearing in Eq. (2) can be approximated by treating the interactions between the electrons as a perturbation, and using the formalism of many-body perturbation theory. With this formalism the terms of perturbation theory are represented by Feynman-type graphs. To first order in the effective interaction the graphs which contribute are those shown in Fig. 1. Time proceeds upward in these graphs. The downward-directed double lines represent the core bound-state which is unoccupied before the x-ray emission (the bound state also appears at the top of the graph because Eq. (1) contains the complex conjugate matrix-element which can be symbolically regarded as carrying the system from the final state back to the initial state). The other directed-lines represent the valence states. The wavy line is the interaction with the (x-ray) radiation field, and the line of bubbles is the dynamic effective-interaction between electrons $v(\mathbf{k})/\epsilon(\mathbf{k}, \omega)$ where $v(\mathbf{k})$ is the bare Coulomb interaction and $\epsilon(\mathbf{k}, \omega)$ is the dielectric constant of the valence electrons taken in the Lindhard random-phase approximation (RPA).

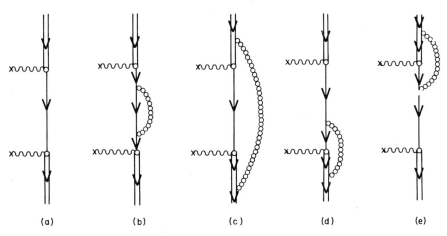

FIG. 1. Basic Feynman-type diagrams contributing to the x-ray emission intensity. (a) Contribution without electron–electron interactions; (b)–(e) graphs of first order in the effective interaction between electrons.

When the emission intensity is calculated to first order (Glick and Longe, 1965) one finds that terms (b), (c), (d) and (e) in Fig. 1 contribute in the tailing and satellite regions of the spectrum ($\omega < |E_B|$, where E_B is the energy of the bound state below the valence band). Terms (d) and (e) act as interference terms which strongly reduce the emission intensity below that of the parent band, and bring it into agreement with the experimental results of Rooke (1963). This calculation is discussed further in this Volume by Brouers and Longe, p. 329.

3. RE-NORMALIZED THEORY

A. Divergences and Re-normalizations

The above theory fails, however, when applied to the region of the parent emission band. First-order perturbation theory gives a divergent result for the intensity throughout the region $|E_B| < \omega < E_F + |E_B|$. To overcome these difficulties we have gone to a higher-order theory which includes certain re-normalizations to account for the fact that we have a strongly interacting system, and that there are charge clouds around each of the particles. The charge clouds cannot be divorced from the particles or holes in the system but must be included consistently in the theory in order to obtain sensible results.

The first type of correction that must be made is the correction for the shifts in energy of the valence electrons due to interaction self-energy. Although these shifts are small they cause difficulty in a perturbation-type treatment. One can gain insight into this difficulty by considering the δ-function of Eq. 1, which can be written

$$\delta(\omega + \epsilon_f - \epsilon_i + \Delta)$$

where, ϵ_f and ϵ_i respectively are the energies of the final and initial states of an electron falling from the valence band into the core state; and Δ represents the correction for the energy-shifts and depends on v, the strength of the interaction between particles. In effect, perturbation theory is an expansion in powers of the coupling constant. Thus perturbation theory would expand the δ-function as a Taylor series

$$\delta(\omega + \epsilon_f - \epsilon_i + \Delta) = \delta(\omega + \epsilon_f - \epsilon_i) + \Delta\delta'(\omega + \epsilon_f - \epsilon_i) + \frac{\Delta^2}{2}\delta''(\omega + \epsilon_f - \epsilon_i) + \dots$$

Such an expansion is mathematically meaningless, and introduces highly singular terms into succeeding approximations. These terms cause greatest difficulty near the edges of the main band where the derivatives of the δ-function provide a very poor representation of a slightly shifted threshold. Fortunately there is a well-prescribed procedure for eliminating this difficulty. The energy shifts are incorporated in perturbation theory, approximately to all orders, by introducing re-normalized electron propagators in which the energy of the valence electron undergoing the transition is adjusted to include interaction self-energy. The resulting contribution to the emission intensity is represented diagrammatically in Fig. 2(a), where the heavy line represents the re-normalized electron as indicated in Fig. 3. It can be seen that this term includes the contributions of Fig. 1(a), (b), as well as certain higher-order terms.

It is interesting to compare this result with that obtained earlier by Landsberg (1949), who calculated the spectrum taking into account the broadening of the valence-electron levels due to interactions. He treated the interaction as a static screened Coulomb-force and adjusted the screening length to fit the experimental shape. Figure 2(a) contains the same physical effect as considered by Landsberg; however, the present calculation is more general since it includes the dynamic aspects of the effective interaction between electrons. Over the parent-band region both approaches give similar contributions. However, the dynamic interaction introduces into the tail a new feature—the plasmon satellite. A collective oscillation of the

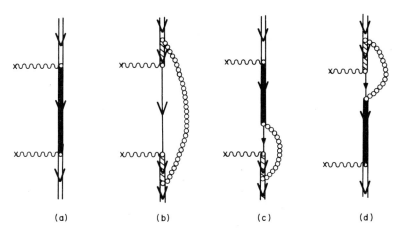

FIG. 2. Basic diagrams of the re-normalized theory. The graphical definitions of the re-normalized lines are given in Figs. 3 and 4.

FIG. 3. Graphical representation of the integral equation satisfied by the re-normalized conduction-electron propagator. $\Sigma(k, \omega)$ is the electron self-energy.

valence electrons in a metal can be set up during the radiation process. Since it requires energy to create the plasmon, the x-ray photon is emitted with reduced energy, giving rise to a satellite band with energy shifted downwards from the main emission band.

These features were already present for the analogous term in the first-order theory, Fig. 1(b). But now a new feature is introduced. Hedin et al. (1967) have pointed out that the real part of the electron self-energy could play an important role in modifying the single-particle spectrum of an electron gas. Hedin (1967; this Volume, p. 337) applied this result to soft x-ray emission and suggested that additional structure could be expected in the tail of the spectrum, near the low-energy end of the plasmon satellite. They identified this structure with an excitation which they call the plasmaron. Such structure appears quite markedly in the contribution from term (a) of Fig. 2. However, as discussed below, it is reduced in magnitude by other terms, and it appears that this effect will cause only a small change in the actual emission-spectrum. In the first-order theory it was found that terms (c), (d) and (e) of Fig. 1 contribute strongly in the tailing region. Their contribution to the higher-order theory must also be considered before drawing conclusions about the true shape of the spectrum.

The divergence of the term (c) in Fig. 1 has a different origin, and is interesting because it has not previously to our knowledge been noticed to play a role in these problems. Normally, the interactions between the valence electrons in a metal cause them to scatter in such a way that they conserve momentum. However, in the present case the missing bound electron acts, in effect, as a localized charged impurity and breaks up the translational symmetry experienced by the valence electrons. As a consequence the impurity-potential scatters electrons out of the Fermi sea without momentum restrictions. Among the forms of scattering which can occur, are those that carry electrons from *occupied* states very close to the Fermi surface to *unoccupied* states very close to the surface. The energy of excitation is then arbitrarily small, though the momentum change of the electrons can be as large as $2k_F$. In this way a large number of electron–hole pairs can be created at very little cost of energy. When considered in conjunction with the x-ray emission process, this effect appears to give rise to an infrared type of divergence. It differs from the ordinary infrared divergence in that the low-energy particles are not photons, but rather Fermion particle–hole pairs.[†]

This process should, in fact, be self-limiting. As the pairs are excited, they should form a screening cloud which inhibits further pair excitation. The theory is modified to include this effect by re-normalizing the propagator of the bound-hole state. The state must always be considered together with its interaction "self-energy" which is the result of three types of processes as indicated in Fig. 4. In the last self-energy correction, on the right side of Fig. 4(b), the dashed line represents a radiative correction while the other two terms represent electron excitations. The term on the left contributes only to the real part of the self-energy which cancels out of the calculation. The most important term is the middle one, which gives rise to the Auger-width of the bound state. In effect it is this width of the level which removes the divergence and makes the contribution finite. Then the graphical contribution of term (c) of Fig. 1 to the x-ray emission spectrum is replaced by that of term (b) of Fig. 2.

The interference terms (d) and (e) of Fig. 1 are also affected by the infrared process. These terms diverge at the upper edge of the spectrum at $\hbar\omega = |E_B| + E_F$. In the re-normalized theory the valence-electron lines and the bound-state lines are re-normalized as indicated in (c) and (d) of Fig. 2. Though the divergences are eliminated by this procedure, some effects of the infra-red process seem to persist in the x-ray emission spectrum. The calculation leads to a peak in intensity, rising above the $E^{1/2}$-behaviour, near the high-

---

[†] Recently, Anderson (1967) has written about an infrared catastrophe in the Fermi gases with local scattering potentials. While these two effects may be related the precise nature of the relationship is not apparent to the authors at this time.

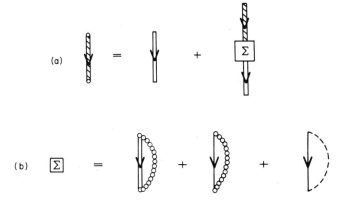

FIG. 4. Graphical representation of the integral equation satisfied by the bound-state hole propagator. The hole self-energy, $\Sigma_B(\omega)$, has a radiative part and an electron-interaction part, and the intermediate hole can be in the core or conduction band.

energy edge as indicated for the main emission band shown in Fig. 5. This result is particularly interesting because it may offer an explanation for the peak observed by Skinner (1940) in sodium (see the dashed line in Fig. 5). As reported in this Volume by Rooke (see p. 3), Crisp and Williams also observe this peak; although it is not present in the spectrum of Cady and Tomboulian (1941). It would be of interest to have additional experimental verification of this feature and information about its temperature dependence. The fact that Crisp and Williams found it to persist at different electron densities indicates that it is not due to band structure, and could indeed originate from the infrared effect that we have found.

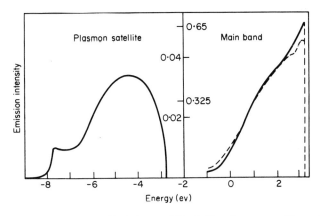

FIG. 5. Results for the re-normalized theory of Fig. 2. The plasmon satellite on the left shows some residual plasmaron structure. The main band on the right shows a high-energy peak due to the fermion "infrared" effect.

B. Results

Figure 5 shows the results obtained for the terms represented in Fig. 2. The spectrum differs in several ways from those expected from a simple density-of-states theory. The $E^{1/2}$ behaviour seems to persist only in the middle portion of the main band. A low-energy tail is added with a plasmon-satellite band and possibly a plasmaron structure. At the high-energy edge there is a peak due to the particle–hole infrared effect. Although the spectrum depends on the structure of the density of states of the electrons, the relationship is very complicated, and it seems unlikely that one could work backwards from the experiments to obtain a reliable density of states.

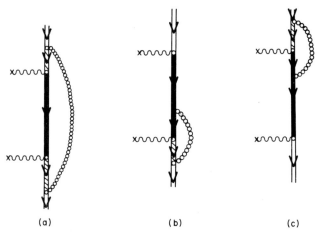

FIG. 6. Basic diagrams of a "thoroughly re-normalized theory" for which calculations are at present in progress.

As mentioned above, Hedin has interpreted the spectrum in terms of a generalized-interaction density of states, but even this picture is open to question as it neglects terms such as (b) of Fig. 2, and incorporates only a portion of the interference terms (c) and (d). With these additional terms one finds that the plasmaron effect is severely reduced. Its only surviving effect, seen in Fig. 5, is the small ridge on the left-hand side of the plasmon-satellite band.

Another interesting feature of the results is shown by the comparison of the absolute magnitude of the intensity with that of a one-electron model, for the main band and for the first-order plasmon satellite. In first order it was found, after the cancellation due to the interference terms, that the peak of the plasmon-satellite band amounted to $\sim 2.5\%$ of the peak-intensity of the main band (as given by the one-electron model). This result is significant because the experimental results of Rooke (1963) show a similar relationship of magnitudes. The re-normalized theory differs markedly in magnitude from the simple theory, in that the absolute magnitude of the main band is enhanced by a factor of about three due to electron interactions. Unfortunately the experiments only provide relative intensities and the validity of the theory cannot be checked on this basis. However, the plasmon satellite is even further enhanced, and reaches almost 8% of the corrected main-band peak. Such a strong satellite could appear as a result of the neglect of related terms which should be included in the present approximations. The delicate cancellation, found in the first-order theory, could be lost as a result of neglecting these terms. The additional corrections can be incorporated into the theory by replacing terms (b), (c) and (d) of Fig. 2 by, respectively, (a), (b) and (c) of Fig. 6. It is very difficult to calculate the graphs of Fig. 6 and they are still being studied. Preliminary results are encouraging; they indicate that the changes will be small in the main-band region, while the intensity of the tailing and plasmon satellite will be somewhat reduced and therefore more in accordance with the experimental magnitude.

4. CONCLUDING REMARKS

The calculation reported above was carried out for sodium in order to minimize the role of band structure and to bring out most clearly those effects in the soft x-ray emission spectrum which can be attributed to the interactions between the electrons in the metal. It was found that interactions do indeed play a major role. They change the absolute intensity of emission by large factors, and also introduce qualitatively new features which cannot be attributed to a one-electron density of states.

Since these effects are so marked in the theory, the study of the soft x-ray emission spectrum of simple metals provides an important method for experimentally testing the theoretical approximations, as well as for gaining new insight into the role of electron correlations in metals. In particular, the following features of the spectrum of sodium deserve further study and verification.

(1) The structure of the main band. Is there a peak at the high-energy edge? What is its magnitude and width? How does it depend on temperature and electron density?

(2) The structure of the plasmon satellite band. Is there any residual plasmaron structure, or is it completely removed by higher-order cancellation?

(3) The energy-dependence of the satellite band near its high-energy edge; the cancellation between graphs results in a $(E_0 - E')^{3/2}$-behavior immediately below the threshold, where the threshold energy $E_0 = |E_B| + E_F - \hbar\omega_p$ (i.e. the main-band threshold minus the plasmon energy). In fact this behavior may not be observable since the numerical results shown in Fig. 5 appear to give a more rapid rise, even quite close to the emission edge.

(4) The relative magnitude of the plasmon satellite band and the main band. Is there a temperature or density dependence?

(5) The strength and fall-off of the low-energy tail of the main band, and the separation of the tail due to electron interactions from the experimental background.

A thorough understanding of the spectrum of sodium and probably lithium would provide a necessary prerequisite to the effective application of soft x-ray spectrometry in the determination of the properties of more complicated metals and alloys. The present theory is an attempt in this direction, but there is also an urgent need for additional experimental information.

Acknowledgement

This work was supported in part by the U.S. Air Force Office of Scientific Research. Computer time for the work was assisted by Grant no. NsG398 from the National Aeronautics and Space Administration to the Computer Science Center of the University of Maryland.

References

Anderson, P. W. (1967). *Phys. Rev. Letters* **18**, 1049.
Cady, W. M. and Tomboulian, D. H. (1941). *Phys. Rev.* **59**, 381.
Glick, A. J. and Longe, P. (1965). *Phys. Rev. Letters* **15**, 589
Hedin, L., Lundqvist, B. I. and Lundqvist, S. (1967). *Solid State Commun.* **5**, 237.
Hedin, L. (1967). *Solid State Commun.* **5**, 451.
Landsberg, P. T. (1949). *Proc. Phys. Soc.* **A62**, 806.
Pirenne, J. and Longe, P. (1964). *Physica* **30**, 277.
Rooke, G. A. (1963). *Phys. Letters* **3**, 234.

A New Perturbative Interpretation of the Satellite Plasmon Emission Band

F. Brouers[†] and P. Longe[‡]

Institut de Physique, Université de Liège, Belgium

ABSTRACT

Two successful theoretical interpretations of the plasmon satellite are compared. The first due to Brouers uses canonical transformations. The second due to Longe and Glick uses the dielectric constant method. A better physical understanding of the satellite emission is obtained where the interference of the proper fields of the quasi-particles is emphasized. A new perturbative approach to this problem is foreseen where the interaction of the electron with the whole electronic medium is treated in the same manner as the interaction with the radiation field.

1. INTRODUCTION

In recent years improvements in the sensitivity of soft x-ray spectrometers (Fisher *et al.*, 1958; Watson *et al.*, 1967) have permitted the detection of low-intensity features in soft x-ray spectra reflecting the collective behaviour of the conduction or valence electrons.

One of the most interesting features, due to the many-body character of the electron gas, is the detection (Rooke, 1963; Watson, Dimond and Fabian, this Volume, p. 45) in the $L_{2,3}$-emission spectra of Al, Mg and Na and in the K-emission spectrum of Be, of a satellite band on the low-energy side of the parent emission band at an energy separation of the order of the plasmon energy $\hbar\omega_p$. The relative intensity of these satellites, with respect to the parent band, is of the order of 1 or 2%. They do not reflect exactly the profile of the parent emission band. For the $L_{2,3}$ satellites, instead of the sharp edge characteristic of an L-emission band, a gradual rise in intensity is observed.

The physical nature of these satellite bands is now quite well established (Brouers, 1964, 1965; Glick and Longe, 1965). They are due to the simultaneous emission of an x-ray photon and a plasmon. In this process the x-ray photon being deprived of an amount of energy $\hbar\omega_p$ gives rise to a satellite shifted to the low-energy side of the emission band. The shape of the satellite is a consequence of slight frequency dispersion of the plasmon. Its weak intensity is due to a strong destructive interference between effects which we shall especially emphasize in this paper.

[†] Present address: H. H. Wills Physics Laboratory, University of Bristol, England.
[‡] Chercheur I.I.S.N.

2. THEORETICAL FORMULATION

Using the procedure adopted by Bohm and Pines, Brouers (1964, 1965) obtained a theoretical shape of the satellite band in good agreement with observed spectra (Rooke, 1963). On the other hand, the expression obtained by Brouers is a good approximation of that obtained by Glick and Longe (1965) who by the use of a different but more general method have calculated not only the satellite but also the tailing of the main emission band. This is the subject of the foregoing paper, presented by Glick, Longe and Bose. It is interesting to compare both methods. This will lead to a better physical understanding of the emission process, concerning not only the satellite but also the parent band.

A. The Canonical Transformation Method

Brouers applied two canonical transformations to the Hamiltonian of a model consisting, in its initial state, of N valence electrons together with a particle at rest having an infinite mass and a charge $+e$ which represents the core hole responsible for the x-ray transition. This x-ray transition is treated as a hole–electron annihilation, which gives rise to a final state containing $N-1$ valence electrons.

The Hamiltonian is

$$H = H_0 + EM \tag{1}$$

where EM is the electromagnetic term, and

$$H_0 = \frac{1}{2m} \sum_{p} p^2 C_p^\dagger C_p + \frac{1}{2} \sum_{\substack{p, p' \\ k \neq 0}} v(k)(C_{p+k} C_{p'-k} C_{p'} C_p - N) -$$

$$\sum_{k, p} v(k) C_{p+k}^\dagger C_p b^\dagger(R) b(R) e^{-ik \cdot R} \tag{2}$$

where, $v(k)$ is the Coulomb interaction $4\pi e^2/k^2$, $C_p^\dagger$ and C_p are respectively the creation and annihilation operators of a valence electron of momentum p and $b^\dagger(R)$ and $b(R)$ are respectively the creation and annihilation operators of a core vacancy in an ion localized at R.

Then $N(R) = b^\dagger(R) b(R)$ is equal to unity when there is a core vacancy and to zero when there is none. Hereafter R will be the origin of the co-ordinates. The two first terms are the electronic part of H_0, the last term is due to the presence of a core vacancy.

The term EM of H (Eq. 1) describes the x-ray emission and may be written:

$$EM = \sum_{p} D(p) \alpha^\dagger C_p b(R) \tag{3}$$

where $\alpha^\dagger$ is the creation operator of an x-ray photon, and where $D(p)$ is proportional to the matrix element corresponding to the transition of a valence electron, of momentum p, to the core vacancy.

Using the Bohm–Pines formalism (see for example, Pines, 1964), new terms coupling the electrons and the ion to a plasmon field are introduced in the Hamiltonian (Eq. 1). The first Bohm–Pines canonical transformation is extended to take account also of the charge density around the core vacancy. The transformation $H' = e^{is_1} He^{-is_1}$ is generated by the operator

$$S_1 = \frac{1}{i\Omega} \sum_{k(<k_c)} [v(k)]^{1/2} Q_k (\sum_p C^\dagger_{p+k} C_p - b^\dagger b) \qquad (4)$$

where Ω is the volume of the metal and k_c is the cut-off wave vector separating the collective behavior from the individual behavior of the electron gas. The collective co-ordinate Q_k is related to the creation and annihilation plasmon operators, respectively $A^\dagger_k$ and A_k, by

$$Q_k = (2\omega_p)^{-1/2} (A_k - A^\dagger_{-k}) \qquad (5)$$

This canonical transformation surrounds every electron and core vacancy with a polarization charge-cloud, and gives rise to collective oscillations, of frequency ω_p, weakly coupled to these quasi-electrons. A consequence of this is the screening of the electron–electron and electron–vacancy interactions. Moreover, the transformed Hamiltonian contains an additional term

$$\frac{1}{\Omega} \sum_{\substack{p \\ k(<k_c)}} \left[\frac{\tilde{v}(k)}{2\omega_p}\right]^{1/2} [D(p) - D(p-k)] a^\dagger A^\dagger_k C_p b \qquad (6)$$

which describes the simultaneous emission of an x-ray photon and plasmon; the latter being emitted either by the vacancy or by the electron falling into it.

This term appears in addition to the normal emission term (3). It is equal to zero if the matrix element $D(p)$ does not depend on p, which is almost the case for an L-transition. Physically, this means that at the moment of the x-ray emission, the charge clouds of the two quasi-particles (electron and core vacancy) are exactly superimposed. Because of their opposite charges a destructive interference occurs. Note, however, that until now the plasmon-satellite bands have been observed mostly in L-emission spectra.

The cancellation of the term (6) appears to be complete because the individual motion of the electrons has been neglected. Indeed the quasi-electrons are not proper states of the Hamiltonian (2). Consequently the second Bohm–Pines transformation must be used. This canonical transformation eliminates the weak coupling, which exists between the collective oscillations and the quasi-electron due to their random motions. The transformed Hamiltonian

$$H'' = e^{is_2} H' e^{-is_2}$$

gives rise, on the one hand, to independent modified collective oscillations with a slight dispersion (real plasmons), and on the other hand, to new quasi-electrons whose charge clouds (virtual plasmons) are velocity dependent.

The term (6) describing the simultaneous emission of an x-ray photon and a plasmon becomes

$$\frac{1}{\Omega} \sum_{\substack{p \\ k(<k_c)}} \left[\frac{v(k)}{2\omega p}\right]^{1/2} D(p-k) \frac{2k \cdot p - k^2}{2m\omega_p - 2k \cdot p + k^2} \alpha^\dagger A_k^\dagger C_p b \qquad (7)$$

Now this term is no longer equal to zero, even in the case of an L-transition, and there is always a finite probability of a plasmon being generated during the x-ray emission. One could say that this term is not zero due to the finite mass of the electron; that is the electron can undergo a recoil when the plasmon is emitted, in contrast to the ion containing the core vacancy whose mass is considered as infinite. This interpretation will be more clearly understood when we resort to the results obtained by Glick and Longe.

The intensity of emission of a photon of energy ω may be written in the general form:

$$I(\omega) = \omega \sum_f \delta(\omega + E_f^0 - E_i^0)|\langle \Psi_f^0|EM|\Psi_i^0\rangle|^2 \qquad (8)$$

where Ψ^0 and E^0 denote respectively, exact eigenstates and eigenvalues of the Hamiltonian H_0. After the two above canonical transformations, this intensity may be written as the sum of two terms:

firstly,
$$I_{MB}(\omega) = D^2\omega \sum_{p(<p_F)} \delta(\omega - E_p + E_B) \qquad (9)$$

which comes from perturbation (3) and describes the main emission band, and secondly,

$$I_{SB}(\omega) = \frac{D^2\omega}{2\Omega\omega_p} \sum_p \sum_{p(<p_F)\,k<k_c} \delta(\omega + \omega_p - E_p + E_B) \left[\frac{2k \cdot p - k^2}{2m\omega_p - 2k \cdot p + k^2}\right]^2 \qquad (10)$$

which comes from perturbation (7) and describes the satellite band. In these expressions, $E_p = p^2/2m$ is the energy of a valence electron, and E_B is the energy of the vacant bound level.

B. The Dielectric Constant Method

The calculations performed by Glick and Longe (1965) are also discussed in detail in this Volume (p. 319). These investigators use the formalism of a frequency and wavenumber dependent dielectric constant $\epsilon(k, \omega)$ (Nozières and Pines, 1958). This dielectric constant is computed in the Lindhard approximation, equivalent to the random phase approximation used in the Bohm–Pines canonical transformations. The starting point is expression (8), which may be rewritten in the form

$$I(\omega) = \frac{\omega}{\pi l} \sum_i \mathcal{R} \int_0^\infty dt \, e^{-i\omega t} \langle \Psi_i^0 | (EM)_t (EM)_0 | \Psi_i^0 \rangle \tag{11}$$

where the terms $(EM)_t$ are time-dependent Heisenberg operators. In this expression only the initial states appear explicitly. Note that, to be more exhaustive, we have introduced an average $\frac{1}{l} \sum_i$ over the l possible degenerate initial states. For instance, for an L-transition this would be the three vacant $2p$ bound-levels. In expression (11) the final states are included implicitly in the operator $(EM)_t(EM)_0$ of which l diagonal elements must be computed. These l diagonal elements may be expanded in powers of the effective interaction $v(k)/\epsilon(k, \omega)$ and we shall retain only the zero-order and first-order terms. These terms are related to the Feynman-type graphs (a, b, c, d, e) used by Glick, Longe and Bose (p. 319); their main contributions take the form:

for graph (a), $I_0(\omega) = \omega \sum_{p(<p_f)} \delta(\omega - E_p + E_b) \frac{1}{l} \sum_i |\langle p| - i n \cdot \nabla |B_i\rangle|^2 \tag{12}$

and for graphs (b, c, d and e),

$$I_1(\omega) = \frac{\omega}{\pi \Omega l} \sum_{\substack{p(<p_F) \\ i,k}} \mathcal{I} \left[\frac{v(k)}{\epsilon(k, \omega - E_p + E_B)} \right]_- \times$$

$$\sum_{p'} \frac{\langle p| e^{i k \cdot r} |p'\rangle \langle p'| - i n \cdot \nabla |B_i\rangle}{\omega - E_{p'} + E_B} - \sum \frac{\langle p| - i n \cdot \nabla |B_j\rangle \langle B_j| e^{i k \cdot r} |B_i\rangle}{\omega - E_p + E_B} \Bigg|^2 \tag{13}^\dagger$$

where $|p\rangle$ and $|B_i\rangle$ represent one-electron states of energies respectively E_p and E_B, and n is the photon polarization vector. Term (12) is immediately identified to (9) by writing

$$D^2 = \frac{1}{l} \sum_i |\langle p| - i n \cdot \nabla |B_i\rangle|^2 \tag{14}$$

and concerns the main emission band. Expression (13) describes not only the satellite band but also the tailing of the main emission band; moreover it introduces a contribution in the region of the emission band which must be added to $I_0(\omega)$. However, note that in this region the squared modulus form given to $I_1(\omega)$, expression (13), is not entirely correct; problems concerning principal values must be considered as well as some finer points (divergences) discussed by Glick *et al.* in the foregoing paper, and solved by S. M. Bose (thesis, unpublished). The dielectric-constant method is thus more powerful than the canonical-transformation method. Its physical interpretation is less obvious, however. The form of $I_1(\omega)$ indicated in expression (13) permits such

---

† In this paper we use only the part of $[\epsilon(k, \omega)]^{-1}$ that is analytic in the lower half of the complex ω-plane.

an interpretation. Indeed the terms within the modulus bars of expressions (12) and (13) may be related to the open Feynman-type graphs of Fig. 1, obtained by cutting the closed graphs (a–e) used by Glick, Longe and Bose (p. 321). It can be seen immediately that on squaring the following relationships exist between these open graphs (a′, b′, c′), and the closed ones (a, b, c, d, e):

$$(\text{graph } a')^2 \rightarrow \text{graph } a$$
$$(\text{graph } b')^2 \rightarrow \text{graph } b$$
$$(\text{graph } c')^2 \rightarrow \text{graph } c$$
$$2 \times (\text{graph } b') \times (\text{graph } c') \rightarrow \text{graph } d + \text{graph } e$$

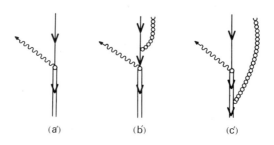

(a′)   (b′)   (c′)

FIG. 1. Open Feynman-type graphs which contribute to the x-ray emission. The double straight line represents the core vacancy; the single straight line, a vacancy in a valence state; the wavy line, a photon; and the bubble line, the plasmon or any excitation in the valence band.

Note, however, that the graphs (b, d and e) may be cut at other points. The terms related to such cuts were neglected in expression (13) for the purposes of the present discussion. Expression (10), computed by Brouers (1967), may be deduced from Eq. (13) by making some quite acceptable approximations. Since we are concerned only with the satellite (and neglect the tailing), we may write

$$\mathscr{I}[\epsilon(k, \omega)]^{-1} \simeq \frac{\pi}{2} \omega_p \delta(\omega + \omega_p)$$

with $k < k_c$. Further, we use Eq. (14) and consider only one bound state— that one for which the dipolar matrix-element is the most important. The wavefunction of this state is considered as highly localized in comparison with k_c^{-1}, and is taken as a δ-function. Thus

$$\langle B | e^{ik \cdot r} | B \rangle \simeq 1$$

These approximations allow us to write

$$I_1(\omega) \simeq \frac{D^2 \omega \omega_p}{2\Omega} \sum_{p(<p_F)} \sum_{k<k_c} \delta(\omega + \omega_p - E_p + E_B) \left| \frac{1}{\omega_p - E_p + E_{|p-k|}} - \frac{1}{\omega_p} \right|^2 \quad (15)$$

which is the expression (10) established by Brouers. The two terms in the modulus now permit a much more complete interpretation of the plasmon generation. These terms are respectively related to graphs (b′) and (c′). Just before the x-ray emission, the emitting electron and the core vacancy may be considered to be at the same place, each surrounded by its polarization charge-cloud. These clouds interfere, and after the emission the small residue of their interference is liberated as a real plasmon. Graphs (b′) and (c′) represent the liberation respectively of the charge-clouds of the electron and of the core vacancy. The first term of the modulus shows clearly that the recoil of the electron is possible only in process (b′). Without this recoil the contribution of processes (b′) and (c′) would cancel each other precisely and no plasmon would be generated. We may consider that the two terms in the modulus of expression (15) represent the form factors of the charge-clouds superimposed during the emission. The disassociation of process (b′) and (c′) is not emphasized in the derivation of expression (10) by the canonical transformation method. However, this disassociation is important for it yields a good physical insight of the plasmon generation process and explains its weak intensity.

C. The Perturbative Interpretation

We shall write expression (13) in another interesting form by using the expression

$$\mathscr{I}\left[\epsilon(k, \omega)\right]^{-1} = \frac{\pi}{\Omega} v(k) \sum_n \delta(\omega + \Delta E_n) |p_{n,0}(k)|^2 \tag{16}$$

where $\rho(k)$ is the Fourier transform of the electron density, and $\rho_{n,0}(k)$ represents its matrix elements between the ground state and an excited state n of the whole electron gas. The energy difference of these two states is ΔE_n. Expression (13) becomes

$$I_1(\omega) - \frac{\omega}{l} \sum_{p(<p_F)} \sum_{k,n} \delta(\omega + \Delta E_n - E_p + E_B) \times$$

$$\left| \sum_{p'} \frac{\langle P|\tau_n(k)|p'\rangle\langle p'|-i n \cdot \mathbf{V}|B_i\rangle}{E_B - (\omega - E_{p'})} + \sum_j \frac{\langle p|-i n \cdot \mathbf{V}|B_j\rangle\langle B_j|\tau_n(k)B_i\rangle}{E_B - (E_B + \Delta E_n)} \right|^2 \tag{17}$$

with

$$\tau_n(k) = \frac{1}{\Omega} v(k)\, \rho_{n,0}(k)$$

This appears to be a second-order perturbation expression, where the perturbative interactions are the interactions of the valence electrons firstly with the radiation field, and secondly with the Coulomb field produced by the $\rho_{n,0}(k)$ charge distribution. This expression, like (13), is applicable not only in the region of the satellite but also in the whole spectrum. It emphasizes a point to which we should like to draw attention. It is only in a first approximation that the x-ray emission of metals may be considered as due to transitions between one-electron states.

To give a complete interpretation of the experimental results, it is desirable not to restrict the discussion to the one-electron model. The discrepancy between observed spectra and spectra computed by means of this model is significant for even the simplest metals like the alkali metals. More elaborate theories (Landsberg, 1949; Pirenne and Longe, 1964) which take into account the instability of the one-electron states, merely introduce a finite level width. It appears now that the levels have not only to be broadened, but that this broadening must have a shape which is sometimes very different from a simple Gaussian profile. Strong changes in absolute intensity may also be expected, but have never been measured. The concept of a one-electron level, it appears from recent papers (Hedin, 1967), is becoming less meaningful. In fact we could say that in many x-ray transitions, the electronic medium tends to participate as a whole in the process. A similar situation is known to exist in nuclear spectroscopy.

Acknowledgement

We should like to thank Professor J. Pirenne, of the University of Liège, and the Solid State Group of the University of Maryland for helpful discussions.

References

Brouers, F. (1964). *Phys. Letters* **11**, 297.
Brouers, F. (1965). *Physica Stat. Sol.* **11**, K25.
Brouers, F. (1967). *Physica Stat. Sol.* **22**, 213.
Fisher, P., Crisp, R. S. and Williams, S. E. (1958). *Optica Acta* **5**, 31.
Glick, A. J. and Longe, P. (1965). *Phys. Rev. Letters* **15**, 589.
Hedin, L. (1967). *Solid State Commun.* **5**, 451.
Landsberg, P. T. (1949). *Proc. Phys. Soc.* **A62**, 806.
Nozières, P. and Pines, D. (1958). *Nuovo Cimento* **X9**, 470.
Pines, D. (1964). "Elementary Excitations in Solids", p. 104. Benjamin, New York.
Pirenne, J. and Longe, P. (1964). *Physica* **30**, 277.
Rooke, G. A. (1963). *Phys. Letters* **3**, 234.
Watson, L. M., Dimond, R. K. and Fabian, D. J. (1967). *J. Sci. Instrum.* **44**, 506.

Many-body Effects in the Soft X-Ray Emission from Metals

L. HEDIN

Institute of Theoretical Physics, Göteborg, Sweden

ABSTRACT

Using general non-perturbative arguments, the following approximate expression is derived for the x-ray intensity

$$I(\omega) = C\omega \sum_{k} |p_{ck} + p_{ck}^{(+)} + (\omega + \epsilon_c - \epsilon_k^{HF})p_{ck}^{(-)}|^2 A(k, \omega + \epsilon_c)$$

$A(k, \omega)$ is the one-electron spectral function and for small-core metals, is close to the electron-gas result; C is a constant. The quantities $p_{ck}^{(+)}$ and $p_{ck}^{(-)}$ sample matrix-elements p_{ck} with k respectively above and below the Fermi surface. They are a consequence of the presence of a hole in the core.

In the main emission band the quasiparticle contribution to A dominates, and the coefficient of the $p_{ck}^{(-)}$ term is small. The satellite band derives from the structure in $A(k, \omega)$ below the quasiparticle peak. The factor $\omega + \epsilon_c - \epsilon_k^{HF}$ is then approximately constant and equal to $-\hbar\omega_p$, resulting in a reduction in the intensity. Values of k close to the Fermi surface give rise to the high-energy edge of the satellite (the plasmon edge) while small k should give a low-energy edge to the satellite (the plasmaron edge). The intensity of the plasmaron edge is estimated to be quite small; for aluminium less than about 1% of the parent band-intensity (per wavelength unit). The analysis made here applies only when the relaxation of the valence electrons before the emission is complete and when the hole in the core causes no resonant scattering.

1. INTRODUCTION

The simplest approximation for the intensity of soft x-ray emission is given by the well-known expression

$$I(\omega) \simeq \omega \sum_{k} |p_{ck}|^2 \delta(\omega + \epsilon_c - \epsilon_k)$$
$$= \omega N(\epsilon_k)|p_{ck}|^2 \text{ with } \epsilon_k = \epsilon_c + \omega \tag{1}$$

There are two important effects which modify this simple result. These are: polarization of the valence electrons by the presence of the hole in the core; and the dynamical interactions between the valence electrons. The two preceding papers (by Brouers and Longe, and by Glick, Longe and Bose) have elaborated the Bohm–Pines and perturbation theoretical treatments of electron–electron and electron–hole interactions, and three important points

337

were noted. (1) The intensity in the main band is increased relative to the simple result of Eq. (1); this is due to the presence of the hole and the effect has been discussed earlier by Ferrell (1956) and treated theoretically by Shuey (1966). (2) In the low-energy tail a plasmon edge is found; this effect was predicted by Ferrell (1956) and found experimentally by Rooke (1963) and also by Watson, Dimond and Fabian (this Volume, p. 45). (3) When calculating the intensity in the low-energy tail, cancellations were observed due to interference between the polarization introduced by the hole and the emitted plasmon; this made the intensity very small.

Despite the work by Glick *et al.* and by Brouers, there are still many questions to be resolved. I shall discuss here simple many-body effects in an essentially non-perturbative manner, thus avoiding the difficulties of convergence in perturbation expansions and the question of subsidiary conditions in Bohm–Pines theory. The discussion represents an elaboration on previous work (Hedin, 1967). Attention is given to the possibility of observing the "plasmaron" which is a new type of elementary excitation in an electron gas, and was proposed by Hedin *et al.* (1967). The excitation broadly corresponds to the resonance interaction of a plasmon with a hole (Lundqvist, 1967a).

2. THE SATELLITE EMISSION

A. Survey of Relevant Experimental Data

Before proceeding, it is useful to briefly survey the experimental situation and the features we are trying to describe. In Fig. 1 the emission curve for aluminium is plotted from the raw data obtained by R. K. Dimond (unpublished). The raw data give the number of counts during some time unit taken *per wavelength unit*. To convert the raw data to intensity per unit energy (i.e. to the quantity $I(\omega)$ in Eq. (1)) we must multiply by ω^{-2}. However, the quantity most naturally calculated is $I(\omega)/\omega$, and this is what we shall discuss here. This is also the quantity that Brouers and Longe, and Glick *et al.* calculate and give numerical results for, while Rooke (1963) gives the raw-data results.

The curve of Fig. 1 was obtained from the raw data, obtained by Dimond, by subtracting a linear background and then multiplying by ω^{-3}. In the figure additional background has also been drawn, starting discontinuously at the Fermi edge; this background is mostly guesswork. However, the high step at the Fermi edge may not be exaggerated, judging from the results on self-absorption obtained by Dimond (1967). The low-energy part of the curve is enlarged by a factor of 10, and all experimental points are included. Above it the satellite is re-drawn after subtracting the estimated background.

In Fig. 2 the main features of an emission band are shown with the satellite grossly exaggerated. Figure 3 is the density-of-states curve obtained by Lundqvist (1968) for an electron gas at roughly the density for aluminium. This curve would closely represent the experimental curve if the effect of the hole could be neglected, and if the matrix-elements p_{ck} were constant (Hedin, 1967).

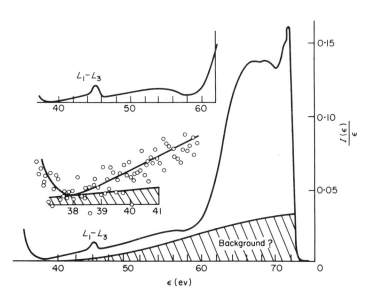

FIG. 1. Soft x-ray emission spectrum for Al. $I(\epsilon)$ is the intensity per unit energy. Inset, with experimental points, is an enlargement of the low-energy part of the curve, scaled by a factor of 10. The upper inset shows the satellite with the crudely estimated background subtracted. The data were obtained by R. K. Dimond (unpublished).

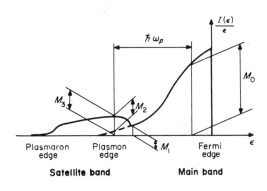

FIG. 2. The main features of a soft x-ray emission curve.

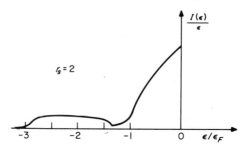

FIG. 3. Density-of-states curve for an electron gas with $r_s = 2$ (reproduced with permission from Lundqvist, 1968).

One characteristic feature of the satellite that has been given much attention is the value of its maximum intensity above the tail of the main band (M_2 in Fig. 2) relative to some intensity in the main band (for which, following Rooke (1963), we choose M_0 in Fig. 2). In Table I we give some theoretical and experimental results for M_2 as a percentage of M_0.

TABLE I

Values for M_2

Author	Al	Mg	Na
Brouers (1965)	1·1	1·3	2
Glick and Longe (1965)			2
Bose (1967)			5
Rooke (1963); Expt.	2·4	2·4	4

If we take M_0 relative to the background which has been drawn in Fig. 1, and use a slightly lower extrapolation of the main emission-band tail than that used by Rooke, we obtain the results given in Table II for M_1, M_2 and M_3 as percentages of M_0. For comparison the electron-gas values of Fig. 3 are also given.

TABLE II

Results for M

	M_1	M_2	M_3
Experiment	4	5	8
Electron gas	4	12	15

The background in Fig. 1 was deliberately drawn to cause the experimental and electron-gas values to agree for M_1 and for M_3 minus M_2. The reason for this will be explained later. It should be emphasized that the raw data obtained by Rooke and by Dimond agree very well. The difference between the M_2 values in the two tables is due only to our method of interpreting the data.

It is clearly difficult to obtain accurate experimental values for M_2, and the results depend very much on the theoretical model chosen; however, we feel that the Bohm–Pines treatment used by Brouers (1965) gives intensity values that are far too low, perhaps by as much as a factor of 5.

While the raw data for aluminium are very good, for sodium they are considerably less accurate. The theoretical calculation by Glick and Longe (1965) seems to give a rather too low intensity while that of Bose (1967) gives a result which is consistent with experiment. Comparing the electron-gas result for M_2 with the experimental value, we see that it is too large by more than a factor of 2. This is due to a cancellation effect which comes from the presence of the hole, but is not nearly as large as that obtained by Brouers (1965) and by Glick and Longe (1965).

B. Background to Theoretical Interpretation

At this point we introduce formulae which enable a comparison between the present work and that of Glick, Longe and Bose.

If we could neglect the effect of the hole and consider only electron–electron interactions, a reasonable approximation for the intensity (Hedin, 1967) is

$$I(\omega) \simeq \omega \sum_k |p_{ck}|^2 A(k, \omega + \epsilon_c) \tag{2}$$

where A is the spectral density-function for one-electron excitations, and is related to the one-electron Green function

$$G(k, \omega) = \frac{1}{\omega - \epsilon_k - \Sigma(k, \omega)} \tag{3}$$

through the relation

$$A(k, \omega) = \frac{1}{\pi} \mathscr{I} G(k, \omega) = \frac{1}{\pi} \frac{\mathscr{I} \Sigma(k, \omega)}{(\omega - \epsilon_k - \mathscr{R} \Sigma)^2 + (\mathscr{I} \Sigma)^2} \tag{4}$$

If the self-energy Σ was just some real function $\Sigma(k)$ plus an infinitesimal imaginary part, Eq. (4) would give a δ-function for A, and would return us to the simple expression of Eq. (1). In fact $A(k, \omega)$ as a function of ω for fixed k has, besides the quasi-particle peak, additional structure which is associated with the presence of plasmons. The quasi-particle peak is reasonably well-defined for all k. The structure follows from very general arguments (Hedin et al., 1967) saying that $\mathscr{I}\Sigma(k, \omega)$ should have peaks at $\omega = E_k \pm \hbar\omega_p$, where E_k is the quasi-particle energy, and $\hbar\omega_p$ is the plasmon energy. The peaks in $\mathscr{I}\Sigma$ automatically lead as a result of the dispersion relation to sharp steps in $\mathscr{R}\Sigma$, which cause new solutions of the Dyson equation.

Extensive calculations of the form of the function $A(k, \omega)$ have been made by Lundqvist, using a simple approximation (1967a,b) and using the full

Lindhard dielectric-function (1968). Results for $I(\omega)/\omega$ using Eq. (2) with a constant matrix-element (Lundqvist, 1968) are shown in Figs. 3 and 4, respectively for $r_s = 2$ (Al density) and for $r_s = 4$ (Na density). The dashed curves of Fig. 4 show the result obtained from the first-order theory of Glick and Longe (1965) according to their Feynman-diagram (b). The curves are plotted from numerical results taken from the work of Bose (1967). The first-order theory is an expansion of G to first order in Σ

$$G(k, \omega) = \frac{1}{\omega - \epsilon_k} + \frac{\Sigma(k, \omega)}{(\omega - \epsilon_k)^2} \qquad (5)$$

and gives for A in the satellite region

$$A(k, \omega) = \frac{1}{\pi} \frac{\mathscr{I} \Sigma(k, \omega)}{(\omega - \epsilon_k)^2} \qquad (6)$$

The difference, shown in Fig. 4, between the respective results for $I(\omega)/\omega$ obtained with Eqs. (4) and (6) is not very sensitive to the chosen approximation for $\Sigma(k, \omega)$ but follows from the structure of the two formulae and from the general nature of $\Sigma(k, \omega)$. The main emission-band is almost unchanged in relation to the Sommerfeld model. This is because the width of the quasi-particle peaks is relatively small, and because the dispersion of the quasi-particle energies E_k is close to that of the free-particle energies ϵ_k.

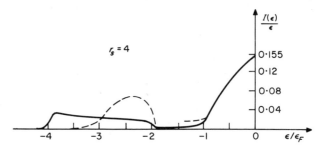

FIG. 4. Density-of-states curve for an electron gas with $r_s = 4$ (reproduced with permission from Lundqvist, 1968). The dashed curve is drawn from calculations by Bose (1967) to give the results of the "first-order theory" of Glick and Longe (1965).

When the effect of the hole was considered by Glick and Longe (1965) they found large cancellations for the satellite intensity. The net effect was essentially a re-scaling of the dashed satellite-band of Fig. 4, without change of form. Later, Bose (1967) made calculations (reported in this Volume by Glick et al.), using the full Green function of Eq. (3) rather than the first-order expression of Eq. (5). For the self-energy, Σ, Bose used the same RPA-expression (random phase approximation) as Lundqvist (1968). However, in this case, the effect of the hole does not seem to give a re-scaling of the satellite; instead the curve obtained by Bose is very similar to the result obtained by Glick and Longe (1965), except for a small second hump on the low-energy side of the main hump. The position of this second hump does not agree with the plasmaron edge of Fig. 4, but comes at $\epsilon/\epsilon_F = -3\cdot2$, which raises serious doubt regarding the accuracy of the calculations by Bose. Bose found that the net effect of the hole was to *increase* the intensity in the satellite by almost a factor 2 and in the main band by a factor 4.

According to theoretical considerations a plasmaron edge should appear in the Al $L_{2,3}$-emission at about 39–40 ev. In earlier work (Hedin et al., 1967) we found not only an edge but also a peak; but this peak should disappear with an improved treatment (Hedin, 1967). The magnitude of the intensity drop at the edge is difficult to estimate. If we neglect the influence of the k-dependence of the matrix-elements, p_{ck}, we should expect the electron-gas result at the plasmaron edge to be reduced by the same factor as M_2, i.e. by about $2\cdot5$. This gives a value of $0\cdot01$ in the units marked on the abscissa of Fig. 1, and 1% of the parent band peak-intensity if we measure intensity per unit wavelength. Now even a small variation of p_{ck} with k can appreciably change this result. However, our analysis indicates that even though the magnitude of the edge might be quite small, its form should not significantly change.

3. THE THEORETICAL ANALYSIS

We consider only cases where the "sudden" approximation is valid; that is the relaxation of the valence electrons before the emission of the x-ray photon is complete and unique. This we expect to be a very good approximation for light elements such as aluminium, magnesium and beryllium. The "energetics" of the problem is then very simple. For the energy E_i of the initial state, we write $E(N-1, c)$ to remind ourselves that one electron is missing in a core level. We neglect the width of the initial state and consider it as a truly stationary state. For the energy E_f of the final state, we write $E(N-1, s)$ where s indicates some quantum state of the valence electrons. The energy of the emitted x-ray photon is

$$\omega = E_i - E_f = E(N-1, c) - E(N-1, s) \qquad (7)$$

The one-electron Green function for the metal has poles at energies

$$\epsilon_s = E(N) - E(N-1, s) \quad \text{and} \quad \epsilon_c = E(N) - E(N-1, c)$$

where $E(N)$ is the ground-state energy of the metal when *all* core states are filled. The x-ray photon energy can thus be written

$$\omega = E_i - E_f = \epsilon_s - \epsilon_c \qquad (8)$$

The largest possible value of ϵ_s is the chemical potential μ, i.e.

$$\omega_{\max} = \mu - \epsilon_c \qquad (9)$$

Thermal excitations of the valence electrons, and the width of the core level, will give rise to some intensity at higher values of ω. The quantity

$\mu - \epsilon_c$ can also be experimentally obtained from photoemission measurements, which can nowadays be made to an accuracy of 0·1 ev. For the light metals the results agree very well with soft x-ray data.

A successful theoretical analysis of the contributions to $\mu - \epsilon_c$ has been made earlier (Hedin, 1965a). The essential contributions to the energy ϵ_c are found to be, (1) the ionization energy of the free ion, (2) the pure coulomb-potential from the valence electrons, calculated with no hole in the ion, (3) a polarization contribution, equal to half the coulomb interaction between the core-electron charge density and the change in the valence-electron density caused by the presence of the hole. These contributions can be understood if we "switch off" the charge of the core electron before we take it out of the metal and then "turn it on" again. The polarization contribution is quite important and in magnitude comparable to the potential from the undisturbed charge distribution, i.e. contributions (2) and (3) are of comparable magnitude. The results indicate an increase in charge density in the unit cell, where the core has a hole, by a factor ~ 2 for monovalent metals and by a rather smaller factor for polyvalent metals such as aluminium. This is what we might expect: the screening distance of an impurity charge is of the order of an inter-atomic distance.

The intensity of x-ray emission is given by the well-known expression

$$I(\omega) \simeq \omega \sum_f \left| \langle \Psi_f | \int \psi^*(r) p \psi(r) dr | \Psi_i \rangle \right|^2 \delta(\omega - E_i + E_f) \qquad (10)$$

where the momentum operator is written in second quantization using the field operator $\psi(r)$. As a first and very plausible approximation we neglect *dynamical* correlations between core and valence electrons and write

$$\begin{aligned} |\Psi_i\rangle &= |\Psi_{\text{ion with hole}}\rangle |N_v^*\rangle \\ |\Psi_f\rangle &\doteq |\Psi_{\text{ion}}\rangle |N_v - 1, s\rangle \end{aligned} \qquad (11)$$

where $|N_v^*\rangle$ is the state vector for the valence electrons in the presence of the hole.

The core hole polarizes the valence electrons appreciably and the difference between $|N_v^*\rangle$ and the state vector for the ground state with no core hole, $|N_v\rangle$, is significant. Using Eq. (11) the expression for the intensity becomes

$$I(\omega) \simeq \omega \sum_s \left| \sum_k p_{ck} \langle N_v - 1, s | a_k | N_v^* \rangle \right|^2 \delta(\omega + \epsilon_c - \epsilon_s) \qquad (12)$$

where p_{ck} is now an ordinary matrix-element, as in Eq. (1). If we replace $|N_v^*\rangle$ by $|N_v\rangle$, Eq. (12) reduces to Eq. (2) which follows from the definition of the spectral function A:

$$A(k, \omega) = \sum_s |\langle N_v - 1, s | a_k | N_v \rangle|^2 \delta(\omega - \epsilon_s) \qquad (13)$$

To approximate $|N_v^*\rangle$ we use the expression

$$|N_v^*\rangle = (\alpha + \sum_{pk} \alpha_p^k a_k^+ a_p)|N_v\rangle \tag{14}$$

The operator creating $|N_v^*\rangle$ from $|N_v\rangle$ is of the same structure as the plasmon-creating operator derived by Bohm and Pines (1953) and by Sawada et al. (1957). To estimate the coefficients α we consider a model problem with a positive point-charge in an electron gas, and minimize the energy using $|N_v^*\rangle$ as a trial function with $|N_v\rangle$ as a single Slater-determinant. The resulting equations for determining α can be written

$$\alpha_p^k = \frac{\alpha}{\Omega} \frac{w_p^k}{\epsilon_k - \epsilon_p + \epsilon_0} \tag{15}$$

$$\epsilon_0 = \frac{2}{\Omega^2} \sum_{pk} \frac{w_p^k v(k-p)}{\epsilon_k - \epsilon_p + \epsilon_0} \tag{16}$$

$$w_p^k = v(k-p)\left[1 - \frac{2}{\Omega}\sum_q \frac{w_q^{k-p+q}}{\epsilon_{k-p+q} - \epsilon_q + \epsilon_0}\right] +$$
$$\frac{1}{\Omega}\sum_q \left[\frac{w_p^q v(q-k)}{\epsilon_q - \epsilon_p + \epsilon_0} - \frac{w_q^k v(q-p)}{\epsilon_k - \epsilon_q + \epsilon_0} + \frac{w_q^{k-p+q} v(q-p)}{\epsilon_{k-p+q} - \epsilon_q + \epsilon_0}\right] \tag{17}$$

$$\alpha^2 + \sum_{kp} |\alpha_p^k|^2 = 1 \quad \text{(normalization condition)}. \tag{18}$$

The main problem is to solve the integral equation (17) for w. Using Eq. (16), ϵ_0 must be adjusted self-consistently; the coefficients α can then be determined from Eqs. (15) and (18). In these equations we have used the notation

Ω = total volume
$v(k) = 4\pi e^2/k^2$
$\epsilon_k = \dfrac{\hbar^2 k^2}{2m} + V_{ex}(k)$ = HF single-particle energy
ϵ_0 = energy gain from presence of point charge ($\epsilon_0 > 0$).

Note that here ϵ_k stands for the HF (Hartree–Fock) energy while earlier we used the symbol ϵ_k for the Hartree or free-particle energy.

When $k \to p$, the dominant contribution to the right-hand side of Eq. (17) comes from the first term. Neglecting the other terms we have an explicit solution for w

$$w_p^k = w(k-p) = \frac{v(k-p)}{1 + v(k-p)\dfrac{2}{\Omega}\sum_q \dfrac{\pi_{p+q}(1 - \pi_{k+q})}{\epsilon_{k+q} - \epsilon_{p+q} + \epsilon_0}} \tag{19}$$

This is almost, but not quite, an RPA-screened potential. Explicit evaluation shows that w is singular for small q

$$\lim_{q \to 0} w(q) \simeq \frac{1}{q} \tag{20}$$

This singularity is of correct kind to make $|N_v^*\rangle$ satisfy the screening condition discussed by Bergersen and Terrell (this Volume, p. 351).

Using the approximation for $|N_v^*\rangle$ given by Eq. (14), we can write the matrix element in the expression Eq. (12) for the intensity in the form

$$\langle N_v - 1, s| \sum_{\kappa'} p_{c\kappa'} a_{\kappa'} |N_v^*\rangle =$$

$$\alpha \left[\left(p_{c\kappa} + \frac{1}{\Omega} \sum_q p_{c,\kappa+q} \frac{w_\kappa^{\kappa+q}}{\epsilon_{\kappa+q} - \epsilon_\kappa + \epsilon_0} \right) \langle N_v - 1, s|a_\kappa|N_v\rangle + \right.$$

$$\left. \frac{1}{\Omega} \sum_{kq} \frac{p_{c,\kappa+q} w_k^{k+q}}{\epsilon_{k+q} - \epsilon_k + \epsilon_0} \langle N_v - 1, s|a_{k+q}^+ a_k a_{\kappa+q}|N_v\rangle \right] \tag{21}$$

where we have taken $|N_v - 1, s\rangle$ to have momentum κ. Note that the quantities w_p^k are zero unless $k > k_F$ and $p < k_F$ (cf. Eq. (14) and the definition of w).

If we could neglect the second term in Eq. (21), that is the triple term containing three a-operators, we would again arrive at Eq. (2), but with p_{ck} replaced by an effective matrix-element involving states above the Fermi surface.

To estimate the triple term we evaluate the commutator $[H, a_\kappa]$ for the model system of a point charge in an electron gas

$$[H, a_\kappa] = -\frac{\hbar^2 \kappa^2}{2m} a_\kappa - \frac{1}{\Omega} \sum_{kq} v(q) a_{k+q}^+ a_k a_{\kappa+q} \tag{22}$$

We take matrix elements of this equation with respect to the states $|N_v - 1, s\rangle$ and $|N_v\rangle$:

$$\left(\epsilon_s - \frac{\hbar^2 \kappa^2}{2m} \right) \langle N_v - 1, s|a_\kappa|N_v\rangle = -\frac{1}{\Omega} \sum_q v(q) \langle N_v - 1, s|a_\kappa \pi_{\kappa+q}|N_v\rangle +$$

$$\frac{1}{\Omega} \sum_{\substack{kq \\ k \neq \kappa}} v(q) \langle N_v - 1, s|a_{k+q}^+ a_k a_{\kappa+q}|N_v\rangle \tag{23}$$

The first term on the right-hand side is not very different from the Hartree-Fock exchange term multiplied by $\langle N_v - 1, s|a_\kappa|N_v\rangle$. This approximation gives

$$(\epsilon_s - \epsilon_\kappa) \langle N_v - 1, s|a_\kappa|N_v\rangle = \frac{1}{\Omega} \sum_{\substack{kq \\ k \neq \kappa}} v(q) \langle N_v - 1, s|a_{k+q}^+ a_k a_{\kappa+q}|N_v\rangle \tag{24}$$

Compare the last terms in Eqs. (21) and (24). The k-dependence of the factor in front of the triple term in Eq. (21) is weak. The q-dependences of the factors in front of the triple terms in Eqs. (21) and (24) are not quite the same, but they are similar. Thus the last two terms in Eqs. (21) and (24) should be roughly proportional. We write for Eq. (21)

$$\langle N_v - 1, s| \sum p_{c\kappa'} a_{\kappa'} |N_v^* \rangle = \alpha [p_{c\kappa} + p_{c\kappa}^{(+)} + (\epsilon_s - \epsilon_\kappa) p_{c\kappa}^{(-)}] \langle N_v - 1, s|a_\kappa|N_v \rangle \quad (25)$$

which should be a reasonable approximation with a proper choice of $p_{c\kappa}^{(+)}$ and $p_{c\kappa}^{(-)}$. Eq. (25) leads us, once again, back to Eq. (2) but now with *an effective matrix-element that depends on energy*. The quantity $p_{c\kappa}^{(+)}$ is an average of matrix-elements p_{ck} with k above the Fermi surface while $p_{c\kappa}^{(-)}$ samples matrix-elements below the Fermi surface.

In the main band the contributions to $\langle N_v - 1, s|a_\kappa|N_v \rangle$ for a given κ essentially come from states with one quasi-particle. We can then replace ϵ_s in Eq. (25) by the quasi-particle energy E_κ. Since $E_\kappa - \epsilon_\kappa$ is negative at the Fermi surface and positive at $\kappa = 0$ (Hedin, 1965b), we expect a decrease in intensity at the Fermi edge and an increase at the bottom of the band, if the matrix elements are constant.

The effects in the satellite from $p_{c\kappa}^{(-)}$ are larger than in the main band. To see this we first consider states with energies just below the plasmon edge. They have κ close to the Fermi surface and energy $\epsilon_s = E_\kappa - \hbar \omega_p$. In this case $\epsilon_s - \epsilon_\kappa = E_\kappa - \epsilon_\kappa - \hbar \omega_p$ is negative and of magnitude much larger than for states in the main band. States which give rise to the plasmaron edge have small κ and energy $\epsilon_s = E_\kappa - \gamma \hbar \omega_p$. The quantity γ varies somewhat with electron density. For aluminium it is $\simeq 1\cdot 4$. The change in $\epsilon_s - \epsilon_\kappa$ going down the satellite band is approximately

$$(\epsilon_s - \epsilon_\kappa)_{plasmon} - (\epsilon_s - \epsilon_\kappa)_{plasmaron} \simeq (\gamma - 1) \hbar \omega_p + V_{ex}(k_F)$$

This is a small number since $V_{ex}(k_F)$ is negative and of about the same size as $(\gamma - 1) \hbar \omega_p$. We find that $\epsilon_s - \epsilon_\kappa$ does not change much over the satellite band which should therefore be rather uniformly scaled-down relative to the main band, when the κ-dependence of $p_{c\kappa}^{(-)}$ is not too different from that of $p_{c\kappa} + p_{c\kappa}^{(+)}$.

We argued earlier that the charge density should be about doubled in the unit cell with a core hole. However, unlike the positron case we cannot expect more than a doubling, say, of the total x-ray intensity. Since most of the intensity comes from the main band this means that the effective matrix-element $\alpha(p_{c\kappa} + p_{c\kappa}^{(+)})$ should be approximately equal to $\sqrt{2} p_{c\kappa}$, which with $\alpha = 0\cdot 9$, say, gives $p_{c\kappa}^{(+)} = 0\cdot 6\, p_{c\kappa}$. For aluminium we must use a smaller value for the density increase, say $4/3$ instead of 2. With $\alpha = 0\cdot 9$ this gives $p_{c\kappa}^{(+)} = 0\cdot 3\, p_{c\kappa}$. These numbers are given only to show that the effect of the hole embodied in $p_{c\kappa}^{(+)}$ need not "drown" the matrix-element $p_{c\kappa}$ of the "simple" theory. On the other hand the states above the Fermi surface certainly have an important influence on the emission spectrum and one cannot expect to get detailed agreement if they are neglected.

The intensity in the main band and in its tail before the plasmon edge should according to our discussion be essentially given by the theoretical curve in Fig. 3, properly modified with the effective matrix-element. The "guessed" background in Fig. 1 has been drawn with this in mind. In particular the ratio between the intensity just before the plasmon edge and the average intensity in the main band are the same in Figs. 1 and 3. The intensity after the plasmon edge in Fig. 1 increases to about 5% of the value in the main band, while in Fig. 3 it increases to 12%. The discrepancy can be accounted for by a proper choice of $p_{c\kappa}^{(-)}$. We have, approximately,

$$\left(\frac{5}{12}\right)^{1/2} (p_{c\kappa} + p_{c\kappa}^{(+)}) = p_{c\kappa} + p_{c\kappa}^{(+)} - \omega_p p_{c\kappa}^{(-)}$$

which gives

$$\omega_p p_{c\kappa}^{(-)} = 0 \cdot 35 (p_{c\kappa} + p_{c\kappa}^{(+)}).$$

The modification by the energy dependence of the effective matrix-element

$$\alpha [p_{c\kappa} + p_{c\kappa}^{(+)} + (\epsilon_s - \epsilon_\kappa) p_{c\kappa}^{(-)}]$$

is therefore relatively small.

Our argument has assumed that the matrix-elements $p_{c\kappa}$ and $p_{c\kappa}^{(\pm)}$ are always real and of the same sign. This is a plausible assumption for $p_{c\kappa}$ and $p_{c\kappa}^{(-)}$, but not necessarily for $p_{c\kappa}^{(+)}$. If $p_{c\kappa}$ and $p_{c\kappa}^{(+)}$ had essentially different signs this would not affect our discussion of the satellite, but would result in a total intensity that was smaller than in the simple theory. This is a quite possible situation since the intensity depends on matrix-elements and not only on charge density.

According to our analysis the shape of the plasmon edge should be given by the expression

$$I(\omega)/\omega \simeq \sum_k |p_{ck} + p_{ck}^{(+)} + (E_k - \epsilon_k - \hbar\omega_p) p_{ck}^{(-)}|^2 A(k, \omega + \epsilon_c) \qquad (27)$$

If the effective matrix-element is constant we obtain the form shown in Fig. 3. This shape is not expected to change much with a more accurate treatment of the electron gas. The edge might, however, be somewhat influenced by the rapid change of the Hartree–Fock energy ϵ_k at the Fermi surface.

The region of energy over which the satellite rises to its peak value (or plateau value, depending on how the background is drawn) has been associated with the dispersion region of the energy of a free plasmon (Rooke, 1963). This interpretation is doubtful on the grounds that the form and approximate size of the plasmon edge are approximately the same whether calculated using an RPA dielectric-function (Lundqvist, 1968) or a model in which the plasmon has no damping whatsoever and carries all the oscillator strength (Lundqvist, 1967a,b). Framed in more physical terms, the results from the electron-gas calculations tell us that a well-defined free plasmon cannot exist in the neighbourhood of a hole, but that the coupled plasmon–hole aggregate, or plasmaron, can.

4. SUMMARY

From fairly general arguments we have arrived at a very simple formula for the x-ray emission intensity (cf. Eqs. (12), (25) and (13))

$$I(\omega) \simeq \omega \sum_{k} |p_{ck} + p_{ck}^{(+)} + (\omega + \epsilon_{c\bar{k}} - \epsilon_{k}^{HF}) p_{ck}^{(-)}|^2 A(k, \omega + \epsilon_c) \qquad (28)$$

In the main band this reduces to essentially the same result as Eq. (1)

$$I(\omega) \simeq \omega \sum_{k} |p_{ck}^{eff}|^2 \delta(\omega + \epsilon_c - E) \qquad (29)$$

except that we have to use an effective matrix-element

$$p_{ck}^{eff} = p_{ck} + p_{ck}^{(+)} + (E_k + \epsilon_c - \epsilon_k^{HF}) p_{ck}^{(-)} \qquad (30)$$

that depends significantly on states above the Fermi surface through $p_{ck}^{(+)}$ and to some extent on states below the Fermi surface through $p_{ck}^{(-)}$.

This qualitative form has some interesting consequences. The quasiparticle density of states comes in as a factor and the van Hove singularities discussed by Rooke should appear in the spectrum. However, they will be smeared due to the width of the quasiparticles. Numerical estimates of these widths may be found in several reports (Hedin, 1965b; Bose *et al.*, 1967; Lunqvist, 1968). Another consequence is the influence of $p_{ck}^{(+)}$ on the lower part of the main band. For K-emission spectra, where $p_{ck} \to 0$ when $k \to 0$, $p_{ck}^{(+)}$ in general will tend to a constant. Because most K-emission bands in fact have something like a $E^{3/2}$ form, $p_c^{(+)}$ cannot completely dominate over p_{ck}, which in turn means that the total intensity cannot be very strongly enhanced.

An approximation for the quantity $p_{ck}^{(+)}$ can be obtained from Eq. (21)

$$p_{ck}^{(+)} = \frac{1}{\Omega} \sum_{q} p_{c,k+q} \frac{w_k^{k+q}}{\epsilon_{k+q} - \epsilon_k + \epsilon_c} \qquad (31)$$

where

$$w_k^{k+q} \simeq w(q) = \frac{v(q)}{1 + v(q)\dfrac{2}{\Omega} \sum_{k} \dfrac{\pi_k(1 - \pi_{k+q})}{\epsilon_{k+q} - \epsilon_k + \epsilon_c}} \qquad (32)$$

These equations are directly useful for crude numerical estimates, but the integral equation (17) for w should also be susceptible to numerical treatment. A suitable check on the quality of the result is to evaluate the charge density at the impurity and compare with positron lifetimes.

The crucial approximation made was the replacement of the last term in Eq. (21) by the $p_{ck}^{(-)}$ term in Eq. (28). To obtain a crude estimate of $p_{ck}^{(-)}$ we can take out the average values of $\langle N_v - 1, s | a_{k+q}^+ a_k a_{\kappa+q} | N_v \rangle$ in Eqs. (21) and (24) and then sum over k and q with the restrictions $k+q > k_F$, $k < k_F$ and $\kappa + q < k_F$. Finally we identify the two averages.

Admittedly, there are uncertainties in the analysis that we have presented, but we believe it to be essentially correct. We feel more sure of the conclusions regarding the main band than regarding the satellite. We look forward with great interest to new results on the plasmaron edge. The very existence of the plasmaron has not yet been rigorously established and the experimental verification of its existence by the observation of a plasmaron edge would be extremely valuable. One could then more confidently proceed with the task of finding its manifestations in other circumstances, for example photo-emission of electrons and tunnelling. Plasmaron effects might also occur in degenerate semiconductors where the energies lie in a more accessible range.

Acknowledgement

I am grateful to R. K. Dimond for communicating his excellent data on Al and his permission to use them here; to B. Bergersen for discussions on consistency requirements in the impurity problem of the core hole; and to J. Krumhansl, J. Wilkins, G. A. Rooke, G. Brogren and S. Hagström for helpful discussions.

My interest in soft x-ray spectra was aroused in connection with the study of the plasmaron problem for an electron gas, carried out together with B. I. Lundqvist and S. Lundqvist. I am grateful to these investigators for fruitful co-operation and for many discussions on the soft x-ray problem.

References

Bohm, D. and Pines, D. (1953). *Phys. Rev.* **92**, 609.
Bose, S. M. (1967). Thesis University of Maryland.
Bose, S. M., Bardasis, A., Glick, A. J., Hone, D. and Longe, P. (1967). *Phys. Rev.* **155**, 379.
Brouers, F. (1965). *Physica Stat. Sol.* **11**, K25.
Dimond, R. K. (1967). *Phil. Mag.* **15**, 631.
Ferrell, R. A. (1956). *Rev. Mod. Phys.* **28**, 308.
Glick, A. J. and Longe, P. (1965). *Phys. Rev. Letters* **15**, 589.
Hedin, L. (1965a). *Arkiv Fysik* **30**, 231.
Hedin, L. (1965b). *Phys. Rev.* **139**, A796.
Hedin, L. (1967). *Solid State Commun.* **5**, 451.
Hedin, L., Lundqvist, B. I. and Lundqvist, S. (1967). *Solid State Commun.* **5**, 237.
Lundqvist, B. I. (1967a). *Phys. kondens. Materie* **6**, 193.
Lundqvist, B. I. (1967b). *Phys. kondens. Materie* **6**, 206.
Lundqvist, B. I. (1968). *Phys. kondens. Materie* **7**, 117.
Rooke, G. A. (1963). *Phys. Letters* **3**, 234.
Sawada, K., Brueckner, K., Fukuda, N. and Brout, R. (1957). *Phys. Rev.* **108**, 507.
Shuey, R. T. (1966). *Phys. kondens. Materie* **5**, 192.

The Use of Consistency Requirements in Calculating the Response to a Point Impurity in an Electron Gas

B. BERGERSEN

Institute of Theoretical Physics, Göteborg, Sweden

J. H. TERRELL

Mithras, A Division of Sanders Associates, Cambridge, Massachusetts, U.S.A.

ABSTRACT

We investigate the electron distribution in the polarization cloud around a charged point-impurity in the light of two physical consistency-requirements of the problem, namely: (1) the impurity charge should be completely screened as seen at large distances from the impurity; (2) the expectation value of the electron density should be non-negative for any sign and reasonable magnitude of the electron charge. The second requirement is used to extend the validity of the linear-response results to obtain a qualitative theory for the positron-annihilation rate in metals. In the case of a heavy impurity it is shown that the consistency requirements offer a convenient tool to check the validity of different approximation methods. In particular we discuss the violation of the requirements in several methods based on approximations to the Hartree self-consistent-field model.

1. INTRODUCTION

We wish to investigate the electron distribution in the polarization cloud around a charged impurity in an electron gas. In positron-annihilation experiments the observed lifetime is inversely proportional to the electron density at the position of the impurity (i.e. the positron). Generally, the electron distribution around an impurity is not measured directly, and one is usually interested in this quantity only within the framework of the Hartree self-consistent-field approximation, where it determines and is determined by an effective potential through the Poisson equation. The effective interaction is then used to calculate further properties of the system. The accuracy with which the effective interaction must be known varies with the problem at hand. This sometimes obscures the fact that a charged impurity causes a disturbance which, at metallic densities, is markedly non-linear in the impurity charge and thus, is poorly approximated by linear

351

response theory. In the positron case, the nonlinear nature of the problem is indicated by an observed electron density at the position of the positron which is from 5 to 50 times the unperturbed density. (For recent experimental results see Weisberg and Berko, 1967.)

In soft x-ray spectrometry the core vacancy can be considered as a positively charged impurity, which polarizes the medium before an electron is captured into the core state. Unlike the positron lifetime, the core vacancy lifetime, which unfortunately cannot be measured at present, is inversely proportional to an average of the electron density over fairly large distances—by atomic standards—from the impurity. The polarization will cause a shortening of the lifetime. However, because of the averaging which takes place, this effect may be fairly independent of the details of the charge distribution, as long as the total polarization charge is calculated correctly.

In positron annihilation experiments one measures, in addition to the lifetime, the angular correlation of the two annihilation photons. These emerge at almost opposite angles, and the deviation is proportional to the centre-of-mass (c.m.) momentum of the annihilating pair (for a description of the experiment see for example Stewart and Roellig, 1967). From the observed positron lifetime one might expect the polarization to have a marked effect on the angular correlation. However, this is not so. The qualitative aspects of the experimental angular correlation curves can be crudely explained by assuming that the dominant part is due to a non-interacting positron, which is annihilated from the zero momentum state, in a non-interacting electron gas (the Sommerfeld model). In addition there is a small tail due to the penetration of the positron wavefunction into the atomic cores. Polarization effects are only important when we come to explain the relative areas of the two parts of the curve (Terrell et al., 1967; Carbotte, 1967). The agreement of the main part of the curve with the simple model is sufficiently striking that one can clearly distinguish effects due to the anisotropy of the Fermi surface from experiments with oriented crystals (Berko, 1962; Stewart et al., 1962).

This apparent contradiction between the lifetime and angular correlation experiments can be explained by assuming a picture in which the positron independently correlates with the different valence electrons in the Fermi sea, starting out approximately from the state of zero momentum. Since the c.m. momentum is conserved in the electron–positron collisions, one would expect an angular correlation which roughly resembles that of the Sommerfeld model, provided the positron has freedom of choice of electrons with which to correlate, and in particular no bound states or resonance occur.

Similar arguments, in the case of soft x-ray spectroscopy indicate that the main emission band is relatively insensitive to the polarization due to the core vacancy, provided that no bound states or resonances with the proper quantum numbers for allowed transitions are formed. The picture is changed if such resonance scattering occurs (see Allotey, 1967). Here a detailed knowledge of the polarization distribution is needed to obtain a correct picture.

In what follows we shall consider the abstract problem of calculating the electron distribution around a point impurity. The point-impurity idealization limits the applicability for atomic impurities, since no orthogonalization to core levels is performed.

In particular we shall discuss the importance of effects which are non-linear in the impurity charge. The criteria by which we choose to analyze the need for non-linear correction terms, are two idealized independent physical consistency-requirements. One of these chiefly relates to long-range properties of the response and the other to short-range properties. The requirements are (1) that the impurity should be completely screened when seen at large distances and (2) that the expectation value of the electron density should be finite and non-negative for all real values of the impurity charge.

Requirement (2) means that if the impurity is negatively charged it must not be allowed to repel more electrons than were actually there to begin with. This requirement is satisfied in any complete calculation with an Hermitian Hamiltonian. Problems will arise from secular terms in a perturbation expansion which is terminated after a finite number of terms. The first requirement is equivalent to the Friedel sum-rule, but is formulated in a different manner. It can be considered to be a stability requirement. As such it is intuitively obvious, since in order that a charged impurity will set up a disturbance that does not extend throughout the system, long-range forces must be compensated, i.e. charge neutrality established.

In Section 3 a short discussion is given of the positron-annihilation rate in simple metals. This is done to show how the requirement of non-negativity of the electron density, together with assumptions regarding the analytic nature of the response with respect to the impurity charge, can be used to give a correct qualitative theory of the annihilation rate. Naturally, more sophisticated theories exist (see Kahana, 1960, 1963; Carbotte and Kahana, 1965; Crowell et al., 1966; Carbotte, 1967), but these are not discussed here. For a recent summary of present knowledge of positron annihilation see Stewart and Roellig (1967).

In the case of a heavy fixed impurity the polarization distribution and the effective potential have been studied by many authors; see for example, Alfred and March (1955), who solved the problem using the Thomas–Fermi approximation and in particular discussed the importance of terms which are non-linear in the impurity charge. The Thomas–Fermi approximation fails to take account of the electrons as dynamic particles—in particular requirement (2) above cannot be satisfied using this approximation. Later, March and Murray (1960) solved the self-consistent problem of a charged impurity in a medium in which a finite number of electrons are in a box of the appropriate volume. In cases where it is possible to obtain a complete solution the number of particles is not large enough to avoid significant surface effects. The same authors, by relating the Fermi–Dirac density distribution to the density function for a system of particles obeying Boltzman statistics (March and Murray, 1960a) obtained expressions for the self-consistent electron density; at first using a linear approximation and later a quadratic approximation (March and Murray, 1961, 1962). The quadratic approximation has also been discussed by Stoltz (1966) and by Jha (1966).

In Section 3 we develop the self-consistent field method in a form which is more suitable for our purposes than the original formulation used by Ehrenreich and Cohen (1959) and by Cohen (1963).

The linear approximation in our formulation is identical to that of Langer and Vosko (1959) and also (although it is not apparent from the formulation) with the results of March and Murray (1961). In this approximation the screening requirement is identically satisfied, but the method suffers from the defect that, for a charge $Z = -1$ or smaller, consistency-requirement (2) is violated at all metallic densities, not only very close to the impurity, but within a sphere of radius $1 \cdot 0 - 1 \cdot 3 k_F^{-1}$.

Similar observations have been made previously by Glick and Ferrell (1960), Hedin (1965) and also Bergersen (1964) in the case of the electron–electron pair correlation function, and by Bergersen (1964) and Carbotte (1967b) in the case of a light impurity. In the latter two cases, although they are equally unsatisfactory from a theoretical point of view, the situation is less serious since the volume in which the positive definite functions can become negative is much smaller and the violation sets in at lower electron densities. The correction terms—which should be included to compensate for this violation of the non-negativity requirement—also affect the calculation of the electron density in the case of an impurity with positive charge, since the density must be expected to be an analytic function of the impurity charge (Kohn and Majumdar, 1965).

For an impurity with charge $Z = -1$, the quadratic approximation over-compensates for the errors introduced by the linear approximation. In fact, March and Murray (1961) were not able to obtain convergent results at an electron density corresponding to that of copper. The expressions in Section 4, for the quadratic response, differ slightly from those obtained by March and Murray (1961); the latter include some higher order effects. We show, that in the limit of small momenta q, the contribution to the Fourier transform of the electron distribution—a quadratic in the impurity charge—can be reduced to a simple integral, which we have evaluated numerically for some values of the electron density. We find that for sufficiently low electron-densities, in the metallic density range, the quadratic contribution to the small q-part of the effective interaction becomes larger than the linear term. It follows that for $Z = +1$, if a quadratic approximation is made for the effective interaction this can finish up with the wrong sign.

We conclude that if the above defects of the theory are to be avoided account must be taken of contributions to all orders in the electron impurity interaction. The "h-approximation" considered by Layzer (1963) is an example of such an approach. This approximation is equivalent to substituting the linearized effective interaction back into the self-consistent-field equation, except that Layzer includes in his formulation the Hartree–Fock self-energy. Since the Hartree–Fock term is not taken into account in the numerical work performed by Layzer, this difference is of no practical importance.

The main interest of Layzer, in this investigation, was to examine the possible formation of bound single-particle states in the presence of a positive charge, and he found that such states exist in the metallic density range. However, he emphasized that, strictly speaking, his approximation is valid only in the high density limit.

We show in Section 4 that the approximation used by Layzer (1963) is associated with an unphysical accumulation of extra charge, i.e. that the screening requirement is violated. Correction terms will weaken the effective potential and make the formation of bound states less likely.

For a discussion of the possible existence of bound plasmons in the presence of an impurity with negative charge, we refer to Sziklas (1965) and Sham (1968).

2. UNITS AND BASIC FORMALISM

We shall, throughout, put $\hbar = 1$. The average electron density is expressed by the dimensionless parameter

$$\beta = \frac{1}{k_F a_0} \tag{1}$$

where a_0 is the Bohr radius, k_F the Fermi momentum, and $1 < \beta < 3$ approximately corresponds to the metallic range of density. The other common electron-density parameter r_s is given by

$$\frac{N}{\Omega} = \frac{3}{4\pi a_0 r_s^3} \tag{2}$$

The relationship between β and r_s is

$$\beta = \left(\frac{4}{9\pi}\right)^{1/3} r_s \tag{3}$$

In order to simplify our formula we shall express energies in units of the kinetic energy at the Fermi surface, momenta in units of the Fermi momentum, lengths in units of the inverse Fermi momentum and charge in units of the positron charge. The average density (Eq. (2)) in these units is $(3\pi^2)^{-1}$. Summations over momenta are transformed, in the limit of infinite volume, into integrals using the rule

$$\frac{8\pi}{3N}\sum_{p} \Leftrightarrow \int d^3p \tag{4}$$

We shall use second-quantized operators in momentum space and let $C_{p\sigma}^{\dagger}$ and $C_{p\sigma}$ respectively create and annihilate an electron with momentum p and spin σ. The complete Hamiltonian, in the case of one fixed heavy impurity located at the origin and interacting with a uniform non-relativistic electron gas and a smeared out positive background, is given by

$$H = \sum_{p,\sigma} \frac{p^2}{2m} C_{p\sigma}^{\dagger} C_{p\sigma} + \frac{2\pi e^2}{\Omega} \sum_{\substack{q \neq 0 \\ p,k\sigma\sigma'}} \frac{1}{q^2} C_{p+q,\sigma}^{\dagger} C_{k-q,\sigma'}^{\dagger} C_{k,\sigma'} C_{p\sigma}$$

$$-\frac{4\pi e^2}{\Omega} Z \sum_{\substack{q \neq 0 \\ k,\sigma}} \frac{1}{q^2} C_{k+q,\sigma}^{\dagger} C_{k,\sigma} \tag{5}$$

In our units and with

$$\rho_q = \sum_{p,\sigma} C_{p+q,\sigma}^{\dagger} C_{p,\sigma} \qquad q \neq 0 \tag{6}$$

Eq. (5) simplifies to

$$H = \sum_{p,\sigma} p^2 C_{p\sigma}^{\dagger} C_{p\sigma} + \frac{\beta}{2\pi^2} \int d^3q \frac{P}{q^2} [\rho_q \rho_{-q} - N] - \frac{\beta Z}{\pi^2} \int d^3q \frac{P}{q^2} \rho_q \tag{7}$$

Here P indicates the principal value. In the case of a free-electron gas, interacting with an external spherically symmetric potential with Fourier transform $v(q)$, the Hamiltonian becomes

$$H = \sum_{p\sigma} p^2 C_{p\sigma}^{\dagger} C_{p\sigma} + \frac{\beta}{\pi^2} \int d^3q \, v(q) \, \rho_q \tag{8}$$

We shall not write down explicitly the corresponding Hamiltonian in the positron case (see for example, Bergersen, 1964). In the case of a fixed heavy impurity, it is convenient to work with the single particle Green function

$$G(p, k, t) = -i\langle T[C_{p\sigma}(t+t')C_{k,\sigma}^{\dagger}(t')]\rangle \tag{9}$$

Here the electron operators are defined in the Heisenberg picture, and we define

$$\delta_p = \begin{cases} +\delta & p > 1 \\ -\delta & p < 1 \end{cases} \qquad \eta(x) = \begin{cases} 1 & x > 0 \\ 0 & x < 0 \end{cases} \tag{10}$$

The frequency Fourier-transform of the non-interacting Green function is

$$G_0(p, k, u) = \frac{\delta_{p,k}}{u - p^2 + i\delta_p} \equiv \delta_{p,k} G_p \tag{11}$$

From Eq. (6) we have

$$\langle \rho_q \rangle = \frac{-i}{\pi} \sum_p \int_{-\infty}^{\infty} du\, e^{iu0^+} G(\mathbf{p}, \mathbf{p}+\mathbf{q}, u) \tag{12}$$

In this expression the summation over spin states is performed, giving a factor of 2.

The consistency-requirements (1) and (2) can be formulated simply in terms of $\langle \rho_q \rangle$, and through $\langle \rho_q \rangle$ in terms of the single-particle Green function. The screening requirement (1) can be written as

$$Z = \lim_{R \to \infty} \mathcal{N}(R) \tag{13}$$

where $\mathcal{N}(R)$ is the expectation value of the extra number of electrons within a sphere of radius R drawn around the impurity, and is given by

$$\mathcal{N}(R) = \sum_\sigma \int_{|r|<R} d^3 r \langle \psi_\sigma^\dagger(\mathbf{r})\psi(\mathbf{r})\rangle - \frac{4}{9\pi}R^3$$

$$= \frac{2}{\pi}\int_0^\infty d(qR)\left[\frac{\sin(qR)}{qR} - \cos(qR)\right]\langle \rho_q \rangle \tag{14}$$

We have

$$\lim_{R \to \infty}\int_0^\infty d(qR)\cos(qR)\langle \rho_q\rangle = 0$$

$$\lim_{R \to \infty}\int_0^\infty \frac{d(qR)}{qR}\sin(qR)\langle \rho_q\rangle = \frac{\pi}{2}\lim_{q \to 0+}\langle \rho_q\rangle \tag{15}$$

Eq. (15) when substituted into (14) and (13) gives

$$Z - \lim_{q \to 0+}\langle \rho_q\rangle \tag{16}$$

In terms of the single-electron Green function the screening requirement becomes

$$Z = \lim_{q \to 0+}\frac{-i}{\pi}\sum_p \int_{-\infty}^{\infty} du\, e^{iu0^+} G(\mathbf{p}, \mathbf{p}+\mathbf{q}, u) \tag{17}$$

Consistency requirement (2) is that the electron density should be everywhere non-negative. We define

$$n(\mathbf{r}, Z) = \langle \psi^\dagger(\mathbf{r})\psi(\mathbf{r})\rangle$$

$$= \frac{1}{3\pi^2}\left(1 + \frac{3}{8\pi}\int d^3 q\, e^{i\mathbf{q}\mathbf{x}}\langle \rho_q\rangle\right) \tag{18}$$

and our requirement is then

$$n(r, Z) \geqslant 0 \tag{19}$$

Equation (19) is most easily violated at the position of the impurity, in which case it becomes

$$-1 < \frac{3}{2} \int_0^\infty q^2 dq \langle \rho_q \rangle \tag{20}$$

From (12), Eq. (20) can be expressed in terms of the single-particle Green function.

We conclude this section by noting that for a system described by the single-particle Hamiltonian (8), the Green function satisfies the integral equation

$$G(p, k, u) = \delta_{p,k} G_p + \frac{\beta}{\pi^2} \int d^3 p' v(p - p') G_p G(p', k, u) \tag{21}$$

For a proof of Eq. (20) see Rickayzen (1961).

3. QUALITATIVE THEORY FOR POSITRON ANNIHILATION IN METALS

It has been shown (for example by Ferrell, 1956) that in the non-relativistic limit the positron-annihilation rate λ, which is due to two-photon decay, is proportional to the electron density n at the position of the positron

$$n = \int d^3 r \langle \phi^\dagger(r)\phi(r)\psi^\dagger(r)\psi(r) \rangle \tag{22}$$

$$\lambda = \pi c a_0^{-1} \beta^{-3} (137)^{-4} n \tag{23}$$

Equation (21) requires that the annihilation can be treated as an instantaneous point process.

We assume zero temperature and also that sufficient time is available for thermalization of the incoming fast positron (the validity of the latter assumption is discussed by Lee Whiting, 1955 and Carbotte and Arora, 1967). Consequently the expectation value in Eq. (1) should be taken with respect to the ground state of the N-electron–1-positron system. The operators $\phi^\dagger(r)$, $\phi(r)$ are respectively the positron creation and annihilation operators in configuration space. The experimental annihilation-rate found by Weisberg and Berko (1967) is plotted in Fig. 1 for aluminium and the alkali metals, against the electron-density parameter β. In the theoretical curves, lattice effects and core annihilation are ignored (for a discussion of this problem, see Terrell et al., 1967, and Carbotte, 1967). The annihilation rate, if the positron were a non-interacting test particle, is plotted as the curve λ_s in Fig. 1 This curve is obtained by substituting the average density $n_0 = (3\pi^2)^{-1}$, in Eq. (23), As seen from the figure, the discrepancy with the observed annihilation rate is quite large. This means that, locally, the electron–positron Coulomb interaction causes a considerable perturbation of the electron gas.

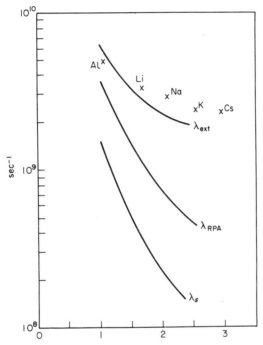

FIG. 1. Experimental annihilation rate for aluminium and the alkali metals (Weisberg and Berko, 1967), plotted against the electron-density function β.

We assume that the electron density at the position of the positron is an analytic function of the positron charge, with the power series expansion

$$n(Z) = n_0(1 + \alpha_1 Z + \alpha_2 Z^2 + ...) \qquad (24)$$

In linear-response theory, we approximate

$$n(Z) \simeq n_0(1 + \alpha_1 Z) \qquad (25)$$

The simplest linear-response theory is the random-phase approximation (RPA) in which

$$\alpha_1 = \frac{3i}{8\pi^2} \int_{-\infty}^{\infty} du \int d^3q S_q(-u) \frac{Q_0(q, u)}{q^2 + Q_0(q, u)} \qquad (26)$$

where

$$S_q(u) = \frac{1}{u - q^2 + i\delta} \qquad (27)$$

$$Q_0(q, u) = \frac{2\beta}{\pi^2} \int d^3k \left[\frac{\eta(1-k)\eta(|k+q|-1)}{u+q^2+2k\cdot q - i\delta} - \frac{\eta(k-1)\eta(1-|k+q|)}{u+q^2+2k\cdot q + i\delta} \right] \quad (28)$$

The integral (26) has been evaluated numerically (see for example, Bergersen, 1964) and the annihilation rate, obtained by substituting Eq. (25) in (23), is plotted as the curve λ_{RPA} in Fig. 1. We note that the random-phase approximation fails to account for the huge enhancement of electron density observed around the positron impurity. The inclusion of exchange corrections to the RPA, according to the approximation of Hubbard (1957), improves the calculation only slightly (see Bergersen, 1964).

We conclude that linear-response theory cannot account for the observed annihilation rate. This can also be seen by noting that $n(Z)$ must be non-negative for all real Z. In the case of a negative charge, with $\alpha_1 Z < -1$, the non-linear terms in Eq. (24) must be large enough to restore the non-negative nature of $n(Z)$. But if the non-linear terms are large for a given negative Z, they must be large also for a positive charge of the same magnitude. We can use the above argument to guess approximate values of the non-linear terms, at least in cases where these are only moderately large. We can do this by reshuffling the terms in Eq. (24), so that $n(Z)$ is positive definite for all real Z in all orders of the approximation. The simplest way to do this is to write

$$n(Z) = n_0 e^{g(z)} \quad (29)$$

where

$$g(Z) = \beta_1 Z + \beta_2 Z^2 + \ldots$$

$$\beta_1 = \alpha_1$$

$$\beta_2 = \alpha_2 - \tfrac{1}{2}\alpha_1^2 \quad (30)$$

Thus an extended linear approximation corresponds to

$$n(Z) \simeq n_0 e^{\alpha_1 z} \quad (31)$$

which reduces to the linear approximation for small Z. However, the density is now positive definite for all real Z and goes, as expected, to zero for large negative Z and to infinity for large positive Z. The resulting annihilation rate, when the RPA value for α_1 and $Z = +1$ is substituted in Eq. (31) is plotted as the curve λ_{ext} in Fig. 1. We see that we now have qualitative agreement with the experimental data.

The second approximation would be

$$n(Z) = n_0 e^{\alpha_1 Z + (\alpha_2 - 1/2\alpha_1)Z} \quad (32)$$

and does not have the desirable property of giving a density which is always an increasing function of Z, going to zero for large negative Z and to infinity for large positive Z. Instead, the density has a non-physical extremum for a finite Z. If the coefficient β_2 is calculated and is found to be small, this would be a valuable check on the validity of the approximation (31). On the other hand, if the coefficient turns out to be large, a different functional ansatz should be used so that the second approximation will also have all the desired properties.

4. THE SELF-CONSISTENT-FIELD APPROXIMATION FOR A HEAVY POINT-IMPURITY

In the self-consistent-field method the Hamiltonian (25) is approximated by

$$h_s = \sum_{p\sigma} p^2 C_{p\sigma}^\dagger C_{p\sigma} - \frac{\beta}{\pi^2} \int d^3q \frac{P}{q^2}(Z-\langle \rho_q \rangle)\rho_q \tag{33}$$

i.e. the total Hamiltonian is replaced by one in which each particle moves in the average field of the others according to a Hartree approximation. The expectation value in Eq. (33) is to be taken with respect to the self-consistent-field ground-state. For a given $\langle \rho_q \rangle$ the Hamiltonian (33) is of the form (28). The Green-function equation is therefore

$$G(p, k, u) = \delta_{p,k} G_p - \frac{\beta}{\pi^2} G_{\bar{p}} \int d^3q \frac{P}{q^2}(Z-\langle \rho_q \rangle) G(p-q, k, u)$$

$$= \delta_{p,k} G_p - \frac{\beta}{\pi^2} G_p \int d^3q \frac{P}{q^2}\left[Z - \frac{-i}{\pi} \sum_{k'} \int_{-\infty}^{\infty} e^{iu'0^+} du' G(k', k'+q, u') \right] \times$$

$$G(p-q, k, u) \tag{34}$$

Some of the formal properties of the solution to this equation can be studied from the following power series expansions

$$G(p, k, u) = G_0(p, k, u) + ZG_1(p, k, u) + Z^2 G_2(p, k, u) + \ldots$$

$$\langle \rho_q \rangle = Z\langle \rho_q^{(1)} \rangle + Z^2 \langle \rho_q^{(2)} \rangle + \ldots \qquad (q \neq 0) \tag{35}$$

The linear terms in Eq. (35) satisfy

$$G_1(p, p+q, u) = -\frac{8\beta}{3\pi N} \frac{P}{q^2}(1-\langle \rho_q^{(1)} \rangle) G_p G_{p+q} \tag{36}$$

Using Eq. (12) and putting

$$\langle \rho_q^{(1)} \rangle = \frac{\beta b_1(q^2)}{q^2 + \beta b_1(q^2)} \tag{38}$$

with

$$b_1(q^2) = \frac{i}{\pi^3} \int d^3p \int_{-\infty}^{\infty} du G_p G_{p+q}$$

$$= \frac{2}{\pi}\left[1 + \frac{1}{q}\left(1 - \frac{q^2}{4}\right) \ln \frac{q+2}{q-2} \right] \tag{37}$$

Equation (36) gives

$$G_1(\boldsymbol{p}, \boldsymbol{p}+\boldsymbol{q}, u) = \frac{8\beta}{3\pi N} \frac{1}{q^2 + \beta b_1(q^2)} G_p G_{\boldsymbol{p}+\boldsymbol{q}} \tag{39}$$

(see Dubois, 1959; Layzer, 1963).

Up to terms that are linear in the impurity charge, the effective interaction becomes

$$\frac{Z}{q^2}(1 - \langle \rho_q^{(1)} \rangle) = \frac{Z}{q^2 + \beta b_1(q^2)} \equiv z v_1(q) \tag{40}$$

We also need the quadratic term $\langle \rho_q^{(2)} \rangle$ and have

$$G_2(\boldsymbol{p}, \boldsymbol{p}+\boldsymbol{q}, u) = +\frac{8\beta}{3\pi N} \frac{P}{q^2} \langle \rho_q^{(2)} \rangle G_p G_{\boldsymbol{p}+\boldsymbol{q}}$$

$$-\frac{8\beta}{3\pi N} \sum_{q_1 \neq 0} v_1(q_1) G_p G_1(\boldsymbol{p}-\boldsymbol{q}_1, \boldsymbol{p}+\boldsymbol{q}, u) \tag{41}$$

which gives

$$\langle \rho_q^{(2)} \rangle = q^2 v_1(q) \frac{-i\beta^2}{\pi^5} \int_{-\infty}^{\infty} du \int d^3 p \int dq_1 v_1(q_1) v_1(q_1-q) G_p G_{\boldsymbol{p}-\boldsymbol{q}_1} G_{\boldsymbol{p}+\boldsymbol{q}} \tag{42}$$

The Fourier transform of Eq. (38)

$$\Delta n = \frac{1}{(2\pi^3)} \int d^3 q \langle \rho_q^{(1)} \rangle \tag{43}$$

has been tabulated by Langer and Vosko (1959). If the electron density is approximated by this quantity, the positive-definiteness requirement becomes

$$Z \Delta n_B \geqslant -\frac{1}{3\pi^2} \tag{44}$$

For physical reasons we must have $|Z| \geqslant 1$. Comparison with the tabulated values of Langer and Vosko shows that for $Z = -1$ Eq. (43) is violated inside a sphere of radius $1 \cdot 0 - 1 \cdot 3 k_F^{-1}$.

In the limit of small q, the integral (42) can be simplified considerably: we perform the integration over u and use

$$\lim_{q \to 0} \eta(1-k)\eta(|\boldsymbol{k}+\boldsymbol{q}|-1) = \boldsymbol{q} \cdot \boldsymbol{k} \eta(\boldsymbol{k} \cdot \boldsymbol{q}) \delta(1-k)$$

$$\lim_{q \to 0} \eta(k-1)\eta(1-|\boldsymbol{k}+\boldsymbol{q}|) = -\boldsymbol{q} \cdot \boldsymbol{k} \eta(-\boldsymbol{k} \cdot \boldsymbol{q}) \delta(1-k) \tag{45}$$

to give

$$b_1(0) \lim_{q \to 0} q^{-2} \langle \rho_q^{(2)} \rangle \equiv I(\beta)$$

$$= \frac{\beta^2}{\pi^4} \int d^3k \int d^3q_1 [v_1(q_1)]^2 \frac{\delta(1-k)P}{[(k-q_1)^2 - 1]}$$

$$= \frac{4\beta^2}{\pi^2} \int_0^\infty q_1 dq_1 [v(q_1)]^2 \ln \left| \frac{q+2}{q-2} \right| \tag{46}$$

Since

$$\lim_{q \to 0+} \langle \rho_q^{(2)} \rangle = 0 \tag{47}$$

the quadratic term does not contribute to the total polarization charge, and it is easy to see that the higher order terms show the same behaviour. Since the linear term exhausts the screening requirement this requires that this requirement is identically satisfied in a complete self-consistent-field calculation. However, the second-order term does contribute to the small q-part of the effective interaction. In the quadratic approximation

$$\lim_{q \to 0} v(q) = \frac{Z}{C_1(0)} [1 - ZI(\beta)] \tag{48}$$

The integral $I(\beta)$ has been evaluated numerically for different values of β; the results are tabulated below

$$\beta = 0 \cdot 5 \quad \beta = 1 \cdot 0 \quad \beta = 2 \cdot 0 \quad \beta = 3 \cdot 0 \tag{49}$$

| $I(\beta)$ | 0·094 | 0·255 | 0·672 | 1·155 |

We note that for $|Z| > 1$ the quadratic term is of the same order of magnitude as the linear term throughout the metallic region, and that even for $Z = +1$ the effective interaction will have the wrong sign at densities corresponding to low-density metals.

This result indicates that it is important to sum contributions to the polarization density over all orders in the effective interaction. One way of doing this is to re-substitute the linear-response results into the self-consistent equations. As noted earlier, this is essentially equivalent to the h-approximation considered by Layzer (1963). For the purpose of investigating the existence of negative-energy bound states, the single-particle wavefunctions constitute a more convenient tool than Green functions. However, our interest is to show only the extent to which the screening requirement is violated in this approximation, and for this the Green-function method appears more convenient. The integral equation for the single-particle Green function becomes

$$G^h(p, k, u) = \delta_{p,k} G_p - \frac{\beta}{\pi^2} G_p \int d^3q Z v_1(q) G^h(p-q, k, u) \tag{50}$$

As before, we expand the solution of Eq. (50) in powers of Z. The linear terms are unchanged and are given by Eqs. (38) and (39). However, the quadratic term is changed and becomes

$$\langle \rho_q^{(2)} \rangle = q^2 v_1(q) \langle \rho_q^{h(2)} \rangle \tag{51}$$

where $\langle \rho_q^{(2)} \rangle$ is given by Eq. (42). In the limit of small q we have

$$\lim_{q \to 0+} \langle \rho_q^{h(2)} \rangle = I(\beta) \tag{52}$$

where $I(\beta)$ is given by Eq. (46). In order to satisfy the screening requirement for all Z it is necessary that

$$\lim_{q \to 0+} \langle \rho_q^{(2)} \rangle = 0 \tag{53}$$

and we observe from Eq. (53) that this requirement is violated in the h-approximation. From the numerical values of $I(\beta)$ we note that the violation can become quite large. The correction terms will weaken the interaction and make the formation of a bound state less likely, as long as one restricts the system to framework of Eq. (33). We wish to point out that this result is model-dependent. If a bound state is formed, all the particles no longer see the same effective impurity potential, and modifications of the Hamiltonian must be made so that the bound-state electron does not repel itself. For an indication of the way such a modification can be made, see Bergersen and Terrell (1964).

5. CONCLUSION

We have shown that the two consistency requirements outlined offer a convenient tool to check the validity of various approximations to the response of a charged point-impurity, when dealing with both long-range and short-range properties. We have also shown that the requirement of non-negativity of the electron density can be used to give a simple, qualitative theory of the positron-annihilation rate in metals. The importance of the non-linear terms in the problem indicates the need for a complete self-consistent numerical calculation of the response to a heavy impurity. The formalism developed appears to be suited to the purpose, and we hope to be able to present the results of such a calculation in the near future.

Acknowledgement

One of the authors (B.B.) wishes to acknowledge financial support from Norges Almenvitenskaplige Forskningsråd and NORDITA. We also thank Dr. B. Stölan and the staff of the Institute of Theoretical Physics in Göteborg, especially Dr. L. Hedin, for many useful discussions.

References

Alfred, L. C. R. and March, N. H. (1955). *Phil. Mag.* **46**, 759.

Allotey, F. K. (1967). *Phys. Rev.* **157**, 467.

Bergersen, B. (1964). Thesis, Brandeis University.

Bergersen, B. and Terrell, J. H. (1964). *In* Bergersen (1964), Chap. V.

Berko, S. (1962). *Phys. Rev.* **128**, 2166.

Carbotte, J. P. and Kahana, S. (1965). *Phys. Rev.* **139**, A213.

Carbotte, J. P. (1967a). *In* "Positron Annihilation", ed. by A. T. Stewart and L. O. Roellig. Academic Press, New York.

Carbotte, J. P. (1967b). *Phys. Rev. Letters* **18**, 837.

Carbotte, J. P. (1967c). *Phys. Rev.* **155**, 197.

Carbotte, J. P. and Arora, H. L. (1967). *Can. J. Phys.* **45**, 387.

Cohen, M. H. (1963). *Phys. Rev.* **130**, 1301.

Crowell, J., Anderson, V. E. and Ritchie, R. H. (1966). *Phys. Rev.* **150**, 243.

Dubois, D. F. (1959). *Ann. Phys. (N.Y.)* **7**, 174; **8**, 24.

Ehrenreich, H. and Cohen, M. H. (1959). *Phys. Rev.* **115**, 786.

Ferrell, R. A. (1956). *Rev. Mod. Phys.* **28**, 308.

Glick, A. J. and Ferrell, R. A. (1960). *Ann. Phys. (N.Y.)* **11**, 359.

Hedin, L. (1965). *Phys. Rev.* **139**, A796.

Hubbard, J. (1958). *Proc. Roy. Soc.* **A243**, 336.

Jha, S. S. (1966). *Phys. Rev.* **150**, 413.

Kahana, S. (1960). *Phys. Rev.* **117**, 123.

Kahana, S. (1963). *Phys. Rev.* **129**, 1622.

Kohn, W. and Majumdar, C. (1965). *Phys. Rev.* **138**, A1617.

Langer, J. S. and Vosko, S. H. (1959). *J. Phys. Chem. Solids* **12**, 196.

Layzer, A. J. (1963). *Phys. Rev.* **129**, 897; **129**, 908.

Lee Whiting, G. E. (1955). *Phys. Rev.* **97**, 1557.

March, N. H. and Murray, A. M. (1960a). *Phys. Rev.* **120**, 830.

March, N. H. and Murray, A. M. (1960b). *Proc. Roy. Soc.* **A256**, 400.

March, N. H. and Murray, A. M. (1961). *Proc. Roy. Soc.* **A261**, 119.

March, N. H. and Murray, A. M. (1962). *Proc. Roy. Soc.* **A266**, 559.

Rickayzen, G. (1961). *In* "Lecture Notes on the Many-Body Problem from the First Bergen International School of Physics", ed. by C. Frönsdal. Benjamin, New York.

Sham, L. J. (1968). *In* "Proceedings of Int. Conf. on Localized Excitations in Solids. Plenum Press, New York.

Stewart, A. T., Shand, J. B., Donaghy, J. J. and Kusmiss, J. H. (1962). *Phys. Rev.* **128**, 118.

Stewart, A. T. and Roellig, L. O. (1967). "Positron Annihilation". Academic Press, New York.

Sziklas, E. A. (1965). *Phys. Rev.* **138**, A1070.

Stoltz, H. (1966). *Physica Status Solidi* **18**, 251.

Terrell, J. H., Weisberg, H. L. and Berko, S. (1967). *In* "Positron Annihilation", ed. by A. T. Stewart and L. O. Roellig. Academic Press, New York.

Weisberg, H. and Berko, S. (1967). *Phys. Rev.* **154**, 249.

A Theoretical Summing-up

Simon L. Altmann

I should like to start with some remarks that follow from the philosophy of Harrison on the relation between theoretical calculation and experimental verification. His story of the theoretician reproducing the experimentalist's curve, whereupon both are forgotten, should always be borne in mind. It cannot be overemphasized in studies of the solid state that, unless a genuinely crucial hypothesis is being tested, the purpose of a theoretical study is not merely to agree with experiment; rather it is the extraction from experiment of information that would not otherwise be available.

One of the items of data that we should like to obtain from x-ray emission studies is the bandwidth or Fermi-energy of a metal. This is an extremely important parameter in theoretical work, for the following reason: at present, the major difficulty in calculating band structures is the choice of potential field, and whereas no serious trouble arises with simple metals, there is still a great deal of uncertainty regarding the field to be used for a transition metal. The Fermi-energy depends sensitively on the potential field, so that if we had reliable values for it we could gauge whether or not a given field is acceptable.

In the old days, alas, one obtained Fermi-energies directly from measured x-ray bandwidths. Admittedly, there was some hesitation regarding the low-energy tail, but any inhibitions people had were easily placated. We now know better. First, we expect the x-ray data to be presented to us with full *instrumental* corrections. I emphasize the word "instrumental" to reinforce the plea—often voiced at the Meeting on which this Volume is based—that half-baked theoretical corrections, such as those arising from assumed variations of the transition probabilities with energy, should not be included in the experimental curves. Whereas plasmon satellites and Auger-tails can safely be subtracted from the experimental curves, extreme care must be taken to ensure that other corrections are really what they purport to be. For example, sharp variations in the transition probabilities could masquerade as edge-widths, so that these should be subtracted only when the strongest guarantees can be given that they represent nothing else.

When a curve with proper instrumental corrections is available, we know how to deal with the low-energy tail and the plasmon satellites, as the theoretical work of Hedin and of Glick *et al.* has shown. However, we also know that the bulk of the curve is not simply proportional to the density of states, as it was accepted to be in the past. In fact, Harrison, and Stott and March, conclusively show that the variation of the oscillator strengths with the energy is at least as important as the density-of-states distribution itself. This means that we shall have a long haul to the reliable determination of Fermi-energies and of densities-of-states from x-ray emission experiments. In principle, we should proceed as follows: first, we choose a potential field for the metal and carry out a band calculation from which the Fermi-energy, the wavefunctions, the density-of-states, the oscillator strengths and the spectral shapes are obtained; second, using the emission intensities as a rough guide only, we identify the van Hove singularities in the spectrum and adjust the potential field until the energies of the various singularities agree with experiment. When this is achieved, the new field should be checked by recalculating the spectral shape: if agreement is obtained with the experimental spectrum, the potential field can be considered reasonably reliable and other metal properties, such as Fermi-surface orbits and density-of-states curves, can be safely derived. However, to obtain such agreement it might be necessary to refine the potential even further.

As Harrison has shown, the programme sketched can be partly simplified for simple metals, where the band calculation can be by-passed in order to go straight from a pseudo-potential to the x-ray intensities. Whatever procedure we use, difficulties may arise along the way. Ashcroft suggests that the effect of the potential-change at the ionized sites must be included; though Stott and March and Harrison are more optimistic in this respect.

This report highlights the substantial advances in experimental technique and also the success of many-body calculations, in particular when dealing with plasmon satellites. At the same time, it has become clear that much work is still needed in order to understand the details of x-ray emission intensities, which are now of sufficient experimental significance to warrant more theoretical effort on their interpretation. I have no doubt that much interesting work will be done along these lines in the next few years, and that much of the impetus for this will stem from the extremely stimulating discussions during the conference that led to this Volume.

Summary Bibliography

GENERAL

Skinner, H. W. B. (1940). Soft X-Ray Spectroscopy of Solids, *Philosophical Transactions of The Royal Society* **A239**, 95.

Parratt, L. G. (1959). *Reviews of Modern Physics* **31**, 616.

Tomboulian, D. H. (1957). Soft X-Ray Spectrometry (*in English*). *In* "Handbuch der Physik", Vol. 30, p. 246, Springer-Verlag, Berlin.

Abelès, F. (1966). "Optical Properties and Electronic Structure of Metals and Alloys". North-Holland Publishing Company, Amsterdam.

Blokhin, M. A. (1957). "The Physics of X-rays". State Publishing House of Technical-Theoretical Literature, Moscow (*in Russian*). English translation: Document No. AEC-tr-4502, Clearing-house for Federal Scientific and Technical Information, Springfield, Virginia.

Faessler, A., Cauchois, Y. and Zemany, P. D. (1963). Proceedings of the 10th Colloquium Spectroscopicum Internationale, 307, 321, 341. Spartan Books, Washington.

Faessler, A. (*in press*). "Compilation of Soft X-Ray Spectra and Binding States", Landolt-Börnstein, New Series. Springer-Verlag, Berlin. (Scheduled for publication in 1969 as Vol. II/5).

Holliday, J. E. (1967). *In* "Handbook of X-rays", chap. 38. McGraw-Hill, New York.

Yakowitz, H. and Cuthill, J. R. (1962). "Annotated Bibliography on Soft X-Ray Spectroscopy". NBS Monograph 52, U.S. Government Printing Office, Washington D.C.

THEORETICAL

Callaway, J. (1964). "Energy Band Theory". Academic Press, New York.

Harrison, W. A. (1966). "Pseudopotentials in the Theory of Metals". W. A. Benjamin, New York.

Hedin, L. (1965). Effect of Electron Correlation on Band Structure of Solids, *Arkiv för Fysik* **30**, 231.

Slater, J. C. (1962). "Quantum Theory of Molecules and Solids", Vol. 2. McGraw-Hill, New York.

Author Index

Subject Index

A

α-brass, *see* Alloy, Cu–Zn
α-particle, *see* d-state, nearly-bound
Absorption spectrograph, 82–83
Alkali metal
  annihilation rate, 359
  band spectra, *see* names of individual
    elements
  halides, band spectra, 38, **41**–43
  optical absorption, 194–199
Alloy
  dilute, 176, 238–243, 306–309
  electronic structure, theory of emission,
    226, 238–244, 297–298, 303–308
  optical absorption, 209–212
  spectra, band and optical absorption
    Al_2Cu, Al L-emission, 179–180
    AlFe, Al L-emission, 180–181
    $AlFe_3$, Al L-emission, 175
    Al–Mg, Al and Mg L-absorption,
      39–41
    Al–Mg, Al and Mg L-emission, **174**–
      175, **178**, 181, 186
    Al_3Mn, Al L-emission, **178**–179
    Al–Ni, band emission, L and M, 160
    $AuCu_3$, Cu M-emission, 175
    Au Ni, optical absorption, 212
    Au–Pd, optical absorption, 212
    Cu–Ge, optical absorption, 211
    Cu_3Si, Si L-emission, 182–183
    Cu–Zn, Zn M-emission, 175
    optical absorption, 209–210
    $Fe–Fe_3C$, Fe L-emission, 125
    In–Na, Na L-emission, 188–189
    Mg_2Ni, Mg and Ni L-emission, 181–
      182
    Mg_2Si, Si L-emission, 182–183
    NiTi, Ni M-emission, 161–162
    Ni–Zn, Ni and Zn M-emission, 177–
      178
Aluminium
  electronic structure, theory of emission,
    236, 260, 339, 343, 359

K-emission, 24, 76
L-absorption, 39
L-emission, 24
optical absorption, 203–204
photo-electron spectrum, 217
plasmon satellite, 7, 24
plasmaron, 158
Annihilation operator, 330
Annihilation rate, *see* Alkali metal, annihilation rate
Auger broadening, *see* Bandwidth *and*
  Spectra, interpretation
Auger transition, 140, **218**–219, 256
Augmented-plane-wave method, 158, 166–
  169, 187, 266, 279

B

β-brass, 187
Background continuum, correction, **48**,
  98, 110, **133**–134, 144–145, 155
Band gap, *see* Energy-band gap *and* Van
  Hove singularities
Band spectra, *see* Spectra
Band structure, *see* electronic structure
  under elements by name
Bandwidths, **12–15**, 36, 51, 110–111, 146,
  179, **277**
  alloy spectra, **179–184**, 185–189
  Auger broadening of, 8, **13–14**, 55, 157
Beryllium
  alloy, Li–Be, 297–298
  electron interaction, 343
  K-absorption, 38–39
  K-emission, 17, **23**, 32, 34–36, **55**–56
  plasmon satellite, 7, 23, **57**
Bohm-Pines theory, 10, **331**, 337–338
Bohr-formula, 89
Bonding, effect on emission band, 94,
  102–103, 116–130, **125–126**
Boron trioxide
  K-absorption, 89
  emission band, 81–91

377

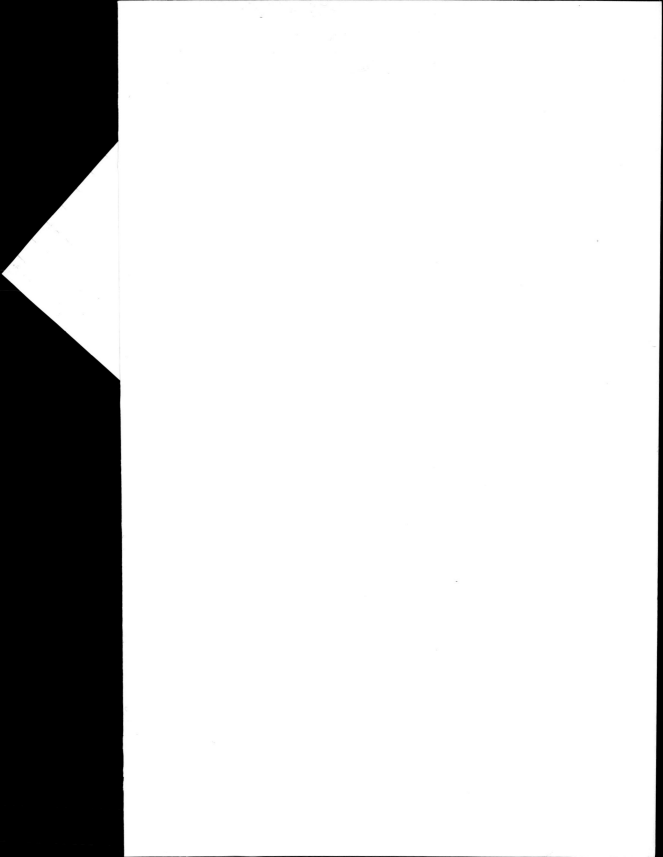

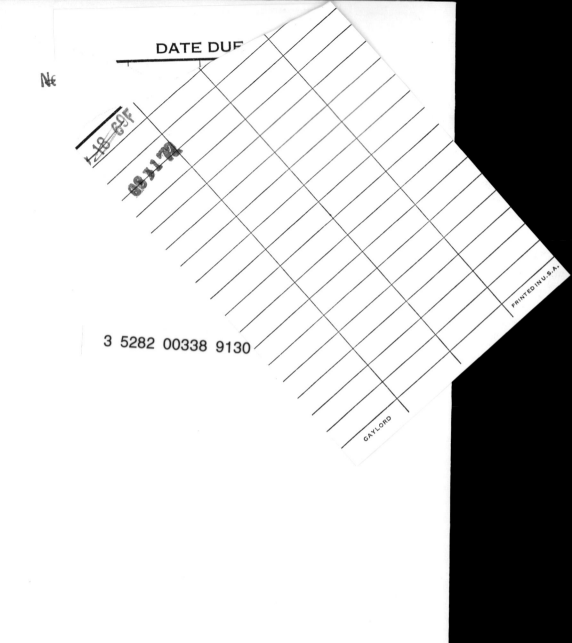